When the Invasion of Land Failed

Critical Moments and Perspectives in Earth History and Paleobiology

CRITICAL MOMENTS AND PERSPECTIVES IN EARTH HISTORY AND PALEOBIOLOGY

David J. Bottjer, Richard K. Bambach, and Hans-Dieter Sues, Editors

Mark A. S. McMenamin and Dianna L. S. McMenamin, *The Emergence of Animals: The Cambrian Breakthrough*

Anthony Hallam, *Phanerozoic Sea-Level Changes*

Douglas H. Erwin, *The Great Paleozoic Crisis: Life and Death in the Permian*

Betsey Dexter Dyer and Robert Alan Obar, *Tracing the History of Eukaryotic Cells: The Enigmatic Smile*

Donald R. Prothero, *The Eocene-Oligocene Transition: Paradise Lost*

George R. McGhee Jr., *The Late Devonian Mass Extinction: The Frasnian/Famennian Crisis*

J. David Archibald, *Dinosaur Extinction and the End of an Era: What the Fossils Say*

Ronald E. Martin, *One Long Experiment: Scale and Process in Earth History*

Judith Totman Parrish, *Interpreting Pre-Quaternary Climate from the Geologic Record*

George R. McGhee Jr., *Theoretical Morphology: The Concept and Its Applications*

Thomas M. Cronin, *Principles of Paleoclimatology*

Andrey Yu. Zhuravlev and Robert Riding, Editors, *The Ecology of the Cambrian Radiation*

Patricia G. Gensel and Dianne Edwards, Editors, *Plants Invade the Land: Evolutionary and Environmental Perspectives*

David J. Bottjer, Walter Etter, James W. Hagadorn, and Carol M. Tang, Editors, *Exceptional Fossil Preservation: A Unique View on the Evolution of Marine Life*

Barry D. Webby, Florentin Paris, Mary L. Droser, and Ian G. Percival, Editors, *The Great Ordovician Biodiversification Event*

Frank K. McKinney, *The Northern Adriatic Ecosystem: Deep Time in a Shallow Sea*

Hans-Dieter Sues and Nicholas D. Fraser, *Triassic Life on Land: The Great Transition*

When the Invasion of Land Failed

The Legacy of the Devonian Extinctions

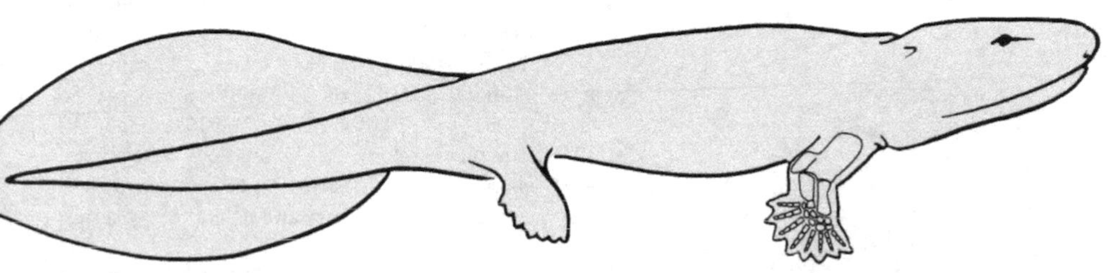

George R. McGhee Jr.

COLUMBIA UNIVERSITY PRESS NEW YORK

Columbia University Press
Publishers Since 1893
New York Chichester, West Sussex

cup.columbia.edu

Library of Congress Cataloging-in-Publication Data
McGhee, George R.
When the invasion of land failed : The legacy of the Devonian extinctions /
George R. McGhee Jr.
p. cm. — (Critical moments and perspectives in earth history
and paleobiology)
Includes bibliographical references and index.
ISBN 978-0-231-16056-8 (cloth : alk. paper) —
ISBN 978-0-231-16057-5 (pbk. : alk. paper) —
ISBN 978-0-231-53636-3 (ebook)
1. Extinction (Biology) 2. Paleontology—Devonian.
3. Catastrophes (Geology) I. Title. II. Series: Critical moments and
perspectives in earth history and paleobiology series.

QE721.2.E97M393 2013
560'.174—dc23

2013011766

∞

Columbia University Press books are printed on permanent and durable acid-free paper.
This book is printed on paper with recycled content.
Printed in the United States of America

COVER ART: Kalliopi Monoyios
COVER DESIGN: Milenda Nan Ok Lee

For Marae

Is gràdhaich leam thu.

Contents

	Preface	ix
CHAPTER 1	The Evolution of Life on Land	1
CHAPTER 2	The Plants Establish a Beachhead	27
CHAPTER 3	The First Animal Invasion	53
CHAPTER 4	The First Catastrophe and Retreat	99
CHAPTER 5	The Second Animal Invasion	159
CHAPTER 6	The Second Catastrophe and Retreat	179
CHAPTER 7	Victory at Last	213
CHAPTER 8	The Legacy of the Devonian Extinctions	263
	Notes	277
	References	295
	Index	313

Color plates insert between pages 98 and 99

Preface

This book is about the profound evolutionary consequences of the twin extinction events that occurred in the Late Devonian period of geologic time, some 375 and 359 million years ago, for the evolution of terrestrial animal life on Earth. Our ancestors, the first vertebrates on land, were dealt a severe blow in their infancy by those catastrophic events, one that almost drove them back into the sea. If those extinctions had been just a bit more severe, spiders and insects might well have become the ecologically dominant forms of animal life on land—we, the vertebrates, would have remained in the oceans.

We know that five catastrophic losses of biological diversity have occurred in the history of life on Earth since the evolution of animals more than 600 million years ago. Each of these biodiversity crises, commonly called the "Big Five," changed the direction and pattern of the evolution of life on Earth and had consequences that extend to our present world.

The most recent of these biodiversity crises is the end-Cretaceous extinction, which occurred 65 million years ago. The environmental catastrophe that triggered the end-Cretaceous extinction destroyed the dinosaurian ecosystem, a highly successful terrestrial ecosystem that had persisted for 150 million years. We mammals owe our current ecological dominance to that event. It was only after the destruction of the dinosaurian ecosystem that mammals, in a classic example of ecological release, evolved explosively and rapidly established the mammalian terrestrial ecosystem that persists to the present day. If that event had not occurred, the dinosaurs might have continued to reign supreme, and we might have continued to be small nocturnal animals that scurried in the underbrush at night, seeking to avoid the dinosaur predators who hunted by day.

The next most recent biodiversity crisis is the end-Triassic extinction, which occurred approximately 200 million years ago. If the end-Cretaceous extinction is seen as beneficial to mammals, the end-Triassic was the opposite. Our ancestors, the synapsid amniotes, were in fact more diverse than the ancestors of the dinosaurs prior to that catastrophe, but that changed after this crisis. In essence, the end-Triassic extinction ushered in the dominance of the dinosaurs, and the end-Cretaceous extinction terminated that dominance. What if the end-Triassic event had never happened? Might we, the mammals, have become the ecologically dominant land animals and have kept the dinosaurs in check, rather than the other way around?

The next most recent biodiversity crisis in geologic time is the catastrophic end-Permian extinction, which nearly ended animal life on the planet Earth about 250 million years ago. All forms of life, both terrestrial and marine, suffered in that global catastrophe. If it had been only a little more severe, it would have erased the previous 350 million years of animal evolution, leaving only the simplest animals, such as jellyfish and sponges, as survivors. As it was, this extinction event triggered a restructuring of the ecosystems of the entire planet, both on the landmasses and in the seas, and is recorded in the fossil record as the largest loss of biodiversity ever seen in geologic time.

This brings us back in time to the two remaining biodiversity crises of the Big Five: the Late Devonian and the end-Ordovician. The end-Ordovician did not have a great impact on life on land simply because there was not much life on land to be affected at that time. For much the same reason, it is often thought that the Late Devonian crisis also did not have much of an effect on terrestrial life. That point of view is profoundly wrong, however, as will be demonstrated in this book.

Our world is the product of the Devonian extinctions. Only one group of four-limbed vertebrates exists on the Earth today: we, the tetrapods.

All of our close relatives, the once diverse lineages of the other tetrapod-like fishes, are extinct. Our more distant relatives, the once numerous lobe-finned fishes, are rare—only three families of air-breathing, lobe-finned lungfishes exist today. Before the Devonian extinctions, the sight of an air-breathing fish crawling along a riverbank—out of water—was not unusual, but the Devonian extinctions eliminated all of our intermediate vertebrate relatives, resulting in a large phylogenetic gap between us, the limbed vertebrates, and our very distant living relatives, the familiar ray-finned fishes that fill our rivers and lakes today. This gap in living intermediate forms is why the very idea of a fish with limbs and feet seems so peculiar to us today; yet such fishes existed in the Late Devonian. If the Devonian extinctions had not happened, such fishes might still have been commonplace today. No one would have conceived of creating a Darwin bumper sticker for their car that shows a fish with feet simply because a fish with feet would not have been unusual—you could see one at any local fish market.

This book is a summary of more than three decades of my research into the Devonian extinctions. I thank Columbia University Press, publishers of my initial book *The Late Devonian Mass Extinction* (1996), for making this second volume possible. I thank Kalliopi Monoyios, scientific artist, for her original illustrations of our ancient vertebrate ancestors and kin. And last, I thank my wife, Marae, for her patient love.

When the Invasion of Land Failed

CHAPTER 1

The Evolution of Life on Land

The Long Beginning

The planet Earth is 4,560 million years old. For over 3,000 million years, life on Earth was confined to the oceans, and the land areas of the planet looked much like the rocky landscapes of Mars today. Complex life on land, and the terrestrial ecosystem, is a relatively new phenomenon in the history of the Earth. The oldest simple terrestrial ecosystems of macroscopic organisms—plants, fungi, and animals—can be traced back only to the middle of the Ordovician Period of geologic time (for the geologic timescale, see table 1.1),[1] some 468 million years ago. That is, only in the past 10 percent of the total age of the Earth has complex life evolved to the point where it could emerge from the oceans and conquer the hostile environment of the terrestrial realm.

It is not easy living on land. Dehydration is a very serious problem for land-dwelling organisms. Many of the morphological and physiological changes that took place in the evolutionary sequence of fish to amphibians to amniotes were adaptations for preventing water loss in the dry environments of the land, where animals are surrounded only by the thin gaseous layer of the atmosphere. Life in the sea is surrounded at all times by water; dehydration is an unknown phenomenon. Land-dwelling organisms must deal with the crushing force of gravity,

TABLE 1.1 The geologic timescale.

Eon	Era	Period	Time at Beginning (Ma BP)
Phanerozoic	Cenozoic	Neogene	23.03
		Paleogene	65.5
	Mesozoic	Cretaceous	145.5
		Jurassic	199.6
		Triassic	251.0
	Paleozoic	Permian	299.0
		Carboniferous	359.2
		Devonian	416.0
		Silurian	443.7
		Ordovician	488.3
		Cambrian	542.0
Proterozoic	Neoproterozoic	Ediacaran	635
		Cryogenian	850
		Tonian	1,000
	Mesoproterozoic	Stenian	1,200
		Ectasian	1,400
		Calymmian	1,600
	Paleoproterozoic	Statherian	1,800
		Orosirian	2,050
		Rhyacian	2,300
		Siderian	2,500
Archaean	Neoarchaean		2,800
	Mesoarchaean		3,200
	Paleoarchaean		3,600
	Eoarchaean		3,850
Hadean			4,560

Source: Modified from Gradstein et al. (2004), Walker and Geissman (2009), and Erwin et al. (2011).
Abbreviations: Ma BP = millions of years before the present.

whereas, for neutrally buoyant swimming fish out in the oceans, gravity is an unknown phenomenon. Life in the oceans experiences very little variation in temperature; in places where temperature does vary, the change is very slow, gently rising and falling with the passing of the four seasons of the year. In contrast, land-dwelling organisms deal with large fluctuations in temperature that occur very rapidly: every 24 hours, temperatures quickly increase as the fireball that is the sun rises above the horizon, and they fall just as quickly as the sun sets into the cold darkness of the night. Even today, under the protective screen of the Earth's ozone layer in the upper atmosphere, land-dwelling organisms are subject to radiation burns from that same fireball of the sun—radiation burns that can lead to fatal skin cancers. For much of the history of the Earth, that ozone layer did not exist at all, and the intense ultraviolet radiation flux present on the continental landmasses was deadly. In contrast, water is a very good radiation shield, and marine organisms located just a few centimeters below the surface of the oceans are protected from ultraviolet radiation from the sun.

Thus land-dwelling organisms have had to evolve complex adaptations to deal with four particularly serious problems that do not exist for organisms in the oceans: dehydration, gravity, temperature fluctuations, and radiation poisoning. The last problem was initially the most serious; it took 3,000 million years of environmental change for the radiation flux present on land to decrease to livable levels and evolutionary change for organisms to adapt to life on land.

Life on Earth Today

Three types of life exist on Earth today: the archaea, the bacteria, and the eukarya (table 1.2). The simplest forms of life are the unicellular; that is, they exist as a single individual cell. There are two types of unicellular life (fig. 1.1): the prokaryotes (simple cells) and the eukaryotes (complex cells). Prokaryotes are the simplest forms of unicellular life: their small cells have no nucleus and no organelles but consist only of a cell membrane within which the chromosomes of the organism are enclosed. The chromosomes contain the DNA of the organism, the coding mechanism of life on Earth. For many years it was thought that there was only one type of prokaryote life on Earth, the bacteria. Then, in 1977, microbiologist Carl Woese discovered that another type of prokaryote life existed—the archaea—a discovery for which he received the Crafoord Prize.[2] The archaea and bacteria look very much like one another when viewed under a microscope, but their genetics are quite different. The

TABLE 1.2 A phylogenetic classification of life.

Archaea (prokaryote cells with ether-bonded lipid membranes)
Bacteria (prokaryote cells with ester-bonded lipid membranes)
Eukarya (eukaryote cells)
– Bikonta
– – Rhizaria
– – Excavobionta
– – Chromoalveolata
– – Green eukaryotes
– – – Glaucophyta
– – – **Metabionta** (unicellular to multicellular bikonta)
– – – – Rhodobionta (red algae)
– – – – Chlorobionta (green algae + land plants)
– Unikonta
– – Amoebozoa
– – Opisthokonta
– – – Fungi
– – – – Microsporidia (unicellular fungi)
– – – – **Eumycetes** (unicellular to multicellular fungi)
– – – Choanozoa
– – – – Choanoflagellata (unicellular choanozoa)
– – – – **Metazoa** (multicellular choanozoa, the animals)

Source: Modified from the phylogenetic classification of Lecointre and Le Guyader (2006).

Note: Multicellular eukaryote clades are marked in bold-faced type.

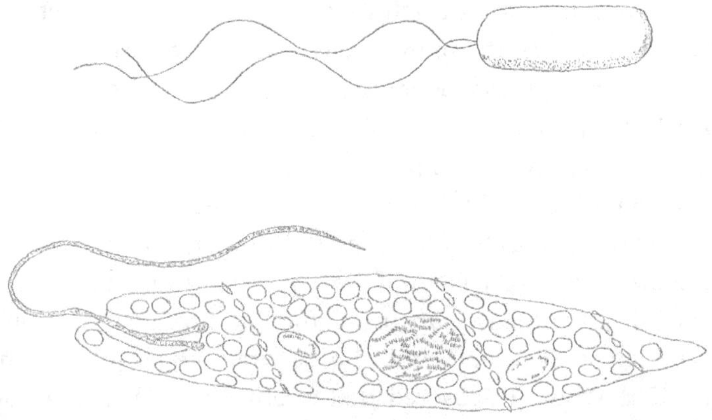

FIGURE 1.1 Top: a prokaryote cell, the bacterium *Rhizobium leguminosarum*. Bottom: a eukaryote cell, the excavobiontan *Euglena spirogyra*. *Credit:* Modified and redrawn from Lecointre and Le Guyader (2006).

easiest way to tell the two types of life apart is by examining their cell membranes: the bacteria have ester-bonded lipid membranes, and the archaea have ether-bonded lipid membranes. Because the possession of a cell membrane is one of the fundamental characteristics of life (the other being the possession of DNA), the difference in cell-membrane construction between bacteria and archaea is of major evolutionary significance.[3]

In our day-to-day lives we are surrounded by a sea of bacteria, but we usually do not notice them because they are so small that they are invisible to the naked eye. Some of these bacteria are very beneficial to us: the *Escherichia coli* ("E. coli") bacteria that live in our intestines help us to digest food (but are beneficial only if they remain in our intestines!). Others can be deadly: *Clostridium botulinum* will kill us with botulism, and its cousin *Clostridium perfringens* will give us gangrene. For the most part, however, the bacteria around us are benign and simply coexist with us.

The archaea usually live in extreme environments; thus, in our normal lives we do not encounter them as often as we do the bacteria.[4] Some, like *Methanopyrus kandleri*, live at the bottom of the oceans in volcanic hydrothermal vents, under incredible pressure and in water as hot as 110°C (230°F). Others, like *Thermoplasma acidophilum*, flourish in the waste pits of coal mines, in sulfuric acid–rich waters with a pH value as low as one.

In contrast to the prokaryotes, the eukaryote unicells are much larger and more complex forms of life (fig. 1.1). Eukaryotes possess a nucleus (which contains the chromosomes) and specialized organelles such as chloroplasts (which can produce food for the cell by photosynthesis), mitochondria (which mediate the energetic enzyme activities of the cell), and Golgi apparati (for processing proteins within the cell). The difference in size between the simple prokaryote unicells and complex eukaryote unicells is striking: the prokaryote bacterium *Rhizobium leguminosarum* illustrated in figure 1.1 is 2 micrometers (0.08 thousandths of an inch) long, and the eukaryote *Euglena spirogyra* is 120 micrometers (4.7 thousandths of an inch) long. That is, the single eukaryote cell is 60 times larger than the prokaryote cell!

It has been proved that the organelles of the eukaryote cell were once separate, independent bacteria. For example, the chloroplasts of a eukaryote cell were once free-living cyanobacteria, and the mitochondria were once free-living purple bacteria.[5] Even today, the mitochondria within the cells in our bodies still possess their own DNA, which is separate from the DNA in the nucleus of our cells that codes for the cell's reproduction. The evolution of the eukaryote cell from the prokaryote cell occurred by the process of biological symbiosis: first as colonies of cooperating independent cells, and then later as a complex single cell of dependent organelles. This evolutionary transition had occurred by 2,100 million years ago, and the geologic evidence for the occurrence of this event will be considered later in this chapter.

The eukarya are the third form of life on Earth (table 1.2), and they all possess eukaryote cells (hence their name). The eukarya split into two separate lineages, the bikonta and the unikonta, early in their evolutionary history (table 1.2). The unicellular forms of these two groups are easy to tell apart: the bikonta have two flagella, and the unikonta have only one flagellum. We are unikonts, and the single flagellum of the human sperm cell is an ancient inherited trait from our unicellular ancestors.

The eukarya include both unicellular and multicellular forms of life (table 1.2). Multicellular organisms evolved in the same fashion as the eukaryote cell itself, i.e., by biological symbiosis, in which free-living eukaryote cells first formed colonies of cooperating independent cells and then multicellular tissues composed of dependent and specialized cells. However, the evolution of multicellularity in the eukarya is convergent—it occurred at least three times independently.[6] In the lineage of the bikonta, multicellularity occurred in the evolution of the metabionta, the red and green algae (table 1.2). In the lineage of the unikonta, multicellularity occurred twice. First, the lineage of the opisthokonta produced the fungi and their sister group the choanozoa. The clade of the fungi is divided into the microsporidia, which remained unicellular, and their sister group, the eumycetes, which evolved multicellularity. Second, the clade of the choanozoa is divided into the choanoflagellata, which remained unicellular, and their sister group the metazoa, which independently evolved multicellularity (table 1.2). Thus the trees, mushrooms, and humans we encounter in our daily lives are all independent inventions of multicellularity by the ancient unicellular eukarya. The bikonta were the first to evolve multicellularity, producing the green algae some 1,200 million years ago, with the unikonta evolving multicellular animals only 780 million years ago, as will be discussed later in this chapter.

The Earliest Life on Earth

It is not known how life originated on Earth. We do know that the most ancient fossils of life yet discovered—simple, prokaryote cells like those found in living archaea and bacteria—are 3,450 million years old. Geochemical evidence suggests that life was present on Earth 3,830 million years ago, though actual fossil cells have yet to be found that are that old.

The first item of geochemical evidence for the presence of life 3,830 million years ago comes from the analysis of carbon isotopes present in the ancient strata. The carbon atom has two stable isotopes, carbon-12 and carbon-13. Carbon-12 possesses six neutrons and six protons in the nucleus of the atom, and the heavier carbon-13 possesses seven neutrons and six protons in its nucleus (carbon has another, heavier isotope, carbon-14, which is unstable and radioactively decays to nitrogen-14). Photoautotrophic organisms prefer the lighter isotope of carbon in photosynthesis; that is, photosynthetic bacteria preferentially extract carbon-12 from the atmosphere to synthesize hydrocarbons and leave the heavier isotope carbon-13 free in the environment. The naturally occurring ratio of carbon-12 to carbon-13 in the atmosphere in the absence of life is well known. Therefore, the isotopic ratio of carbon in ancient strata can be used as a tool to search for the presence of life: if the ratio of carbon-12 to carbon-13 is not the naturally occurring, inorganically-produced ratio in the atmosphere, then some other process is changing that ratio by preferentially removing carbon-12 from the atmosphere. Life is one such process. The carbon-isotope ratios in 3,830 million-year-old strata from Greenland are off; something skewed the ratio of carbon-12 to carbon-13, and this is taken as geochemical evidence that life was present.[7]

The second item of geochemical evidence comes from the presence of banded-iron formations in Archaean strata 3,830 million years old.

The early atmosphere of the Earth was composed mostly of carbon dioxide, much like the present-day atmospheres of Mars and Venus. The present-day atmosphere of the Earth, in contrast, contains about 21 percent molecular oxygen. That oxygen is the product of organic photosynthesis: plants and other aerobic photosynthetic organisms remove carbon dioxide from the atmosphere and combine it with hydrogen from water to produce sugars and other hydrocarbons. In this process, free oxygen is released as a waste product of aerobic photosynthesis. These organisms dump oxygen back into the environment.[8] The earliest organisms to evolve photosynthesis, however, were not aerobic cyanobacteria but anaerobic purple bacteria.[9] Phylogenetic analyses of the five groups of photosynthetic bacteria reveal that the purple bacteria are basal and that they first evolved anaerobic photosynthesis, followed by the evolution of the two types of green bacteria,[10] the heliobacteria,[11] and finally the evolution of aerobic photosynthesis in the cyanobacteria.[12]

Banded-iron formations contain iron that has been oxidized. How could iron be oxidized on an early Earth that had an atmosphere that was anoxic, i.e., was mostly carbon dioxide? One potential way to oxidize iron is in the process of photosynthesis; that is, life. Some of the anaerobic-photosynthetic purple bacteria directly oxidize iron, whereas the aerobic-photosynthetic cyanobacteria produce free oxygen that then secondarily oxidizes iron. Thus the presence of oxidized iron in the banded-iron formations in strata that are 3,830 million years old is taken as geochemical evidence that life was present. Geochemical analyses of the types of iron isotopes present in these most ancient banded-iron formations indicate that anaerobic purple bacteria were responsible for the oxidation of the iron, not aerobic cyanobacteria.[13] Our oldest fossil evidence for the aerobic-photosynthetic cyanobacteria indicates that they evolved later, nearly 400 million years later.[14]

Sometime and somewhere in the 730 million years between the formation of the Earth 4,560 million years ago and our oldest geochemical evidence for the presence of life at 3,830 million years ago, life originated on the planet. Or did it? Some of our modern theories about how life might have originated propose that life may in fact have originated on Mars first, and then was seeded on Earth by Martian meteorites—rocks that had been blasted off the surface of Mars by asteroid impacts, which then fell to the Earth's surface, bringing life with them. If that scenario is ever proved, then we are all in fact Martians. We have found quite a few Martian meteorites here on Earth, and the most ancient of these—the famous 4,510 million-year-old ALH84001 meteorite—contains tiny cylindrical structures that look very much like fossil nanobacteria, embedded in carbonate deposits within the meteorite that are 3,900 million years old (fig. 1.2).[15] These structures, if fossil bacteria, are smaller than any nanobacteria living on Earth today; for example, the one shown in figure 1.2 is only 480 nanometers (0.019 millionths of an inch) long. Alternatively, it has been suggested that these tiny structures may be fossil precursors of prokaryote cellular life, or protobacteria—actual preserved evidence of one of the steps in the evolution of cellular life.[16] In that case, these structures could yield critical evidence about the origin of life itself, and their existence would be even more compelling evidence to go to Mars to find more of them.

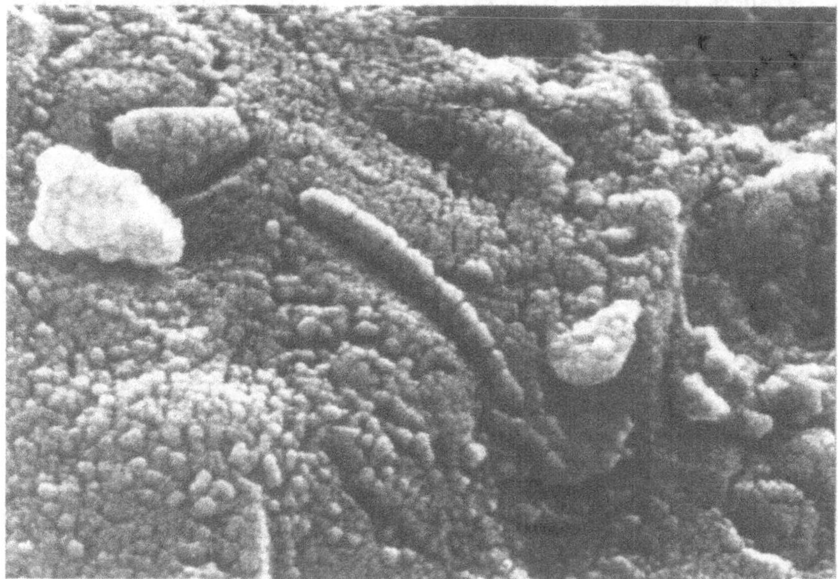

FIGURE 1.2 One of numerous enigmatic cylindrical structures that look very similar to fossil prokaryote cells, found in the Martian meteorite ALH84001. *Credit:* Photography courtesy of NASA.

Scenarios for the origin of life, whether on Earth or on Mars, all consider life to have originated in water. Earlier scenarios considered it probable that life originated in shallow marine waters, and that the energy source necessary for assembling and rearranging molecules was provided by electric discharges in lightning or by the high ultraviolet radiation flux at the surface of the water, or both. More recent scenarios consider it more likely that life originated in deep marine waters around volcanic hydrothermal vents, and that the energy source was both thermal and chemical.[17] Even today, most of the microbial forms of life found around deep-water hydrothermal vents are archaea, an ancient lineage of life that we considered in the previous section of the chapter. We may in fact have to go to Mars to hunt for the earliest evidence for the evolution of life. The ancient strata that we need to examine simply do not exist on Earth any longer. The surface of the Earth is a very dynamic environment: the gigantic plates of the Earth's crust are constantly shifting around, spreading apart at the mid-oceanic ridges and grinding together to form mountain chains in other areas of the planet. The atmosphere is constantly convecting; water evaporates into the atmosphere in some regions of the Earth and condenses in torrential rainfall in other regions. Many of the Earth's dynamic processes destroy sedimentary rocks: they are crushed and metamorphosed in uplifted mountain chains, they are subducted in the oceanic trenches and melted into lava, they are eroded by the rainfall and washed away in the rivers. Not surprisingly, the further we go back in geologic time, the fewer strata are preserved in the rock record. The oldest surviving sedimentary rocks on Earth that could possibly contain evidence of life are 3,830 to 3,850 million years old.[18] We know older sedimentary rocks must have existed on Earth, but they are now long gone, destroyed by Earth's dynamic surface processes. How can we answer the question of how life originated when no evidence remains to be examined on Earth in the critical time interval from 4,560 to 3,850 million years ago?

In contrast to the Earth, we know that 4,510 million-year-old crustal rocks still exist on Mars: we actually have a piece of them, the Martian meteorite ALH84001.[19] The surface of Mars is much less dynamic than that of the Earth; the plates of the Martian crust stopped moving long ago, and most of the Martian volcanoes have ceased erupting molten rock. Mars is a smaller world than the Earth, and it lost its heat at a much faster rate than the Earth. The only major destroyer of sedimentary rocks on Mars today are the winds, which continue to etch away the ancient strata. Did life originate on Mars? If it did, ancient sedimentary rocks that could hold their fossils still exist on Mars.

It is important to realize that the origin of life and the evolution of life are two distinctly separate questions. If we ask the question, "How did life originate?", the answer is: we do not know. We do have some good ideas for how life originated, but we honestly do not know for certain how life originated. However, if we ask the question, "Does life evolve?", the answer is: *yes.* That is an empirical observation; it is a fact. There is nothing theoretical about the fact that life evolves. Theory only enters the equation when we ask the additional question, "Why does life evolve?" Charles Darwin's theory of natural selection is one theory that has been proposed to explain why life evolves; another theory is that of neutral change and genetic drift, and so on. Continued debate about potential causes of evolution does not bring into question the fundamental empirical observation that life evolves.

Evolution in the Archaean and Proterozoic Eons

What are the facts, the empirical observations, about the early evolution of life on Earth? Our oldest (yet) discovered microfossils of prokaryote cells and macrofossil stromatolites are 3,450 million years old.[20] Stromatolites are conical, mound-like layered structures of sediment created by masses of filamentous bacteria that secrete mucilage slime (fig. 1.3). Modern-day cyanobacteria typically form stromatolites if they are allowed to grow undisturbed by grazing marine animals, such as snails, who consume the bacterial mats as food. Thus stromatolites today are found only in extreme marine environments, such as hypersaline lagoons where the salinity of the sea water is so high that it is lethal to complex animals. In the Archaean, however, there were no animals, and stromatolites formed in a variety of normal marine environments.

It is not known with certainty what type of prokaryotes produced the Archaean stromatolites that are found in more than 30 known sites

FIGURE 1.3 A scene from the Archaean Earth, with mound-like stromatolites in the tidal zone. Stromatolites are produced by filamentous cyanobacteria that secrete mucilage slime; they are rare today but were abundant and the only visible form of life on the Earth in the Archaean. *Credit*: From *Evolution: What the Fossils Say and Why It Matters*, by D. R. Prothero, copyright © 2007 Columbia University Press; figure by Carl Buell. Reprinted with permission.

around the Earth. It is generally assumed that at least some of these stromatolites, if not all of them, were formed by cyanobacteria because these bacteria continue to form stromatolites today.[21] Cyanobacteria are aerobic photosynthesizers; that is, they produce free oxygen as a metabolic waste product when they form sugars by photosynthesis, and during the Archaean this process had a profound impact on future evolution of life on Earth.

Molecular evidence from Australia has been used to argue that the cyanobacteria were present in the Earth's oceans 2,700 million years ago and—more controversial—that the eukarya were present as well.[22] This evidence is the presence of distinctive cyanobacterial biomarker chemicals in ancient sediments that are argued to have undergone very little metamorphism for 2,700 million years (metamorphism will destroy organic molecules); these biomarkers have been called "molecular fossils" because actual cellular fossils are absent. These fragile biomarker chemicals are surprising in that they have survived for 2,700 million years, as the previously discussed 3,450 million-year-old stromatolite fossils are generally considered to have been produced by cyanobacteria (table 1.3). What is more surprising is the argued ancient presence of other distinctive biomarkers that are only found in the cell membranes of the eukarya. Our oldest eukaryote cellular fossils are 2,100 million years old—thus, the presence of these molecular fossils would mean that the symbiotic evolution of eukaryote cells actually occurred some 600 million years earlier (table 1.3). If the eukarya had evolved by 2,700 million years ago, then their presence would have important implications for the oxygen content of the Earth's atmosphere. Modern eukaryote cells need an ambient oxygen concentration of 0.042 percent to 0.21 percent in the atmosphere[23] to synthesize these biomarker molecules, which are components of their cell membranes. After 750 million years, since the evolution of the cyanobacteria some 3,450 million years ago, oxygen had finally reached high enough concentrations in sea water that it began to diffuse into the atmosphere.

The atmosphere of the Earth has been radically transformed by the presence of life. Both anaerobic and aerobic photosynthesizers actively remove carbon dioxide from the atmosphere and use the carbon to form complex hydrocarbons for food. The original atmosphere of the Earth

TABLE 1.3 Major biotic and geologic events that occurred in the Archaean and Proterozoic Eons.

Time (Ma BP)	Biotic and Geologic Events
780	← marine animals evolve
	(420 million years of time pass by)
1,200	← marine plants evolve; definite microbial life on land
	(800 million years of time pass by)
2,000	← oxidized redbed strata on land
	(100 million years of time pass by)
2,100	← oldest fossil eukaryote cells
	(100 million years of time pass by)
2,200	← possible microbial life on land?
	(100 million years of time pass by)
2,300	← Snowball-Earth glaciation
	(200 million years of time pass by)
2,500	← GOE: "great oxygenation event"
	(100 million years of time pass by)
2,600	← possible microbial life on land?
	(100 million years of time pass by)
2,700	← molecular evidence suggests eukaryote cells have evolved
	(750 million years of time pass by)
3,450	← oldest stromatolite fossils; life is definitely present in the oceans
	(380 million years of time pass by)
3,830	← geochemical evidence suggests life is present on Earth
	(730 million years of time pass by)
4,560	← formation of the Earth

Abbreviations: Ma BP = millions of years before the present.

was probably composed mostly of carbon dioxide, very similar to that of its rocky volcanic neighbors, Mars and Venus. The atmosphere of Mars today is 95 percent carbon dioxide, and the atmosphere of Venus is 97 percent carbon dioxide, whereas the atmosphere of the preindustrial-age Earth was only 0.03 percent carbon dioxide.[24] On Earth, life has been removing carbon dioxide from the atmosphere for the last 3,830 million years and has contributed to the transformation of the Earth's atmosphere to its present carbon dioxide–depleted state.

About 2,500 million years ago, the oxygen-producing activity of the cyanobacteria had its first major impact on the atmosphere of the Earth: the GOE, or "great oxygenation event" (table 1.3), marked the end of the

Archaean Eon and the beginning of the Proterozoic (table 1.1). For the future evolution of complex—and large—life forms with aerobic metabolism, the GOE was good news indeed. For the ancient anaerobic life forms—the original inhabitants of Earth—the GOE was a disaster. The first mass extinction in the history of life on Earth probably occurred 2,500 million years ago, when vast unknown numbers of species of anaerobic bacteria and archaea perished by oxygen poisoning.

Diffusion of oxygen into the atmosphere was most probably the trigger of the first Snowball-Earth glaciation seen in the Proterozoic, which occurred about 200 million years after the GOE.[25] Snowball-Earth glaciations were much more extensive than any seen in the Phanerozoic, in that ice extended from the poles of the planet all the way down to the equator. From space, the entire planet may have looked like one giant snowball, hence the name "Snowball-Earth." The steady downdraw of carbon dioxide, due to the photosynthetic activity of life, was already reducing the greenhouse effect of this gas in the atmosphere. The presence of free oxygen in the atmosphere may then have begun to remove an even more powerful greenhouse gas—methane—via oxidation.[26] The resulting sharp drop in the greenhouse capacity of the atmosphere of the Earth may have triggered the first great ice age in Earth's history.

Some 300 million years later, oxygen concentrations in the atmosphere reached high enough levels to begin oxidizing iron on land, and the first redbed strata appeared in the terrestrial rock record (table 1.3). Redbeds are layered beds of red sandstones and shales in which the iron has been oxidized, giving the strata a characteristic rusty-red color. By this time, the aerobic photosynthetic activity of cyanobacteria had raised the amount of oxygen in the atmosphere to something between 1 percent and 4 percent of the atmosphere.[27]

The Earliest Invaders of Land

It is not known with certainty when life first ventured out of the oceans and onto dry land, but we do know that the life forms at that time were microbial. Even today, many bacteria can survive extreme dehydration as well as high radiation fluxes, which is probably an ancient adaptation acquired early in their evolution, as the Earth in the Archaean Eon had no protective ozone layer in the upper atmosphere to shield the surface of the planet from high-energy solar radiation. For example, the bacterium *Deinococcus radiodurans* can survive exposure to a radiation flux of 30 to 50 *thousand* sieverts (three to five *million* rem),[28] whereas exposure to only eight sieverts (800 rem) is fatal to a human. Modern marginal-marine cyanobacteria form microbial biofilms that survive in dry air during times of low tide each day. In the eukarya, the chlorobiontan chlorokybophyte and klebsormidiophyte green algae (table 1.2) can live in dry air. Perhaps the most complex algal forms of life on land today are the lichens. Lichens look somewhat like plants but are actually symbiotic associations of eumycete fungi (table 1.2) with green algae or, more anciently, with cyanobacteria.

Definitive proof exists that microbial life was present on land 1,200 million years ago (table 1.3). Numerous occurrences of fossil microbial crusts are present in the Torridon sandstones in northwest Scotland, microbial mats that lived under dry air in sites well away from the nearest shorelines and major river channels, and thus survived on moisture provided by rainfall alone.[29] Ancient fossil soil horizons, presumably produced by microbial activity, and the presence of carbon isotope anomalies suggest that microbial life may have existed on land 2,200 million years ago in South Africa, and 2,600 million-year-old altered rock surfaces hint at the presence of subaerial microbial life in South Africa even in the Archaean.[30]

These tiny microbial invaders of land in the pre-Phanerozoic were important in that they started a biogeochemical process that assisted the eventual invasion of land by multicellular plants in the Phanerozoic: the formation of soil. In the Archaean and most of the Paleoproterozoic, microbial life on land consisted only of archaeal and bacterial films and mats, simply because the only forms of life on Earth were archaea and bacteria. With the evolution of the eukarya, more complex microbial associations became possible. Microfossils interpreted as probable microsporidian fungi indicate that the fungi had evolved around 720 to 800 million years ago,[31] and lichen-like fossils demonstrate the evolution of the eumycete fungi by 600 million years ago.[32] Fossils of fungi are very rare, thus both of these groups may have evolved earlier than those dates.

A lichen symbiont is composed of two components: an eumycete fungal species, the mycobiont, and a photobiont, a species that has the capability to photosynthesize food. Even in modern terrestrial lichens, the photobiont can either be a cyanobacterium or a green alga. The oldest yet-discovered lichen fossils come from China, are about 600 million years old, and consist of phosphatized filaments of fungal hyphae and of coccoidal cells.[33] Unfortunately, it is not possible to prove, simply on the basis of the morphology of the coccoidal cells, whether the photobiont in the lichen was a sheathed cyanobacterium or a colonial green alga. Molecular phylogenetic analyses of the fungi suggest that the major lineages of terrestrial eumycete fungi, the Ascomycota, Basidiomycota, and Glomales, all evolved around 600 to 620 million years ago, although the oldest terrestrial fossils of glomalean fungi yet discovered are 460 million years old, in the middle of the Ordovician (table 1.1).[34]

Lichens can colonize very hostile environments and are found today in habitats as extreme as arid deserts to freezing subpolar tundra. The fungus provides a tough protective microhabitat for the photobiont, shielding the photosynthetic bacteria or algae from dehydration and providing nutrients by dissolving the rock upon which the mycobiont is growing. In turn, the photobiont provides hydrocarbons for the fungus to metabolize. The process of etching and breaking down rock by lichens, and the subsequent addition of organic debris to the rock debris, combine to produce soil. The existence of thin soil horizons and scattered soil pockets provides a foothold upon which future land plants can seize and grow.

The lineage of the plants, the chlorobionta (table 1.2), evolved as marine green algae 1,200 million years ago (table 1.3).[35] Animals, our ancestors, were latecomers: the first marine animals appeared in the fossil record 400 million years after the evolution of the plants (table 1.3). However, it took the plants 732 million years to leave the oceans and to finally invade the land, as will be discussed in the next section of this chapter.

Evolution in the Early Phanerozoic Eon

Animals, the metazoa (table 1.2), evolved 780 million years ago[36] in the Cryogenian Period of the Neoproterozoic (table 1.1). The earliest large animals evolved in the Ediacaran Period of the Neoproterozoic (table 1.1), a geologic period named after the Ediacaran animals, which are themselves named after the Ediacara hills in Australia where their fossils are abundant. These peculiar first animals had no mineralized tissues; that is, they had no skeletons. They were all soft-bodied, and moreover their bodies were often peculiarly pleated and puffy, much like the structure of a modern air mattress that we use to sleep on when on a camping trip. Some of these pleated structures were clearly held vertical in life, strange fronds up to one meter (3.3 feet) in height, looking more like a marine alga than an animal (color plate 1). Some paleontologists have suggested that these multicellular animals may have contained chloroplasts like plants and thus were autotrophic as well as heterotrophic like some unicellular eukaryotes. Others suggest these animals were chemoautotrophic instead of photoautotrophic.[37] Fossils of more familiar forms of animal life, such as the very primitive sponges and the cnidarians, the soft-bodied jellyfishes, and bilaterian burrowing worms are also found with some Ediacarans.

It took another 238 million years for the first animals with mineralized tissues—skeletons—to evolve; this occurred around 542 million years ago (table 1.4). This evolutionary event marks the beginning of the Phanerozoic Eon, the geologic eon of the "visible animals."[38] The event is also often called the "Cambrian Explosion," in recognition of the explosive diversification of animal life that occurred at the beginning of the Cambrian Period of geologic time (table 1.1). Large macroscopic animals with skeletons made of calcium, silica, and phosphorous compounds left fossils in rocks that before contained only the faint impressions made by a few types of the soft-bodied Ediacara animals. Among the numerous animal groups that appear at the beginning of the Phanerozoic are the lineages of the arthropods (fig. 1.4) and of the chordates (fig. 1.5), our ancestors. Both of these marine animal lineages produced descendants who left the oceans and invaded the land.

TABLE 1.4 Major biotic and geologic events that occurred in the late Proterozoic Eon and early Phanerozoic Eon.

Time (Ma BP)	Biotic and Geologic Events
468	← land plants definitely present
	(20 million years of time pass by)
488	← animal trackways on land; beginning of the Ordovician Period
	(25 million years of time pass by)
513	← land plants evolve?
	(29 million years of time pass by)
542	← marine animals with skeletons evolve; beginning of the Cambrian Period
	(38 million years of time pass by)
580	← Gaskiers Snowball-Earth glaciaton
	(60 million years of time pass by)
640	← Marinoan Snowball-Earth glaciaton
	(80 million years of time pass by)
720	← Sturtian Snowball-Earth glaciation
	(60 million years of time pass by)
780	← soft-tissued marine animals evolve

Abbreviations: Ma BP = millions of years before the present.

Why did so many different types of large animals with skeletons appear so suddenly in the fossil record? A clue may be seen in some modern day animals, such as bivalve molluscs that have the capability to switch from aerobic metabolism to anaerobic metabolism when the oxygen content in water becomes depleted. While in the anaerobic-respiration mode, these bivalves also produce acid metabolites as a byproduct, and these acids begin to etch and dissolve the calcium-carbonate shell of the animal.[39] Therefore it has been proposed that the ability to grow and maintain mineralized skeletal tissues is a trait that is found in organisms with aerobic respiration, meaning there has to be enough oxygen present in the environment to allow organisms to respire aerobically. Atmospheric modeling of the evolution of the Earth's atmosphere indicates that, 540 million years ago, in the Early Cambrian,

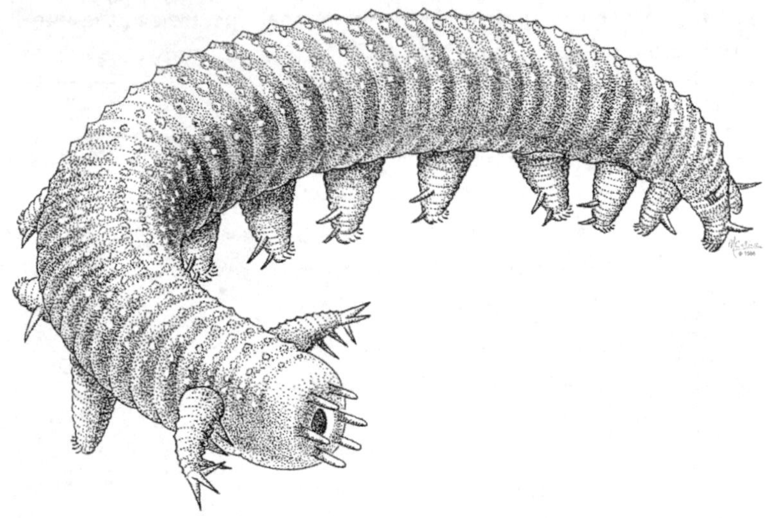

FIGURE 1.4 The Cambrian marine species *Aysheaia pedunculata*, an ancient basal member of the clade of panarthropod animals. The panarthropods were the first major clade of animals to leave the seas and invade the land, and they possessed a key anatomical trait necessary for that invasion: legs. *Credit*: Illustration © Marianne Collins, artist. By permission of W. W. Norton and Company, Inc. (Gould, 1989).

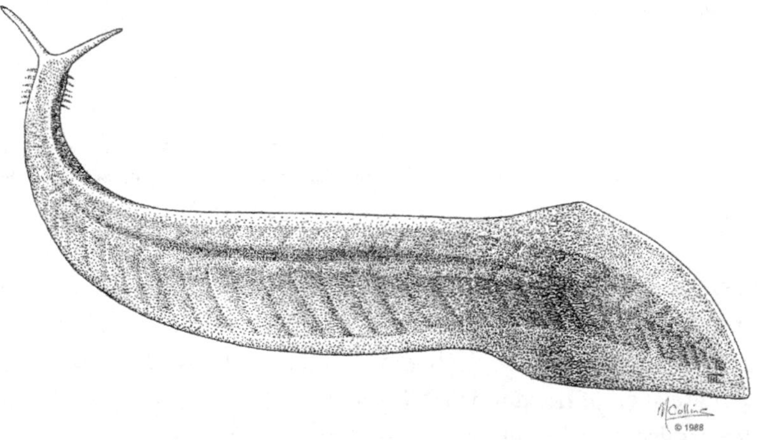

FIGURE 1.5 The Cambrian marine species *Pikaia gracilens*, an ancient basal member of the clade of chordate animals. The chordates were the second major clade of animals to leave the seas and invade the land. The ancient chordates were swimmers, not walkers like the panarthropods, and their fish descendants would later evolve legs from fins to leave the water. *Credit*: Illustration © Marianne Collins, artist. By permission of W. W. Norton and Company, Inc. (Gould, 1989).

oxygen levels in the atmosphere had finally risen to around 16 percent to 18 percent,[40] as compared to present-day levels of 21 percent. Atmospheric oxygen levels of 16 percent to 18 percent may have been the final trigger for animals belonging to numerous disparate phylogenetic lineages to simultaneously achieve sustained aerobic respiration, to increase in size, and to secrete mineralized skeletal tissues of several different chemical compositions in the different lineages. The fact that the planet was hit by at least three more Snowball-Earth glaciations in the latest Neoproterozoic (table 1.4), immediately before the Cambrian animal diversification, may also be evidence of the continued drawdown of the greenhouse gas carbon dioxide from the atmosphere—and the injection of oxygen into the atmosphere—by aerobic photosynthetic life.

Another consequence of the increased oxygen content of the atmosphere is the production of ozone. High above the surface of the Earth, the ultraviolet radiation flux from the sun is so energetic that it will split apart the two oxygen atoms present in an oxygen molecule. A free oxygen atom will then attach to another oxygen molecule, producing a new oxygen molecule consisting of three oxygen atoms bound together, known as ozone.[41] Instead of splitting apart when hit by a photon of ultraviolet radiation, the ozone molecule will instead absorb the ultraviolet photon and stop it from traveling downwards to the surface of the Earth. Thus the ozone radiation shield of the Earth began to form, and the ultraviolet radiation flux present at the surface of the Earth began to decrease. This process may have begun 1,800 million years ago, after the cessation of the formation of banded-iron formations in the Earth's oceans and more oxygen began to diffuse into the atmosphere as it was no longer being bound by iron in the seas. By the early Phanerozoic, the ultraviolet radiation flux present on land may no longer have been strong enough to cause serious radiation poisoning. Complex macroscopic life could venture out from the oceans onto the land without being fried.

Ample evidence exists to prove that land plants had evolved 468 million years ago, in the middle of the Ordovician, and will be discussed in detail in chapter 2. However, some paleontologists argue that land plants had in fact evolved 45 million years earlier, around 513 million years ago in the middle of the Cambrian (table 1.4). The evidence comes in the presence of spores in terrestrial sediments, spores that resemble land-plant spores.[42] Others argue that the spores are from freshwater green algae, not land plants. If land plants had evolved by the middle Cambrian, why did they not proliferate and spread? Instead, there is a strange absence of evidence of even microbial life on land—microbial life that is definitely known to have invaded land as early as 1,200 million years ago (table 1.3)—for the entire 90 million-year period of time from 558 to 468 million years ago. This absence of evidence is variously called the "Cambrian to mid-Ordovician minimum" of microbial-life anomaly.[43]

However, whether tiny land plants were present on the land in the middle to late Cambrian or not, we do know that animals were—at least occasionally. About 488 million years ago, in strata dated to the Cambrian-Ordovician boundary, fossil trackways left by a multi-legged marine arthropod on dry land have been discovered. The animal that made the trackways had at least eight pairs of walking legs and possessed a telson, or tail spine. The trackways are of several different sizes, from eight to thirteen centimeters (three to five inches) wide, indicating the presence of several individuals on land, not just one.[44] These unknown myriapod-like arthropods probably ventured out of the sea onto the land for limited periods of time—protected by their tough chitinous exoskeletons from dehydration in the dry air and from potential radiation burns from the sun. They may have even been engaged in mating swarms, much like modern horseshoe crabs who leave the sea to mate on dry land and then return to the safety of seawater.

The invasion of land began in earnest in the Ordovician: by the middle of the Ordovician, the land plants had definitely evolved from freshwater green algae. By the Late Ordovician, the first land animals, the millipede arthropods, had evolved. In the gradual assembly of what would become the terrestrial ecosystem, the autotrophic organisms—the plants—were the base of the trophic pyramid of life, and their invasion of the hostile realm of dry land will be considered in detail in the next chapter.

Why Such a Long Beginning?

If we accept the geochemical evidence that life was present on Earth 3,830 million years ago (table 1.3), then it took 1,130 million years for life to evolve from a simple prokaryote cell to a more complex eukaryote cell. It took another 1,920 million years for life to evolve from single eukaryote cells to a multicellular animal. All in all, it took life 3,050 million years to evolve from a unicellular form of life to a multicellular animal. Why did it take so long for animals to evolve? The astrobiologist David Grinspoon, in his musings about the possible evolution of life on other planets in the universe, comments on this long beginning on Earth: "Three billion years is a long wait by any standard—human, geological, even cosmological. It is more than half the age of the Earth and probably about a quarter the age of the universe . . . I'll admit I find this disturbing. If life always self-organizes into more complex entities, why did it get stuck?"[45]

Massive deposits of iron oxide (rust) abound in Archaean and Proterozoic strata (color plate 2). These are the banded-iron formations that we considered at the beginning of the chapter as evidence that life, probably anaerobic photosynthetic purple bacteria, had evolved about 3,830 million years ago. The banded-iron formations continued to form throughout the Archaean Eon into the Proterozoic, ending about 1,800 million years ago. Almost all of the iron that is commercially mined today for modern industry comes from these ancient sedimentary strata (color plate 3).[46] The iron oxide molecule is composed of two atoms of iron and three atoms of oxygen—Fe_2O_3—thus the banded-iron formations contain an enormous amount of oxygen, which had been produced by cyanobacteria in aerobic photosynthesis starting about 3,450 million years ago. That oxygen did not go into the Earth's atmosphere in the Archaean—it was immediately bonded by iron and deposited on the seafloor as iron oxide. The size of those iron-oxide deposits is astounding, even today after many of the ancient deposits must have been long ago destroyed by erosion and are no long present in the rock record. The surviving fraction of the banded iron formations that remain in the fossil record show us that at least 640 billion tonnes (705 billion tons) of these strata were deposited on the seafloor in the half-billion-year interval from 2,500 to 2,000 million years ago alone,[47] and that was after the great oxygenation event. As these strata can be composed of up to 55 percent iron in content, they contain an enormous amount of bound oxygen.

Only at the end of the Archaean, evidenced by the great oxygenation event 2,500 million years ago, did the production of oxygen by cyanobacteria finally exceed the rate of bonding by iron in the environment, which allowed the oxygen to escape into the atmosphere of the Earth. It still took another 700 million years before the formation of banded-iron formations finally ceased. At this time, the ozone radiation shield of the Earth may have begun to form in the upper atmosphere. Six hundred million years after the formation of banded-iron formations ceased, the first marine plants evolved. Four hundred million years after these plants, the first animals evolved. The rapid diversification of animal life that occurred at the beginning of the Phanerozoic Eon—the Cambrian Explosion—can be argued as evidence that oxygen in the atmosphere had finally reached levels that would allow animals to achieve sustained aerobic respiration, to become larger, and to grow mineralized skeletons.

In summary, the astrobiologist David Grinspoon engages in a thought experiment in imagining a hypothetical evolutionary scenario where life originates on an Earth-like world located farther out toward the rim of our galaxy than the position of the Earth:

> Picture a planet that is similar to the Earth in many ways except that it formed originally with much less iron in the mix (perhaps orbiting one of those stars farther out in the galaxy having less of the heavy elements with which to make planets). Remember that on Earth iron hogged all the oxygen for the longest time and kept it out of the atmosphere, stunting our growth for billions of years. On this iron-depleted world, all other things being equal, oxygen would build up much faster in the air.[48]

On such a hypothetical iron-depleted world, would multicellular plants and animals have evolved much more quickly? Would the invasion of land have occurred much earlier? Would the long beginning have been a short beginning?

It should be noted that a similar scenario has already been proposed for a member of our own solar system: the planet Mars.[49] Our robot probes to Mars have revealed that Mars was much more Earth-like in the first billion years of its existence, from 4,560 to about 3,500 million years ago. The planet had a magnetic field, its atmosphere was much thicker, surface temperatures were much warmer, and liquid water was everywhere—there was even an ocean covering much of the Northern Hemisphere of the planet. Even with these similarities, Mars is a smaller planet with a smaller mantle and core, and the movements of the plates of its crust were not as dynamic as those seen on Earth. As a result, most of the volcanoes on Mars were created by mantle plumes, not by the melting of plate boundaries in subduction zones like on Earth (for example, the "ring of fire" distribution of volcanoes surrounding the Pacific Ocean).

Because the Martian crustal plates were not melted and recycled as vigorously as those on Earth, much less iron was brought to the surface of the planet—oxygen could have accumulated in the atmosphere of the planet much more rapidly than it did on Earth. That is, life could have evolved at a much more rapid pace on early Mars than it did on Earth. But eventually the core of Mars cooled, the magnetic field collapsed, and much of the atmosphere of the planet was lost to space. Life, if it ever existed on Mars, either became extinct or was driven deep underground where water is still liquid around volcanic hot spots.

With much less iron being brought to its surface by plate-tectonic forces, could Mars have experienced a *short* beginning in the evolution of life in its first billion years? Could eukaryote complex cells and perhaps even multicellular life forms have evolved on Mars from 4,560 to about 3,500 million years ago? We will never know until we travel to Mars and hunt for fossils in its ancient strata.

CHAPTER 2

The Plants Establish a Beachhead

Building the Foundation of the Terrestrial Ecosystem

Plants are photoautotrophic organisms; that is, they have the ability to synthesize complex hydrocarbons like sugars—food—from simple molecules and water by using energy obtained from sunlight. Autotrophic organisms form the base of the trophic pyramid in any ecosystem, whether it be marine or terrestrial, and are themselves used as a source of food by the heterotrophic organisms in the ecosystem—the animals—who cannot synthesize food for themselves. Thus, in order for animals to leave the oceans and invade the terrestrial realm, plants must first establish a beachhead on dry land.

Land plants, the Embryophyta (table 2.1), evolved from ancient freshwater filamentous green algae, and modern molecular analyses indicate that the freshwater coleochaetophyte algae are the closest living relatives to these first land plants. For plants to leave the protection of the water and spread onto dry land, the radiation flux on land had to be low enough so that they did not suffer radiation burns. As discussed in the previous chapter, the oxygen concentration in the atmosphere of the Earth—itself a product of the photosynthetic activity of the plants—had increased by the early Phanerozoic to the point that the ozone shield of the planet had at least partially formed. The first filamentous green

TABLE 2.1 Phylogenetic classification of plants.

Chlorobionta (unicellular to multicellular green algae and plants; from table 1.2)
– Ulvophyta
– Prasinophyta
– Streptophyta
– – Chlorokybophyta
– – Klebsormidiophyta
– – Phragmoplastophyta
– – – Zygnematophyta
– – – Plasmodesmophyta
– – – – Chaetosphaeridiophyta
– – – – Charophyta
– – – – Parenchymophyta
– – – – – Coleochaetophyta
– – – – – **Embryophyta** (land plants)
– – – – – – Marchantiophyta (liverworts)
– – – – – – Stomatophyta (stomate plants)
– – – – – – – Anthocerophyta (hornworts)
– – – – – – – Hemitracheophyta
– – – – – – – – Bryophyta *sensu stricto* (mosses)
– – – – – – – – Polysporangiophyta (branched-sporophyte plants)
– – – – – – – – – Horneophyta†
– – – – – – – – – **Tracheophyta** (vascular plants)
– – – – – – – – – – Rhyniopsida†
– – – – – – – – – – Eutracheophyta
– – – – – – – – – – – Lycophyta (microphyll-leafed plants)
– – – – – – – – – – – – Zosterophyllopsida†
– – – – – – – – – – – – Asteroxylales
– – – – – – – – – – – – – Drepanophycales†
– – – – – – – – – – – – – Lycopodiales (club mosses)
– – – – – – – – – – – – – – Protolepidodendrales†
– – – – – – – – – – – – – – Selaginellales (spike mosses)
– – – – – – – – – – – – – – – Lepidodendrales† (scale trees)
– – – – – – – – – – – – – – – Isoetales (quillworts)
– – – – – – – – – – – Euphyllophyta (megaphyll-leafed plants)
– – – – – – – – – – – – Moniliformopses
– – – – – – – – – – – – – Cladoxylopsida†
– – – – – – – – – – – – – Equisetophyta (horsetails)
– – – – – – – – – – – – – Filicophyta (ferns)
– – – – – – – – – – – – **Lignophyta** (woody plants)
– – – – – – – – – – – – – Aneurophytales†
– – – – – – – – – – – – – Archaeopteridales†
– – – – – – – – – – – – – **Spermatophyta** (seed plants)
– – – – – – – – – – – – – – Lyginopteridales†
– – – – – – – – – – – – – – Medullosales†

TABLE 2.1 (*continued*)

- - - - - - - - - - - - - - - Gigantopteridales†
- - - - - - - - - - - - - - - Core seed plants
- - - - - - - - - - - - - - - - Pinophyta
- - - - - - - - - - - - - - - - Ginkgophyta
- - - - - - - - - - - - - - - - Cycadophyta
- - - - - - - - - - - - - - - - Gnetophyta
- - - - - - - - - - - - - - - - Angiospermae (flowering plants)

Source: Living plant groups follow the phylogenetic classification of Lecointre and Le Guyader (2006), and extinct plant groups follow the phylogenetic classifications of Kenrick and Crane (1997a, 1997b) and Donoghue (2005).

Note: The evolutionary position of the Embryophyta (land plants), Tracheophyta (vascular plants), Lignophyta (woody plants), and Spermatophyta (seed plants) are marked in bold-faced type.

† extinct taxa.

strands of these new land plants then faced the problem of dehydration, i.e., how to prevent losing their water by evaporation into the dry air surrounding them. Land plants evolved a series of adaptations to prevent or at least slow the rate of water loss, such as specialized pore regions in their external tissues, which they could open to exchange carbon dioxide and oxygen with the atmosphere when needed and then close to prevent further water loss to the atmosphere when gas exchange was not needed.

Filamentous plant growth in a two-dimensional plane soon led to crowding and overgrowth, where one plant strand crossed and shaded another plant strand in the competition for light from the sun, the energy source for the survival of the plant. Just as in human cities, when crowding occurs in the two-dimensional plane of the land surface, the solution is to move into the third dimension above the land surface—to construct vertical structures (houses with more than one floor, in the case of humans). To grow into the third dimension of height, a new problem had to be overcome—the force of gravity. First, land plants had to evolve tissue structures that were strong enough to resist the force of gravity to allow the growth of a stem that could rise vertically from the land surface. This vertical growth led to the next challenge: the vertical stem of the plant needs water to live, but it is surrounded by air, not water, so absorbing moisture directly into the vertical tissue was not possible. The plants then evolved systems to transport water up the vertical stems, against the force of gravity, from the moist soil below.

The early land plants that grew in three dimensions were small vertical stems with no leaves. Photosynthesis took place in tissues in the outer surface area of the green stem. In terms of geometry, a stem is a small cylinder, having a fairly large surface area where photosynthesis can take place, and a small volume of internal tissue that is in need of food. The key word in the last sentence is *small*: the larger the cylinder, the smaller its surface area relative to its volume, as surface area increases as a square function of linear dimension, whereas volume increases as a cubic function. Thus, the larger the stem, the smaller its surface area, where photosynthesis occurs, relative to its volume of internal tissue, where food is needed. A large single-stemmed plant thus might not be able to photosynthesize enough food in its surface-area tissues to meet the needs of the tissues in its internal volume.

One solution to this classic area–volume problem is to take the single stem and branch it into multiple stems, each with its own photosynthetic surface area. Thus larger plants quickly evolved branched-stem structures. Another solution to the area–volume problem is to somehow increase the photosynthetic region of the plant from a simple cylinder to a structure with a larger surface area: that is, to evolve leaves. Interestingly, at least nine different phylogenetic lineages of early land plants independently, convergently, evolved leaves.[1]

Larger plants require more water and nutrients, and on land the only source for these is in the soil below the vertical plant. Plants had to evolve solutions to the area–volume problem here as well, to evolve structures that would increase the surface area at the opposite end of the plant that could be used to absorb water and nutrients: that is, to evolve roots. The early land plants had no roots; they were small and did not need them. Larger plants quickly evolved roots, and even larger plants evolved complex, branched roots with large surface areas for the uptake of nutrients and water.

The Mysterious Ordovician Land Plants

As discussed above, we know that the embryophytes, the land plants, evolved from freshwater filamentous green algae. The open question is, however, *when* did the embryophytes evolve from freshwater green algae?

At present, the first appearance of tetrad spores—spores grouped into fours after meiosis—in the fossil record is taken to be evidence that the first of the land plants had evolved.[2] Undisputed tetrad spores attributed to land plants are known from the fossil record in the Middle Ordovician, yet enigmatic tetrad-like spores have been found in river deposits as old as the Middle Cambrian in Australia, and some paleobotanists believe that tiny land plants had evolved by that early period,[3] as mentioned in chapter 1.

In the previous chapter, the finest time divisions of the geologic timescale that we considered were the Periods of the Phanerozoic (table 1.1). To follow the pattern of land-plant evolution in this chapter, we will use the finer time division of the Periods into Epochs and Ages within the critical interval during which the plant invasion of land took place (table 2.2).[4]

The oldest known tetrad spores and cuticular fragments of land plants are found in strata dated to the Dapingian Age of the Middle Ordovician,[5] around 470 million years ago (table 2.2).[6] The type of plant that produced these spores is debated—some paleobotanists believe they are marchantiophyte spores, or liverwort spores, which would make sense as the liverworts are the most plesiomorphic of living land plants (table 2.1). Others believe they are anthocerophyte spores, or hornwort spores, and thus were produced by more advanced stomatophyte land plants. As the phylogenetic affinities of these early spores remains controversial, they are often termed "cryptospores" in acknowledgment of their cryptic origin.[7] Regardless of their affinity, these oldest-known land-plant spores were found in Argentina, thus the land plants appear to have evolved on what is known today as the supercontinent of Gondwana.[8]

The arrangement of the continental landmasses in the Ordovician world was very different from our present-day Earth. At the beginning of the Middle Ordovician, some 472 million years ago, the current separated continents of South America, Africa, India, Antarctica, and Australia were united together in a single supercontinent, Gondwana, that was positioned over the South Pole of the Earth (color plate 4). Three smaller continents were located to the north of Gondwana: present-day North America was located in the west, along the equator, and is named Laurentia after the Laurentian Shield of Canada, the ancient core of the North American continent. Present-day northern Europe was located southeast of Laurentia and is named Baltica after the Baltic region of Europe. Present-day middle Europe was located on a much smaller

TABLE 2.2 Finer time divisions (Epochs and Ages) of the geologic timescale in the critical time interval during which the plant invasion of land occurred.

Period	Epoch	Age	Time at Beginning (Ma BP)
Carboniferous	Late (Pennsylvanian)	Gzhelian	303.9
		Kasimovian	306.5
		Moscovian	311.7
		Bashkirian	318.1
	Early (Mississippian)	Serpukhovian	326.4
		Visean	345.3
		Tournaisian	359.2
Devonian	Late	Famennian	374.5
		Frasnian	385.3
	Middle	Givetian	391.8
		Eifelian	397.5
	Early	Emsian	407.0
		Pragian	411.2
		Lochkovian	416.0
Silurian	Late	Pridolian	418.7
		Ludfordian	421.3
		Gorstian	422.9
	Middle	Homerian	426.2
		Sheinwoodian	428.2
	Early	Telychian	436.0
		Aeronian	439.0
		Rhuddanian	443.7
Ordovician	Late	Hirnantian	445.6
		Katian	455.8
		Sandbian	460.9
	Middle	Darriwilian	468.1
		Dapingian	471.8
	Early	Floian	478.6
		Tremadocian	488.3

Source: Modified from the international geologic timescales of Gradstein et al. (2004) and Walker and Geissman (2009).

Abbreviations: Ma BP = millions of years before the present.

continental sliver and island chain directly south of Baltica, named Avalonia. Last, present-day Siberia was located northeast of Laurentia and north of Baltica, and is named simply Siberia.

Definitive proof that the liverworts had evolved about 15 million years later (about 455 million years ago) comes from fossils from the Katian Age of the Late Ordovician[9] (table 2.2), which were found in present-day Oman,[10] thus also on Gondwana (color plate 4). These fossils include not just isolated spore tetrads but also tissue fragments of the sporangia that held the spores in life. Some of these fossil sporangia hold as many as 7,450 tetrads together, and the shape of the spherical sporangia fragments suggest that they could have held as many as 95,000 spore tetrads in the living plant. The ability to produce large numbers of spores is considered to be an adaptation to living in harsh terrestrial environments and is evidence that these fossils were indeed from land plants. Finally, the microscopic ultrastructure of the walls of the spores reveals parallel-arranged lamellae of a type that is found only in the spores of living liverworts, providing further evidence that the plants that produced them were liverworts (fig. 2.1), the

FIGURE 2.1 The living liverwort *Marchantia polymorpha*. This true plant, with its lichen-like appearance, has no roots, no vascular system, no true stomata, and is small, about five centimeters (two inches) wide. *Credit*: Modified and redrawn from Lecointre and Le Guyader (2006).

most plesiomorphic of the land plants.[11] Today, these simple land plants resemble lichens, the terrestrial algal-fungal symbionts, and, like lichens, can encrust rocks or tree trunks.

The primitive liverworts had no roots, no vascular system, no true stomata, and were confined to humid environments. The next step in the evolution of land plants was the appearance of the stomatophytes (see table 2.1), plants so named because they possess stomata, openings in the surface tissues of the plant, which can be opened and closed by two guard cells and allow the plant to regulate gas and water vapor exchange with the surrounding atmosphere. The most plesiomorphic of the stomatophytes are the anthocerophytes, or hornworts (fig. 2.2). As mentioned previously, some paleobotanists consider some of the Middle Ordovician cryptospores to be hornwort spores, which would mean that the stomatophyte plants had evolved by that point in time. Others consider the first hornwort spores to have appeared in the Late Ordovician, the *Stomatophytes* spores that appeared about 455 million years ago[12] in the Katian Age.[13] The bryophytes, or mosses, are more advanced stomatophytes of the hemitracheophyte clade (table 2.1). Both the hornworts and the mosses (fig. 2.3) have upright sporophytes; that is, the sporangium is held up in the air on a thin vertical stem (fig. 2.2). Upright sporophytes are another adaptation to life in dry air, but it is debated whether this adaptation evolved originally in the hornworts and was simply inherited by the mosses, or the mosses independently evolved upright sporophytes and this adaptation is convergent in the two groups.[14]

In summary, the mysterious Middle Ordovician tetrad cryptospores are evidence that land plants had evolved by 470 million years ago, and the Late Ordovician fossil sporangia are evidence that the simplest of the land plants, the liverworts, were present on land 460 million years ago. Land plants appear to have evolved in Gondwana (color plate 4) and then spread north to Avalonia and Baltica.[15] Whether the stomatophyte plants, the hornworts and more advanced mosses, had evolved by the Middle Ordovician remains controversial, but the *Stomatophytes* cryptospores are taken as evidence that the stomatophyte plants had evolved by the middle of the Late Ordovician. All of these plant types are small, and the Ordovician is sometimes referred to as the "Lilliputian plant world" in recognition of the minute stature of the early land flora.

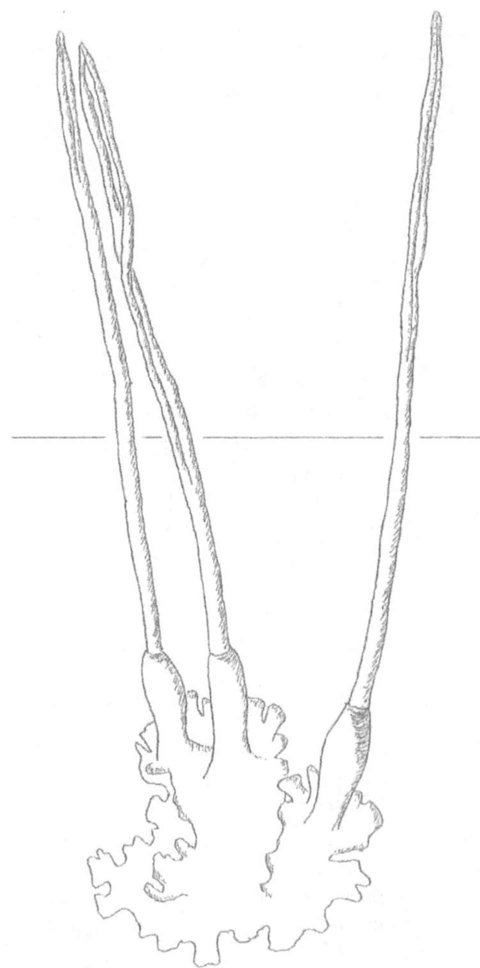

FIGURE 2.2 The living hornwort *Anthoceros fusiformis*. The upright sporophytes of the plant are about three centimeters (a little over an inch) tall, and the plant possesses stomata to regulate gas exchange in dry air. *Credit*: Modified and redrawn from Lecointre and Le Guyader (2006).

A major mystery is the absence of macrofossils of these early plants in the fossil record, aside from the fragmentary liverwort sporangia traced to the Late Ordovician. The microfossils, the cryptospores, prove that land plants had evolved in the Middle Ordovician. The first macrofossils, stems and other plant body fragments, do not appear in the fossil record until the mid-Silurian, some 50 million years later.

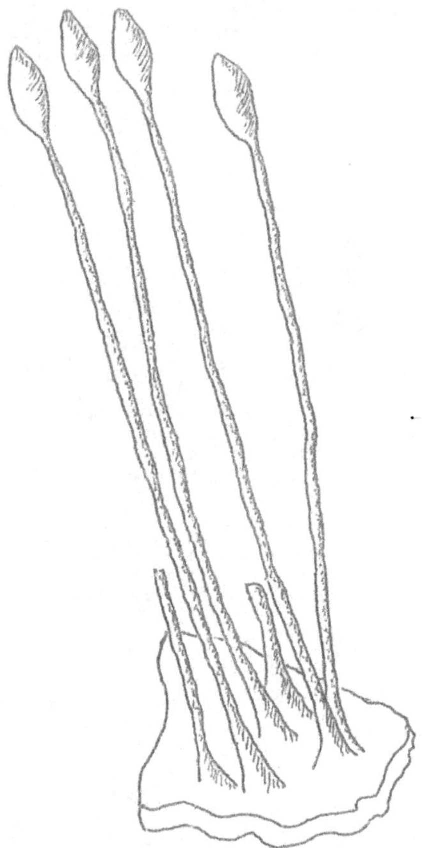

FIGURE 2.3 The fossil bryophyte *Sporogonites exuberans*, an ancient moss species. The numerous sporophytes were about twenty millimeters (0.8 inch) tall. *Credit*: Modified and redrawn from Andrews (1960).

In addition to the fragmentary fossil evidence for the first land plants that is found in the Ordovician, there exists equally sparse fossil evidence for another important terrestrial invader: the fungi. As mentioned in chapter 1, fossil hyphae and spores of eumycete glomalean fungi have been found in 460 million-year-old Ordovician strata. These fungi are particularly important in that they form "arbuscular mycorrhizal symbioses" with land plants;[16] that is, they grow both on the roots and within the root cells of land plants in a symbiotic relationship with the land plant, similar to that seen in the lichens that we considered in chapter 1—the fungus provides nutrients to the plant by dissolving rock upon which the plant is growing, and the plant provides hydrocarbons for the fungus to metabolize. The surface area of the mass of tiny hyphae of the fungus extending from the plant roots also help the plant absorb more water from the soil than would be possible by using the surface area of the roots alone. Almost all living vascular land plants have endomycorrhizal symbionts in their roots, and they are found in the tissues of many of the nonvascular bryophytes as well.[17]

The liverworts and hornworts do not have roots, thus could not have formed fungus–root symbioses (mycorrhizae) in the Ordovician. Thus it is not clear if the symbiotic relationship between the glomalean fungi and the early land plants had yet formed in the Ordovician, although it is known that some modern hornworts do form symbiotic associations with fungi that grow in the thallus of the plant.[18] At present, the oldest definitive fossil evidence for the existence of endomycorrhizal symbionts in land plants is found in the Early Devonian.[19]

Land Plants Emerge from the Mists: Silurian Macrofossils

Why did it take so long for macrofossils of the earliest land plants to appear in the strata of the Earth? The pattern of land plant diversification seen in the macrofossil record is more of an ecological phenomenon than an evolutionary one. We know this to be true because macrofossils of more derived species appear in the fossil record before fossils of the more plesiomorphic species. Phylogenetic analyses of land plant evolution reveal that simple land plants with a single vertical stem (Anthocerophyta and Hemitracheophyta; table 2.1) evolved to more complex land plants with multiple-branched stems, the Polysporangiophyta, which in turn gave rise to plants that evolved water-conducting tracheids, the Tracheophyta, for water transport up the stem and these in turn produced the first plants with simple leaves, the Eutracheophyta (table 2.1). In the fossil record, however, the eutracheophytes appear first, not last.

In the Homerian Age of the Middle Silurian (table 2.2), fossils of early eutracheophytes are represented by the basal lycophyte species

Cooksonia pertoni and *Baragwanathia longifolia*, and species of the more derived zosterophyllopsid lycophytes appear in the Pridolian Age of the Late Silurian.[20] Even this early in time, the floras of the Earth appear to have differentiated into four geographic groups, with the zosterophyllopsids being most abundant in the equatorial regions.[21] Because fossils of eutracheophyte species are present in Homerian strata, we know that the evolution of the polysporangiophytes and tracheophytes (table 2.1) must have occurred *before* the Homerian in the Middle Silurian. Yet the oldest yet discovered macrofossils of simple horneophyte species, which are plesiomorphic polysporangiophytes, do not appear until the very end of the Silurian (in the Pridolian), and the oldest yet discovered macrofossils of simple tracheophytes, the rhyniopsids, do not show up until much later in the Pragian Age of the Early Devonian. Why do fossils of descendant species (eutracheophyte lycophyte species) appear in the fossil record before their ancestors (polysporangiophytes and tracheophytes)?

We know that the fossil record is not an unbiased sample of past life on Earth; Charles Darwin devoted an entire chapter in his famous book *On the Origin of Species* to the subject of "On the Imperfection of the Geological Record."[22] The probability of a species making it into the fossil record is a function of three main factors: the species' anatomy, its population size, and its habitat. Let us consider each of these factors with respect to land plants. Plants have no hard, crystalline skeletal parts like the shells of marine animals or the bones of land animals. Animals, or at least the mineralized tissues of animals, have a high probability of preservation in the fossil record because they are not sources of food for other organisms and are resistant to chemical destruction. Anatomically, plant stems are made of cellulose and, as such, they are food for fungi. Thus plant tissues have a low probability of preservation in the fossil record. In addition, smaller land plants can be consumed by fungi much more quickly than larger plants, hence the fossil record is biased in favor of preserving large plants.

Next, even if a species possesses anatomical features that give it a high probability of preservation, it still will not make it into the fossil record if it has very small population sizes; that is, if it is rare in nature. This bias in the fossil record is purely a sampling phenomenon; that is, the more numerous the number of individuals a species has, the higher the probability that one of these individuals—and hence the presence of the species itself—will be preserved in the fossil record.

Last, the type of habitat in which the species lives is of major importance. Even if a species has anatomical characteristics that give it a very low probability of preservation in the fossil record, and even if it is rare, it still can be preserved if it lives in just the right environment, or if it is transported into that environment after death. A highly unusual environment that sometimes even allows the preservation of soft-tissue details in the fossil record is called a *Lagerstätte.* The analysis of unusual preservation in the fossil record was developed by German paleontologists, hence the German name by which it is known. The word may be roughly translated into English as a "preservation site," from the German verb *lagern,* to store up or preserve, and the noun *Stätte,* a place or a site (the plural form of the word is *Lagerstätten*).

Much of what we know about the delicate details of the evolution of early land plant tissues is due solely to their chance preservation in a few *Lagerstätten* randomly scattered in geologic time: the Middle Silurian Stonehaven Group in Scotland, the Late-Silurian-Pridolian Ludford Lane Bone Bed in Wales, the Early-Devonian-Pragian Rhynie Chert in Scotland, the Middle-Devonian-Givetian Gilboa Shale in New York State, and the Early-Carboniferous-Visean East Kirkton Limestone in Scotland. The Scottish Rhynie Chert and East Kirkton Limestone *Lagerstätten* in particular are hot spring deposits formed near active volcanic fumaroles, and they have exquisite preservation of delicate plant structures in their silica deposits and volcanic ashes. In summary, in the Silurian, land plant species of much larger size began to evolve, their populations became more numerous, and their geographic dispersal became much wider. As a direct result, numerous stems and body fragments of these species begin to appear in the sediments of Silurian strata. However, these stems and body fragments came from the larger and anatomically more robust lycophyte species, rather than the small and more fragile stems of the little horneophytes and rhyniopsids. Fortunately for us, horneophyte and rhyniopsid species are excellently preserved in the unusual Rhynie Chert *Lagerstätte* in the Early Devonian.

The Dawn of the Vascular Land Plants

The hornworts and mosses have simple, upright sporophytes. The next step in the evolution of land plants was the evolution of plants with sporophytes that were branched; that is, vertical stems that supported more than one sporangium. These plants are the polysporangiophytes, the most plesiomorphic of which are the now extinct horneophytes (table 2.1). We know from phylogenetic analyses that the horneophytes must have evolved in the early Silurian because more derived lycophyte eutracheophyte plants are present in the fossil record in the middle Silurian, as discussed in the previous section of the chapter. Yet the first glimpse we have of the horneophytes (fig. 2.4) comes some 25 million years later at the end of the Silurian, in the Pridolian, when the horneophyte species *Tortilicaulis transwalliensis, Caia langii,* and *Cooksonia hemisphaerica* appear in the fossil record at Dyfed, Wales.[23] The ancient horneophytes persisted in time until the Emsian Age of the Early Devonian, where the species *Horneophyton lignieri* and *Aglaophyton major* have been found.

After the evolution of erect, multi-branched stems (the Polysporangiophyta), the next step in the evolution of land plants was the evolution of the vascular plants, the Tracheophyta; that is, plants possessing water-conducting tracheids to transport water up the stem against the force of gravity. The most plesiomorphic of the tracheophytes are another extinct group, the rhyniopsids (table 2.1). As for the horneophytes, we know that the rhyniopsids must have evolved in the early Silurian before the appearance of the mid-Silurian-aged lycophyte fossils, but our first glimpse of the rhyniopsids (fig. 2.5) comes much later in the Pragian Age of the Early Devonian, where the group's namesake species, *Rhynia gwynne-vaughanii,* is preserved in the famous Rhynie Chert *Lagerstätte* in Scotland. Thus the first rhyniopsid macrofossils are found in the fossil record some 30 million years after the group evolved. This ancient group persisted even longer in time than the horneophytes as some rhyniopsid species, like *Stockmansella remyi,* have been discovered in strata as young as the Eifelian Age of the Middle Devonian.[24] And last, the excellent preservation of delicate biotic structures found in the Rhynie Chert *Lagerstätte* reveals to us that the symbiotic relationship between glomalean fungi and land plants had evolved, in that fossil

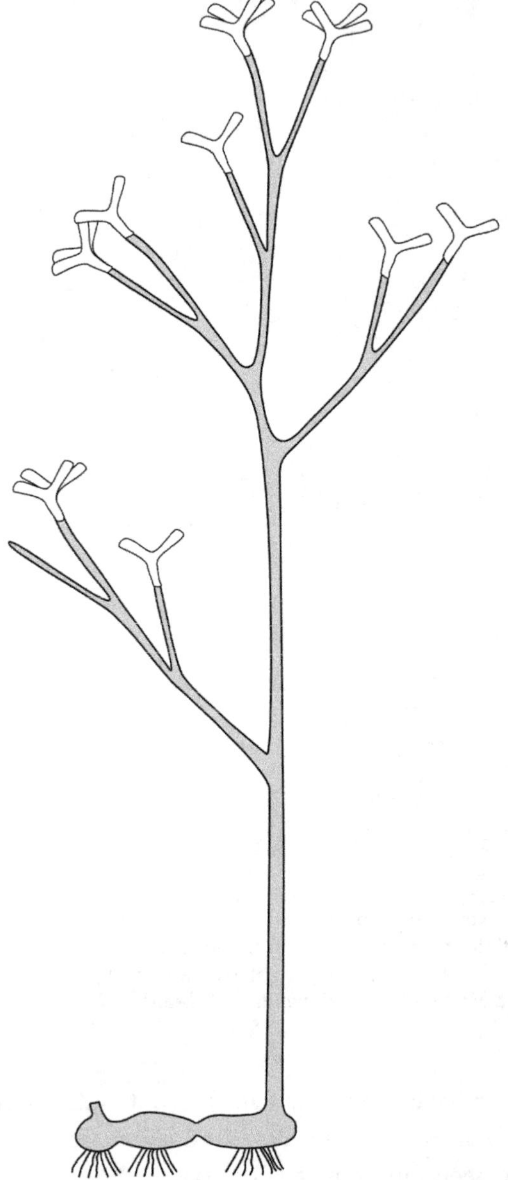

FIGURE 2.4 The extinct horneophyte *Horneophyton lignieri* possessed branched terminal sporangia on naked stems that grew from bulbous, corm-like structures. The living plant was about fifteen to twenty centimeters (six to eight inches) tall. *Credit*: Illustration by Peter Coxhead.

FIGURE 2.5 The extinct rhyniopsid *Rhynia gwynne-vaughanii* had numerous adventitious branches with irregular surface proturberances and possessed tracheids for fluid transport within the plant. The living plant was about eighteen centimeters (seven inches) tall. *Credit*: Modified and redrawn from Edwards (1980).

endomycorrhizae have been found in the root systems of the horneophyte *Aglaophyton major*.[25] As discussed above, fossils of the glomalean fungi have been found in strata as old as the Ordovician, thus the symbiotic association of these fungi with land plants probably occurred much earlier in time, in the Silurian or perhaps even in the Ordovician.

The basal, leafless tracheophytes gave rise to the more derived eutracheophytes, the first land plants to evolve leaf-like structures. The split

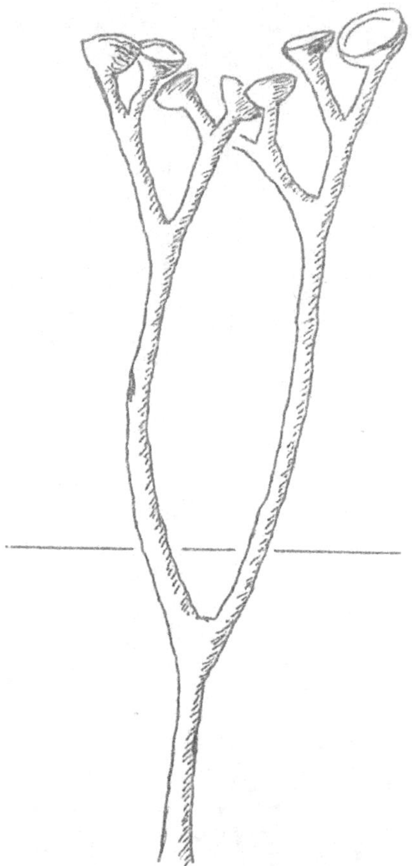

FIGURE 2.6 The basal lycophyte *Cooksonia pertoni*, an extinct species in the clade of the microphyll-leafed plants. The living plant possessed simple, bifurcating stems with disc-like terminal sporangia and was a little over six centimeters (two and a half inches) tall. *Credit*: Modified from *C. caledonica* and redrawn from Edwards (1970).

between the two major types of eutracheophytes, the lycophytes and the euphyllophytes (table 2.1), must have occurred by the Homerian Age of the Middle Silurian because fossils of basal lycophyte species are found in strata of that age. The microphyll-leafed vascular plants, the lycophytes, are abundantly represented in the Middle Silurian, as discussed in the previous section of the chapter. The basal lycophyte species *Cooksonia pertoni* (fig. 2.6) and *Baragwanathia longifolia* appear in

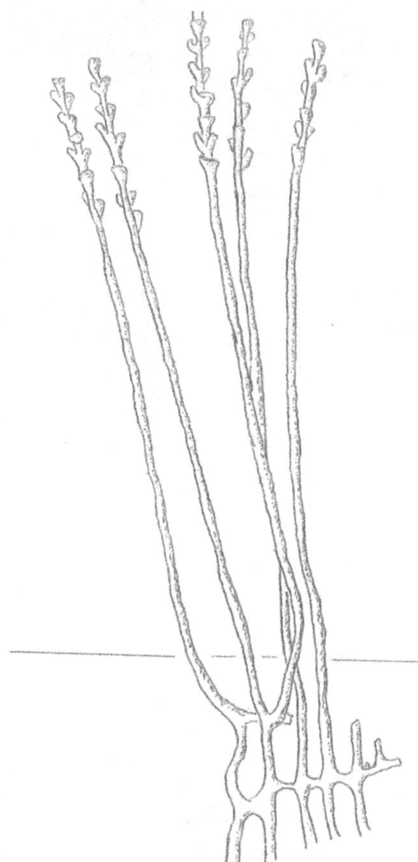

FIGURE 2.7 The extinct lycophyte *Zosterophyllum myretonianum*, a more derived species in the clade of the microphyll-leafed plants. It had numerous lateral sporangia on each vertical stem instead of a single terminal one. The living plant was about fifteen centimeters (six inches) tall. *Credit*: Modified and redrawn from Walton (1964).

the Homerian, and *C. cambrensis* appear in the Pridolian.[26] Fragments of an unidentified species of the more derived zosterophyllopsid lycophtes (table 2.1) have also been traced to the Pridolian, a species from the group's namesake genus *Zosterophyllum*.[27] More numerous zosterophyllopsid species (fig. 2.7) appear immediately later in the Lochkovian Age of the Early Devonian: *Zosterophyllum myretonianum*, *Z. rhenanum*, *Deheubarthia splendens*, and *Gosslingia breconenesis*.[28]

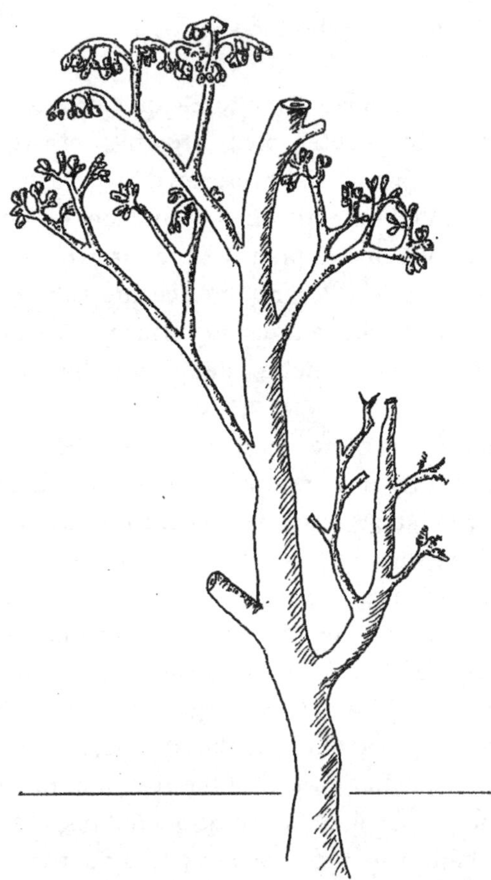

FIGURE 2.8 The basal euphyllophyte *Psilophyton dawsoni*, an extinct species in the clade of the macrophyll-leafed plants. It had a large, central axis with smaller lateral branches that terminated in clusters of sporangia. The living plant was about one meter (3.3 feet) tall. *Credit:* Modified and redrawn from Banks et al. (1975).

The megaphyll-leafed vascular plants, the euphyllophytes, must have evolved in the Homerian Age of the Middle Silurian at the same time as the first lycophytes,[29] but the first euphyllophyte macrofossils are not found (as yet) in the fossil record until the Early Devonian Pragian. These are the basal euphyllophyte species *Eophyllophyton bellum*, *Psilophyton crenulatum*, and *P. dawsonii* (fig. 2.8) from the Rhynie Chert *Lagerstätte* in Scotland.

The Evolution of Trees and the First Forests

Trees are a characteristic adaptation to life on land. As we considered at the beginning of the chapter, plant growth in a two-dimensional plane soon leads to crowding and overgrowth, where one plant shades out another plant in the competition for light from the sun, the energy source for the survival of the plant. When crowding occurs in the two-dimensional plane of the land surface the solution is to move into the third dimension above the land surface, to evolve plants with single vertical stems, then branched stems, and then branched stems with leaves.

Crowding will again become a problem when these types of plants cover the surface of the land. In human cities, when crowding occurs in a region of low buildings, the solution is to move even higher into the third dimension above the land surface—to construct towering skyscraper buildings; in the case of plants, the solution is to evolve trees. The force of gravity is still to be reckoned with, and new support structures had to be evolved to create a massive central structure, the tree trunk, that rises vertically from the land surface. At some distance above the ground, branches extend out from the tree trunk to capture as much sunlight as possible for the survival of the tree and, to ensure the survival of the species, to facilitate fertilization and dispersal of the tree's offspring. All of these structures are heavy, and the tree trunk must be strong enough to support them without breaking or bending. Thus it is astonishing that it can be proved that nine separate phylogenetic lineages of plants independently and convergently evolved the tree form, each with its unique solution to the problem of growing a sufficiently strong central trunk structure.[30]

In land plant evolution, the first trees appeared in the Givetian Age of the Middle Devonian, around 390 million years ago. The trees in those forests were very different from those that constitute the forests of Earth today. The famous Gilboa fossil-forest *Lagerstätte* in New York State gives us a glimpse of the earliest forests on Earth. Numerous upright trunks of trees are preserved as sandstone casts, each one broken off at about one meter (3.3 feet) above the ancient ground surface. The trunks range in size from about a half-meter to a meter in diameter with a wider, bulbous base with numerous small anchoring roots, revealing that these earliest trees had no taproot system for anchoring. The morphology of the upper portions of the trees was completely unknown until the spectacular discovery in 2007 of fossils of the complete tree, logs that had fallen over and were preserved horizontally, in rock outcrops only 13 kilometers (eight miles) east of the Gilboa fossil forest.[31] Some of these trees were over eight meters (26 feet) tall and, most surprisingly, had no laminar leaves or horizontal branches. Instead, they were topped with a crown resembling the hairs in a shaving brush, numerous offshoots that gradually opened outward and were held at an acute angle to the axis of the trunk stem (fig. 2.9). The complete tree *Wattieza (Eospermatopteris) erianus* belongs to an ancient group of fern-like plants that is now extinct, the Cladoxylopsida (table 2.1). *Wattieza (Eospermatopteris) erianus* had no woody tissue, unlike our modern lignophyte trees, and the tree was probably hollow and reed-like and may have grown rapidly, like a weed.[32]

The extinct cladoxylopsids belong to the euphyllophyte clade of the Moniliformopses (table 2.1), a clade that includes the living horsetails (equisetophytes) and ferns (filicophytes).[33] Both the horsetails and the ferns independently and convergently evolved trees, but not until the Carboniferous. The other branch of the eutracheophyte clade, the lycophytes, also convergently evolved trees (color plate 5). The oldest yet known lycophyte tree is also found in the Givetian Gilboa fossil forest *Lagerstätte*,[34] and they eventually evolved towering scale trees in the Carboniferous. Most of these ancient tree groups are now extinct; among the moniliformopses, only the ferns still produce trees, such as the Australian tree fern *Cyathea cooperi*.[35]

In the modern world, what we normally think of as a "tree" is a woody seed plant, a spermatophyte lignophyte (table 2.1). The euphyllophyte clade of the lignophytes evolved in the Emsian Age of the Early Devonian, yet the first of the woody plants still reproduced with spores, not seeds. These early lignophytes are the aneurophytaleans and the archaeopteridaleans, both groups of which are now extinct (table 2.1). The oldest aneurophytalean lignophyte, *Rellimia* sp., comes from Morocco, which indicates that the lignophytes, like the land plants themselves, appear to have evolved first in Gondwana.[36]

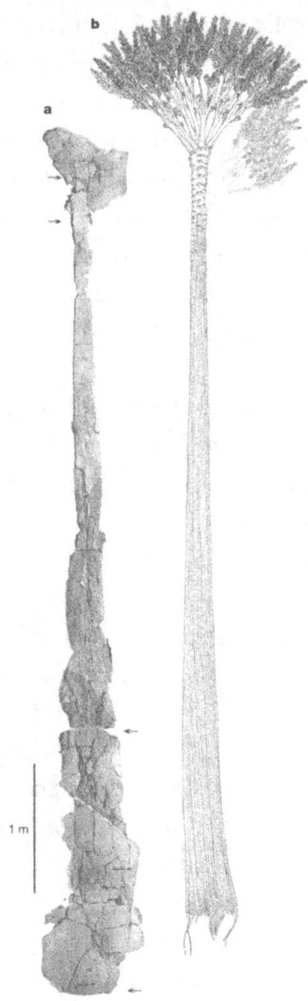

FIGURE 2.9 The extinct cladoxylopsid fern-like tree *Wattieza (Eospermatopteris) erianus*: (a) the reassembled tree fossils, and (b) a reconstruction of the living tree. The living tree was about eight to ten meters (26 to 33 feet) tall. *Credit*: Macmillan Publishers Ltd: *Nature* (Stein et al., 2007), copyright © 2007. Reprinted with permission.

The archaeopteridalean lignophytes in particular produced large trees in the Frasnian Age[37] of the Late Devonian (table 2.2), such as *Archaeopteris hibernica*, which towered 30 meters (100 feet) into the sky. The giant *Archaeopteris* trees (color plate 5) still reproduced with spores, however, and were generally confined to wet coastal areas and lowland areas with meandering rivers and scattered lakes. In these areas they rapidly spread, and by the middle of the Frasnian vast areas of the Earth were covered with *Archaeopteris* forests.

The Final Conquest of Land: Evolution of Seed Plants

The end-Frasnian extinction in the Late Devonian triggered a severe loss of diversity in land plants, resulting in a minimum diversity level that persisted for some seven million years into the Famennian Age (table 2.2). These facts will be considered in detail in chapter 4; of note here is the fact that the land plants recovered in diversity in the late Famennian, and they also produced new adaptive innovations, the most important of which was the evolution of the seed plants, the spermatophytes (table 2.1). The key adaptation of the seed plants is their freedom from needing water in reproduction, unlike the spore-reproducing plants, and the spermatophytes could now colonize the dry highlands and mountains that were out of reach for non-spermatophyte plants. The evolution of seed plants was the ecological equivalent of the evolution of the amniote animals (modern reptiles, birds, and mammals), which also evolved adaptations that freed them from needing water in reproduction. It is these two groups, the seed plants and the amniote animals, that became the victorious plant and vertebrate invaders of the dry land.

The most ancient, undisputedly dated seeds come from late Famennian strata[38] in the Appalachian mountains of West Virginia, the species *Elkinsia polymorpha* (fig. 2.10), and in the Ardennes in Belgium, the species *Moresnetia zalesskyi* and *Dorinnotheca streeli*.[39] A single occurrence of the seed species *Moresnetia zalesskyi* has been reported from strata in Russia that have been variously dated as middle to late Frasnian,[40] however, so seed plants may have evolved before the end-Frasnian extinction. If so, they did not proliferate until long after the catastrophic effects of that extinction had waned—the passing of the Famennian Gap in plant evolution, an interval of time that we will consider in detail in chapter 4. Unlike the earliest land plants and the earliest lignophytes, which appeared first in Gondwana (as discussed above),

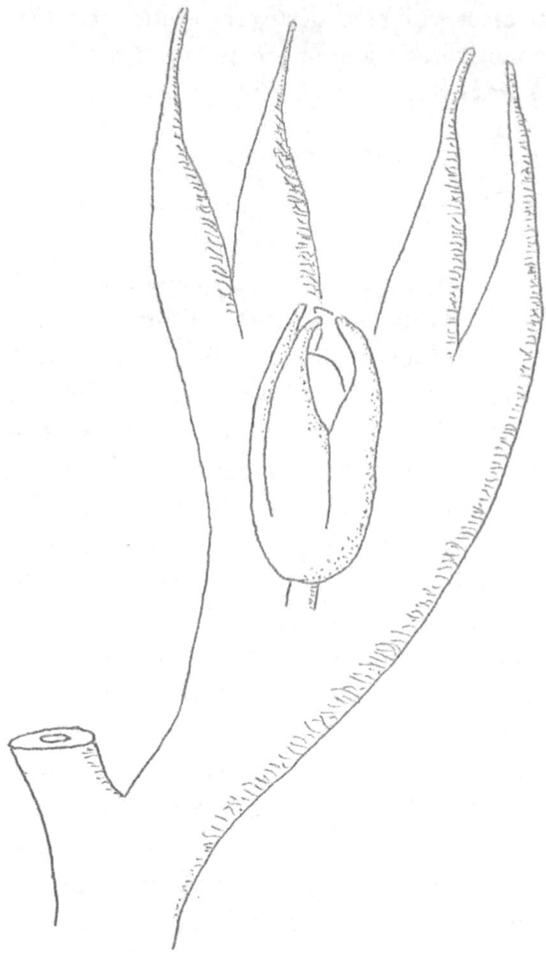

FIGURE 2.10 Branch-tip cupule of the Famennian spermatophyte *Elkinsia polymorpha*. The cupule is about 1.5 centimeters (a little over a half of an inch) long, and the central ovule seed, containing the megaspore, is about 6.5 millimeters (a quarter of an inch) long. *Credit*: Modified and redrawn from Rothwell et al. (1989).

the earliest seed plants appear to have evolved north of Gondwana on the Laurussian continent, a new continental landmass that had been formed in the Silurian by the collision and fusion of Laurentia, Baltica, and Avalonia (color plate 4).

Seed plants evolved rapidly in the late Famennian, and fifteen species of spermatophyte seeds are now known, divided into five different morphotypes: the *Moresnetia, Aglosperma, Dorinnotheca, Condrusia,* and *Warsteinia* types. These early seed plants have been proposed to have been opportunistic species, proliferating in disturbed habitats much like modern ferns do, particularly in the understory of the towering non-spermatophyte lignophyte *Archaeopteris* forests.[41]

The first peat swamps also appeared in the Famennian and would be dominated by *Lepidodendropsis* lycophyte trees later in the Early Carboniferous. Outside of the Famennian peat swamps, in coastal and lowland areas, forests were still dominated by species of non-spermatophyte lignophyte trees: *Archaeopteris hibernica, A. halliana, A. macilenta, A. obtusa, A. sphenophyllifolia,* and the understory was dominated by nearly monospecific stands of the filicophyte fern *Rhacophyton ceratangium.* Ground cover was provided by newly evolved equisetophytes that had adopted a vining habit, *Sphenophyllum subtenerrimum.* Only about 20 percent of the flora were the small, newly evolved spermatophytes in the late Famennian—only in Ireland were the flora already dominated by species of seed plants.[42] The great *Archaeopteris* trees would not survive the end-Famennian extinction, and the seed plants now dominate all the forests of Earth.

The Stage Is Set: The Dawn of the Animal Invasion

By the end of the Late Devonian, the plant invasion of dry land was complete: the seed plants had evolved. The first land plants evolved some 468 million years ago, in the Middle Ordovician. The first seed plants appeared about 364 million years ago, near the end of the Late Devonian; thus, it took plants about 104 million years to finally conquer the terrestrial realm. The plants set the stage for the animal invasion by providing a reliable source of food in the hostile world of dry air; they contributed the first step in the assembly of the trophic pyramid of the terrestrial ecosystem that was to evolve.

The amniote animals, the victorious vertebrate invaders of dry land, and the winged insects, the victorious arthropod invaders of dry land, would not evolve until the end of the Early Carboniferous, as will be discussed in chapter 7. The long saga of the animal invasion of land will be examined in the next four chapters of the book. Leaving the sea and invading the land was not easy for the animals, and twice the vertebrates, our ancestors, almost did not make it.

CHAPTER 3

The First Animal Invasion

Constructing the Higher Levels of the Terrestrial Trophic Pyramid

In the previous chapter, we considered what we know about the land plants that built the foundation of the trophic pyramid of the terrestrial ecosystem. In this chapter, we will consider what we know about the construction of the higher levels of the trophic pyramid: the evolution of the terrestrial herbivorous animals, which live by eating the plants in the foundation of the trophic pyramid, and the evolution of the terrestrial carnivorous animals, which live by eating the herbivores.

At first glance, the construction of the terrestrial trophic pyramid would seem to be a straightforward evolutionary progression: the plants evolve to invade the land, then the herbivorous animals evolve to invade the land when the plants, their source of food, have established a beachhead; then the carnivorous animals invade the land when the herbivores, their source of food, have established themselves. However, as we shall see in this chapter, the terrestrial ecosystem did not evolve in that way. Most surprisingly, the earliest land animals generally did not eat the living plants because many of the earliest animal invaders were carnivores.

Animals Emerge from the Seas

In our day-to-day lives, almost all of the many animal species we see around us belong to only two phylogenetic lineages: the vertebrates and the arthropods. Frogs croaking in the pond, lizards scurrying across the rock, horses galloping in the meadow, dogs barking down the road, and birds soaring through the sky all are descendants of a single clade of vertebrate invaders. Likewise, centipedes zipping across the cellar floor, spiders constructing their intricate webs in the hedgerow, ants traveling single-file along the sidewalk, and mosquitoes humming around our heads in the evening all are descendants of a single clade of arthropod invaders.

There are more clades of terrestrial animal invaders than just these two, but they are small and usually cryptic, so we do not often see them or are aware that they are there. In the oceans, the sponges are the simplest forms of animals[1] (table 3.1) that still survive from the origin of the metazoa in the Neoproterozoic, which we considered in chapter 1. The comb-jellies and jellyfishes are structurally more complex, but it is the clade of the bilaterian animals—animals that evolved a bilaterally symmetric axis, with a distinct anterior and posterior end—that has produced the species richness that we see in the animal world today (table 3.1). The bilaterians were present in the Ediacaran fauna and split into two clades distinguished by their developmental type: the protostomes have protostomous development, and the deuterostomes have deuterostomous development. Curiously, in their long evolutionary history, the deuterostomes have produced only one clade of terrestrial invaders: the vertebrates (table 3.1).

The protostomes produced six clades of animals with species that left the seas and invaded the land (table 3.1), and most of these are worms or worm-like animals: roundworms, hairworms, flatworms, ribbon worms, earthworms, and rotifers. All of these are generally small; only the earthworms attain lengths up to 20 centimeters (about eight inches) and are noticeable to us. Worms, being small and cylindrical in geometry, have very large surface areas and small internal volumes and thus can breathe air directly across their skin membranes without the need for more complicated gas-exchange structures like lungs (which we will

TABLE 3.1 Phylogenetic classification of animals.

Metazoa (multicellular choanozoa, the animals; from table 1.2)

– Placozoa

– Demospongiae (demosponges)

– Hexactinellida (siliceous sponges)

– Calcarea (calcareous sponges)

– Eumetazoa

– – Ctenophora (comb-jellies)

– – Cnidaria (jellyfish, corals, and kin)

– – Bilateria

– – – Protostomia

– – – – Lophotrochozoa

– – – – – Lophophorata (moss-animals and lampshells)

– – – – – Eutrochozoa

– – – – – – Syndermata (terrestrial rotifers and acanthocephalans)

– – – – – – Spiralia

– – – – – – – Entoprocta (kamptozoans)

– – – – – – – Annelida (terrestrial earthworms and leeches)

– – – – – – – Sipuncula (bristleworms)

– – – – – – – Mollusca (marginal-marine chitons and land snails)

– – – – – – – Parenchymia (terrestrial flatworms and ribbon worms)

– – – – Cuticulata

– – – – – Gastrotricha (gastrotrichs)

– – – – – Ecdysozoa

– – – – – – Introverta

– – – – – – – Cephalorhyncha (proboscis worms and kin)

– – – – – – – Nematozoa (terrestrial roundworms and hairworms)

– – – – – – **Panarthropoda** (terrestrial velvet worms and true arthropods; see table 3.2)

– – – – Chaetogratha (arrow worms)

– – – Deuterostomia

– – – – Echinodermata (starfish, sand dollars, sea urchins, and kin)

– – – – Pharyngotremata

– – – – – Hemichordata (acorn worms and sea angels)

– – – – – Chordata

– – – – – – Urochordata (sea squirts)

– – – – – – Myomerozoa

– – – – – – – Cephalochordata (lancelets or "headless fish")

– – – – – – – Craniata

– – – – – – – – Conodonta† (conodonts)

– – – – – – – – Myxinoidea (hagfishes)

– – – – – – – – **Vertebrata** (terrestrial vertebrate animals; see table 3.5)

Source: Modified from the 2006 phylogenetic classification of Lecointre and Le Guyader for living animals, and from the 2005 phylogenetic classification of Benton for extinct animals.

Animal clades (Metazoa) that invaded the land are underlined, and the two major groups of terrestrial invaders—the Panarthropoda and Vertebrata—are marked in bold-faced type.

† extinct taxa.

consider later in the chapter). Many of these small animals live in moist soil, thus are not severely confronted with dehydration, a serious problem in the dry air of the terrestrial realm. Even though they are small and cryptic, living underground, annelid earthworms and nematozoan roundworms are significant components of the terrestrial ecosystem in that they play a very important role in soil aeration and nutrient recycling, both of which assist the growth of the land plants we considered in chapter 2. As these terrestrial worms all have soft tissue, their probability of preservation is near zero. Apart from *Lagerstätten* with marine fossils of their ancestors, the terrestrial fossil record of these groups is almost nonexistent. Thus we do not know precisely when these animal groups emerged from the seas.

The protostome clade of the molluscs has produced another group of terrestrial invaders that are larger and dwell on the surface of the land and are thus familiar to us: the land snails (table 3.1). Most of the numerous mollusc species are marine animals but a few, like some species of the bivalve molluscs, were able to leave saltwater and invade freshwater in the Devonian. The polyplacophoran molluscs, the chitons, also can tolerate some exposure to dry air and have invaded the intertidal zones of the oceans, grazing on algae on rocks or preying on worms. If they begin to lose too much water to the dry air, they will clamp their shells tightly to rocks and await the arrival of the high tide when they are again immersed in water. Some snails, like the limpets, also use this mode of life in the intertidal zone. The land snails have left the water entirely, however, and use complex behavioral adaptations—rather than morphological—to minimize the loss of water from their body tissues.[2] They avoid being exposed to the sun and seek to remain in moist, shaded areas. If the climate becomes too hot and dry, they burrow into the soil and aestivate (which our vertebrate cousins, the lungfishes, also do), slowing down their metabolisms and awaiting the return of the rains and cooler temperatures. Some species climb shrubs and trees during the day to avoid higher ground temperatures, and come down to hunt for food in the cooler evening and night.

Land snails with shells convergently evolved lung structures for breathing dry air: they lost the gill structures in the mantle cavity of their shells and instead formed a sac of the mantle tissue with an opening to the outside air. This mantle tissue sac is used for gas exchange and is in essence a functioning lung. Other land snails, like the slugs, have lost the shell entirely and breathe across their skin surface area.

Many land snail species retain their shells and thus have a fairly high probability of preservation in the fossil record. The oldest known fossils of the land snails are only Jurassic in age, however, thus the land snails are latecomers to the terrestrial realm and did not participate in the Paleozoic establishment of the terrestrial ecosystem. Although interesting animals, they play a relatively minor role in the terrestrial ecosystem even today. For humans, some land snails are pests, in that they feed on garden plants, while others are food, in that individuals of the land snail species *Helix pomatia*—escargot—are delicious when cooked in garlic and parsley butter.

Of all the protostome clades that have emerged from the seas and invaded dry land, it is the clade of the panarthropods that has produced the incredible species richness that we see in terrestrial animals in the modern world. There are many, many more species of living terrestrial insects than there are of vertebrate animals. To further add to their evolutionary accomplishments, the panarthropods emerged from the seas first, long before the vertebrates, as we shall see in the next section of the chapter.

The First Major Wave of Animal Invaders

The first major animal group to make a concerted effort to invade the land was the Panarthropoda, ancestors of modern terrestrial velvet worms, millipedes, centipedes, spiders, and insects (table 3.2). Only much later did vertebrate animals, our ancestors, begin that process, and in part our attempt to invade the land was assisted by these arthropod forerunners. They paved the way for us, chiefly by being a source of food!

The earliest arthropod invaders had a major advantage when they first ventured out of the oceans onto the dry land: they had legs. That is, their arthropod ancestors had already evolved walking structures while still living in the marine realm. Thus it was easy (relatively) to emerge from the water and walk about on dry land. The only major difference was that they were now confronted with the force of gravity: they were heavier on dry land, and walking was not as easy as it was in the water. Much later in their evolution on land the arthropods would evolve a radical new method of locomotion: flight. The flying insects took to the skies of the Earth, and left the slow walkers confined to the land surface. We will consider the evolution of this key locomotory adaptation in detail in chapter 7.

The major problem that the new invaders faced on land was how to breathe when surrounded by the thin, dry air of the atmosphere instead of the waters of the oceans. To reduce water loss while breathing, the early arthropod invaders evolved a developmental process of folding their aquatic respiratory surfaces down into their bodies, producing a series of small tubes—the tracheae—connected to pore-like openings—the spiracles—in their exoskeletons.[3] The tracheae allowed gas exchange to occur between the internal tissues of the animal and the atmosphere, to absorb oxygen and shed carbon dioxide. More advanced arthropods evolved a system to close the spiracles to further minimize water loss to the atmosphere, and to open the tracheal tubes only when gas exchange was needed.[4] What is astonishing is that it can be proved that this tracheal system of breathing evolved independently and convergently in five separate phylogenetic lineages of panarthropod invaders.[5]

The most basal organisms of the Panarthropoda are the onychophorans, or velvet worms (table 3.2). Although they have no mineralized tissues, and thus have a very low probability of preservation in the fossil record, we know they were present in the oceans of the Earth in the Cambrian due to their presence in the Burgess Shale *Lagerstätte.* The velvet worms have somewhat stiff, chitinous cuticle and short, stumpy legs (color plate 6). These little, knob-like legs are arranged in two rows on the ventral side of the worm and are unjointed. The true arthropods have exoskeletons of rigid chitinous cuticle and long appendages that are jointed. Arthropods are divided into two major clades, the Mandibulata and the Cheliceriformes, both of which were present in the Cambrian oceans (table 3.2).

Trackways made by small, multi-legged organisms appear in marginal marine strata in the Ordovician, indicating that some type of marine organism ventured out onto tidal flats and dry land. These indistinct

TABLE 3.2 Phylogenetic classification of living panarthropods.

Panarthropoda (from table 3.1)
- Onychophora (velvet worms)
- Arthropoda (true arthropods)
- - Mandibulata
- - - Myriapoda (millipedes and centipedes)
- - - Pancrustacea (crustaceans and hexapods)
- - Cheliceriformes
- - - Pycnogonida (sea spiders)
- - - Chelicerata (horseshoe crabs, scorpions, spiders, and mites)

Source: Modified from the 2006 phylogenetic classification of Lecointre and Le Guyader.

trackways may have been made by the knobby appendages of velvet worms or by the numerous feet of true arthropods, the millipedes. Trackways that are very similar to those made by millipedes are known from the Sandbian Age of the Late Ordovician.[6] These are placed in the ichnospecies *Diplichnites gouldi* and *Diplopodichnus biformis* and are believed to have been made by eoarthropleurid millipedes,[7] a group of early millipedes that are now extinct (table 3.3). These two species are referred to as "ichnospecies," or trace species, because the actual millipede species that made the trackways is unknown. In the Carboniferous, however, it has been proved that the trackway of ichnospecies *Diplichnites cuithensis* was made by the giant arthropleurid millipede *Arthropleura armata*, which sometimes grew to 2.5 meters (eight feet) in length.

It is not until the Middle Silurian, some 33 million years later than the earliest trackways found in the Late Ordovician, that the oldest yet known body fossils of actual millipede species are found.[8] These are the zosterogrammid *Casiogrammus ichthyeros* and the cowiedesmid *Cowiedesmus eroticopus*, both members of millipede groups that are now extinct (table 3.3). Shortly thereafter, in the Pridolian Age of the Late Silurian, the eoanthropleurid millipede *Eoarthropleura ludfordensis* appeared. Fossils of the other major myriapod arthropod group, the centipedes, also appeared in the Pridolian in the scuterigeromorph species *Crussolum crusserratum*.

The millipedes and centipedes belong to the mandibulate clade of the arthropods (table 3.3). The other major clade of arthropods, the cheliceriforms, also invaded the land during the Silurian. The scorpion species

TABLE 3.3 Arthropod invaders in the Silurian and Devonian that are known from the fossil record.

Panarthropoda
- I. Onychophora (velvet worms)
- II. Arthropoda (true arthropods)
  - A. Mandibulata
    - 1. Myriapoda
      - a. Diplopoda (millipedes)
        - (1). Zosterogrammida†
        - (2). Cowiedesmida†
        - (3). Eoarthropleurida†
        - (4). Archidesmida†
        - (5). Juliformia
        - (6). Microdecemplicida†
      - b. Chilopoda (centipedes)
        - (1). Scutigeromorpha
        - (2). Devonobiomorpha†
    - 2. Pancrustacea
      - a. Hexapoda
        - (1). Collembola (springtails)
        - (2). Insecta (insects)
  - B. Cheliceriformes
    - 1. Chelicerata
      - a. Arachnida
        - (1). Scorpiones (scorpions)
        - (2). Trigonotarbida†
        - (3). Opiliones (harvestmen)
        - (4). Phalangiotarbida†
        - (5). Actinotrichida (mites)
        - (6). Pseudoscorpiones (false scorpions)
        - (7). Uraraneida†

Note: The numbered arthropod groups are listed in the order in which they first appear in geologic time.

† extinct taxa.

Dolichophonus loudonensis is traced to the Middle Silurian,[9] and the spider-like trigonotarbid species *Palaeotarbus jerami* appeared in the Pridolian of the Late Silurian.[10]

The arthropods solidified their hold on the land in the Early Devonian. The myriapod arthropods continued to diversify, with the archidesmid millipede species *Archidesmus macnicoli* appearing in the Lochkovian[11] and the juliformian millipede species *Sigmastria dilata* in the Pragian.[12] In the Middle Devonian, both the microdecemplicid millipede species *Microdecemplex rolfei* and the devonobiomorph centipede species *Devonobius delta* appeared in the Givetian.[13]

The arachnids likewise proliferated. The harvestmen species *Eophalangium sheari* and the phalangiotarbid *Devonotarbus hombachensis* appear in the Pragian,[14] as do five separate species of mites:[15] *Protospeleorchestes pseudoprotacarus, Protacarus crani, Pseudoprotacarus scoticus, Palaeotydeus devonicus,* and *Paraprotacarus hirsti.* In the Middle Devonian, the false scorpion *Dracochela deprehendor* and the spider-like uraraneid *Attercopus fimbriunguis* appear in the Givetian.[16]

In the mandibulate clade of the arthropods, two more evolutionary events of major significance occurred in the Early Devonian: the invasion of the land by the pancrustaceans and the evolution of the first hexapods (table 3.3). The springtail species *Rhyniella praecursor* and the oldest known insect, the rhyniognath species *Rhyniogratha hirsti,* appeared in the Pragian.[17] Last, fragments of another insect, the archaeognath bristletail *Gaspea paleoentognathera,* have been reported from the Emsian of Canada.[18]

Thus, by the beginning of the Late Devonian, the three major groups of modern terrestrial arthropods—the myriapods, hexapods, and arachnids—were well established on land (table 3.3). The evolution of six major groups of millipedes, two major groups of centipedes, seven major groups of arachnids, and the appearance of the very first insects—the most successful and diverse group of arthropods ever to evolve in Earth's history—within the 37-million-year span of time from the Middle Silurian to Middle Devonian is impressive indeed. However, it turns out that what the arthropod invaders actually accomplished is even more impressive than what is seen in table 3.3.

If millipede trackways on land first appear in the fossil record of the Late Ordovician, why is that the oldest body fossils of the first millipedes are only found (as yet) in the Middle Silurian, some 33 million years later? As discussed for the land plants in chapter 2, the pattern of diversification seen for the terrestrial arthropods in the fossil record is more of an ecological phenomenon than an evolutionary one. And, just as for the land plants, we know this to be true because fossils of more derived species of arthropods appear in the fossil record before the more plesiomorphic ones!

A phylogenetic analysis of the full spectrum of mandibulate and cheliceriform arthropods participating in the invasion of land is given in table 3.4. In the fossil record, the zosterogrammid and cowiedesmid millipedes appear first, in the Middle Silurian. Yet it can be seen in table 3.4 that the zosterogrammids and cowiedesmids are more derived chilognath millipedes, and that they actually evolved after the more plesiomorphic microdecemplicid millipedes. The microdecemplicid millipedes appear in the fossil record some 37 million years later than the zosterogrammids and cowiedesmids, in the Givetian Age of the Middle Devonian—yet they evolved *before* the zosteogrammids and cowiedesmids! Likewise, in the fossil record, the juliformian millipedes appear in the Pragian Age of the Early Devonian, yet it can be seen in table 3.4 that the juliformian millipedes are the most highly derived millipedes to have evolved up to the dawn of the Late Devonian. That is, the juliformians appear in the fossil record some 20 million years *before* their ancestral cousins the microdecemplicids do! Thus the sequence of appearance of the arthropod invaders seen in the fossil record (table 3.3) does not reflect the actual evolutionary appearance of these arthropods on Earth, as revealed by phylogenetic analyses of the arthropod lineage (table 3.4).

Consider another example: in the fossil record (table 3.3), the scutigeromoph centipedes appear first (in the Pridolian of the Late Silurian), followed by the devonobiomorph centipedes (in the Givetian of the Middle Devonian). It turns out that the scutigeromorphs did indeed evolve first, and the devonobiomorphs later (table 3.4). There is a problem, however; the lithobiomorph and craterostigmomorph centipedes evolved *before* the devonobiomorphs, yet these two centipede groups have not yet been found in the fossil record of the Silurian and Devonian. We refer to such groups as "ghost" taxa; that is, the lithobiomorph and craterostigmomorph centipedes are "ghost" arthropod groups—we know they evolved sometime between the Pridolian of the Late Silurian (when the scutigeromorphs appeared) and the Givetian of the Middle Devonian (when the devonobiomorphs appeared), but we do not know when. They are present somewhere on Earth during this time interval, but we do not see them in the fossil record. They are ghosts.

TABLE 3.4 Phylogenetic classification of the arthropod invaders.

Panarthropoda (from table 3.2)
- Onychophora (velvet worms)
- [Tardigrada]
- Arthropoda (true arthropods)
- - Mandibulata
- - - Myriapoda
- - - - Progoneata
- - - - - [Symphyla]
- - - - - Dignatha
- - - - - - [Pauropoda]
- - - - - - Diplopoda (millipedes)
- - - - - - - Pencillata
- - - - - - - - Arthropleuridea†
- - - - - - - - - **Eoarthropleurida**†
- - - - - - - unnamed clade
- - - - - - - - **Microdecemplicida**†
- - - - - - - - Chilognatha
- - - - - - - - - **Zosterogrammida**†
- - - - - - - - - [Pentazonia]
- - - - - - - - - unnamed clade
- - - - - - - - - - Archipolypoda†
- - - - - - - - - - - **Cowiedesmida**†
- - - - - - - - - - - **Archidesmida**†
- - - - - - - - - - Helminthomorpha
- - - - - - - - - - - [Pleurojulida†]
- - - - - - - - - - - [Colobognatha]
- - - - - - - - - - - Eugnatha
- - - - - - - - - - - - **Juliformia**
- - - - Chilopoda (centipedes)
- - - - - **Scutigeromorpha**
- - - - - Pleurostigmorpha
- - - - - - [Lithobiomorpha]
- - - - - - Phylactometria
- - - - - - - [Craterostigmomorpha]
- - - - - - - **Devonobiomorpha**†
- - - Pancrustacea
- - - - Hexapoda
- - - - - Entognatha
- - - - - - [Diplura]
- - - - - - Elliplura
- - - - - - - [Protura]
- - - - - - - **Collembola** (springtails)
- - - - - **Insecta** (insects)

(*continued*)

TABLE 3.4 *(continued)*

------ Rhyniognatha†
------ Archaeognatha (bristletails)
-- Cheliceriformes
--- Chelicerata
---- Arachnida
----- Stomothecata
------ **Scorpiones** (scorpions)
------ **Opiliones** (harvestmen)
----- **Phalangiotarbida**†
----- Haplocnemata
------ **Pseudoscorpiones** (false scorpions)
----- [Palpigradi]
----- Pantetrapulmonata
------ **Trigonotarbida**†
------ Tetrapulmonata
------- **Uraraneida**†
----- Acaromorpha
------ **Actinotrichida** (mites)

Source: Phylogenetic data modified from Grimaldi and Engel (2005), Lecointre and Le Guyader (2006), Sierwald and Bond (2007), Shear and Edgecombe (2010), and Dunlop (2010).

Note: Fossil arthropod groups listed in table 3.2 are indicated here in bold-faced type. "Ghost" arthropod lineages are enclosed in brackets (e.g., [Tardigrada]). See text for discussion.

† extinct taxa.

Examination of the phylogeny of the arthropods reveals quite a few ghost arthropod lineages, particularly in the lineage leading to the millipedes and within the millipede lineage itself (table 3.4). The living symphylan and pauropod myriapods evolved before the true millipedes (diplopods), yet they have no fossil record in the Silurian and Devonian (they do show up as fossils in some amber deposits in the Cenozoic).[19] We know the pentazonian, pleurojulidan, and colobognath millipedes all evolved before the highly derived juliformian millipedes, which appear in the fossil record in the Pragian Age of the Early Devonian, but we do not know when (pentazonian and pleurojulidan fossil species do appear in the Carboniferous, long after the juliformians who must have evolved after them). Even the hexapod lineage has its ghosts: the springtails were derived entognaths, thus the diplurans and proturans must have evolved before the collembolans (the diplurans appear in the fossil record in the Late Carboniferous).

The arachnids are much better represented in the fossil record in that only one ghost arachnid group exists, the palpigrades (table 3.4). The palpigrades are documented as being some of the most weakly sclerotized arachnids in existence, i.e., anatomically they have a very low probability of being preserved in the fossil record; they do appear in the fossil record, but only about five million years ago in the Cenozoic![20] Otherwise, the fragmentary nature of the fossil record is seen in the sequence of evolution within the arachnid groups. For example, the harvestmen, phalangiotarbids, and false scorpions all evolved before the trigonotarbids (table 3.4), but they all appear in the fossil record *after* the trigonotarbids (table 3.3).

The fossil record does reveal to us that the myriapod and arachnid arthropods were present on land by the Middle Silurian. The fact that highly derived cowiedesmid millipedes (table 3.4) are present in the Middle Silurian means that an explosive diversification of millipede lineages had to have occurred in the Silurian. Indeed, the myriapod specialists William Shear of Hampden-Sydney College and Gregory Edgecombe of the British Museum of Natural History[21] argue that the symphalans and pauropod myriapods, and the pencillatan, microdecemplicid, zosterogrammid, pentazonian, archipolypod, and helminthomorph millipede groups (table 3.4) all originated before the dawn of the Devonian. Likewise, the University of Berlin arachnid specialist Jason Dunlop[22] argues that all of the major groups of the arachnids (Stomothecata, Phalangiotarbida, Haplocnemata, Palpigradi, Pantetrapulmonata, Acaromorpha; table 3.4) originated in the Silurian.

Just as with the land plants in chapter 2, this mismatch in the sequence in which the arthropods evolved (table 3.4) and the sequence in which they appear in the fossil record is a function of the probability of preservation of the arthropod species, not their evolution. Anatomically, the myriapods, hexapods, and arachnids have exoskeletons made of chitin, a tough, horny polysaccharide. Polysaccharides are complex hydrocarbons, and hydrocarbons are food for numerous organisms, particularly bacteria and fungi. Thus the chitinous exoskeletons of arthropods do not have the high probability of preservation in the fossil record that the hard, crystalline calcite exoskeletons of molluscs and other invertebrates do. Much of what we see in the fossil record of the arthropods is due solely to their chance preservation in a few *Lagerstätten* randomly scattered in geologic time: the Middle-Silurian Stonehaven Group in Scotland and Late Silurian–Pridolian Ludford Lane Bone Bed in Wales, the Early Devonian–Pragian Rhynie Chert in Scotland and Emsian Alken-an-der-Mosel Shale in Germany, the Middle Devonian–Givetian Gilboa Shale and South Mountain Shale in New York State.[23] These scattered *Lagerstätten* are extremely important in that they give us hard proof that certain arthropod groups did exist on Earth at definite points in time, and these data points allow us to anchor in time the phylogenetic sequence of arthropod evolution (table 3.4).

Ecology of the Arthropod Invaders

The Rhynie Chert *Lagerstätte* is a hot spring deposit, thus we know that arthropods lived around the thermal springs and volcanic fumaroles that existed in eastern Scotland in the Pragian. Are we to conclude from this fact that arthropods preferred to live around hot springs? Clearly not—this habitat association is a preservational phenomenon, not an ecological one. Arthropods must have flourished in a broad spectrum of different habitats, just as they do today, but we do not find them in these habitats in the fossil record because their exoskeletons were not preserved. We are simply fortunate in that the silica deposits and volcanic ashes of the Rhynie Chert strata have given us a glimpse into the diversity of arthropod species present in this unique window of time in the Pragian, preserved for us by the unusual environmental conditions present in the *Lagerstätte.* Some of the other *Lagerstätten,* such as the Ludford Lane Bone Bed and the Alken-an-der-Mosel Shale, reveal to us that these early arthropods also liked to live in salt marshes, while others, such as the Gilboa Shale, give us a glimpse of arthropod ecosystems around freshwater ponds near rivers.[24]

The trophic structure of early arthropod ecosystems was not like that of the modern day. In modern terrestrial ecosystems, the autotrophic organisms, the plants, comprise the base of the trophic pyramid. These organisms synthesize the hydrocarbons that other organisms use for food. The herbivorous animals constitute the next level up in the trophic pyramid, as they are consumers of plants. As we saw in chapter 2, plants had firmly established a hold on the coastal and river-edge land areas of the Earth by the Silurian. Thus one would assume that the first animal invaders would be herbivorous to take advantage of this new source of food on the land. To the contrary, however, the earliest arthropod invaders appear to have been detritivores instead of herbivores. They did not eat living plants but rather consumed decaying dead plant material. These are the millipedes, which still occupy the detritivore niche today. Thus the numerous types of millipedes we considered in the previous section of the chapter were consuming dead plant debris and litter, not the living plants themselves. The earliest hexapods, the springtails and insects, also were detritivores. Within the cheliceriforms, the other major clade of arthropods, the mites also were detritivores.[25]

These detritivore millipedes, hexapods, and mites were the prey of the early carnivores. These early predators belonged to both of the arthropod clades: the mandibulate centipedes and the cheliceriform scorpions and spider-like trigonotarbids and uraraneids. These predators were at the top of the food chain at the time because the vertebrates had not yet arrived on land. When the vertebrate invasion began, these top predators were soon to become prey.

Thus the structure of the early arthropod food chain produced energy flow from dead plants to detritivores to carnivores, not living plants to herbivores to carnivores as in the modern world. Some plant damage is seen in a few of the plants preserved in the Rhynie Chert *Lagerstätte*, which gives evidence that a few arthropods were experimenting with consuming live stem tissue in the Pragian. However, full-fledged herbivory was not evolved by the arthropods until the Carboniferous, as will be discussed in chapter 7.

In summary, by the beginning of the Late Devonian, the major groups of land-dwelling arthropods are very well established (table 3.4), even more firmly entrenched than what is revealed in the fragmentary fossil record (table 3.3). The arthropods had experienced a seemingly unstoppable winning streak of invasion and diversification on dry land, spanning some 76 million years, from the Late Ordovician to the dawn of the Late Devonian. In the Late Devonian, however, their luck ran out.

The Second Major Wave of Animal Invaders

The second major animal group to make a concerted effort to invade the land was the Vertebrata, ancestors of modern terrestrial amphibians, reptiles, birds, and mammals (table 3.5). Unlike the arthropods, with their leggy marine ancestors, the marine ancestors of the earliest vertebrate invaders had no legs at all: they were fish and possessed only fins. Thus the vertebrates were confronted with a more drastic initial problem than the arthropods: to successfully invade the land, they had to evolve legs, which are walking adaptations, from fins, which are swimming adaptations.

Two phylogenetic lineages of bony fishes had evolved by the Silurian: the ray-finned fishes (Actinopterygii, table 3.5) and the lobe-finned fishes (Sarcopterygii). Both of these groups have produced species that can venture out of the water and crawl around on dry land. We shall consider some of these odd fish species in more detail in chapter 8. Of note here is the fact that the lobe-finned fishes were ancestral to all the modern terrestrial vertebrates. The saga of the evolution of four legs—of tetrapods—in the vertebrates is a long one. We will begin that saga in this chapter, and it will continue in chapters 5 and 7.

Like the arthropods, the early vertebrate invaders were confronted with the problem of breathing when they ventured out into the thin, dry air of the terrestrial realm. Here the vertebrates had a major advantage in that their marine ancestors had already evolved the capability to breathe air; that is, many already possessed simple lung structures. Therefore, it was easier for them to emerge from the water: they needed only to expand and improve already existing lung structures.

Fish primarily use gill structures to absorb oxygen from water and to shed carbon dioxide. They also have developed a secondary mechanism to assist their gills: they gulp air above the surface of the water and absorb oxygen from the air through esophageal pouches, which are simple lungs. This method of air breathing is termed "buccal," or cheek cavity, pumping. Interestingly, air breathing via buccal pumping was independently and convergently evolved by the ancient ray-finned and lobe-finned fishes.[26] Even today, their method of gulping and breathing air is different: the ray-finned fishes use a four-stroke buccal pump

TABLE 3.5 Phylogenetic classification of early vertebrates.

Vertebrata (vertebrate animals; from table 3.1)
– Petromyzontiformes (lampreys; living jawless fishes)
– unnamed clade
– – Pteraspidomorphi†
– – unnamed clade
– – – Anaspida†
– – – unnamed clade
– – – – Thelodonti†
– – – – unnamed clade
– – – – – unnamed clade†
– – – – – – Osteostraci†
– – – – – – Galeaspida†
– – – – – – Pituriaspida†
– – – – – Gnathostomata (jawed vertebrates)
– – – – – – Placodermi† (armored fishes)
– – – – – – – Acanthothoraci†
– – – – – – – unnamed clade†
– – – – – – – – unnamed clade†
– – – – – – – – – Rhenanida†
– – – – – – – – – Antiarchi†
– – – – – – – – unnamed clade†
– – – – – – – – – Arthrodira†
– – – – – – – – – unnamed clade†
– – – – – – – – – – Petalichthyida†
– – – – – – – – – – Ptyctodontida†
– – – – – – unnamed clade
– – – – – – – Chondrichthyes (cartilaginous fishes)
– – – – – – – unnamed clade
– – – – – – – – Acanthodii†
– – – – – – – – Osteichthyes (bony fishes + descendants)
– – – – – – – – – Actinopterygii (ray-finned fishes)
– – – – – – – – – **Sarcopterygii** (lobe-finned fishes + descendants; see table 3.6)
– – – – – – – – – – Crossopterygii
– – – – – – – – – – – Porolepiformes†
– – – – – – – – – – – unnamed clade
– – – – – – – – – – – – Onychodontida†
– – – – – – – – – – – – Actinistia (living coelacanths)
– – – – – – – – – – Dipnoi (living lungfishes)
– – – – – – – – – – **Tetrapodomorpha** (tetrapod-like fishes + descendants)

Source: Modified from the 2006 phylogenetic classification of Lecointre and Le Guyader for living vertebrates, and from the 2005 phylogenetic classification of Benton for extinct vertebrates.

Note: Major clades of vertebrates are given in bold-faced type.

† extinct taxa.

method, and the lobe-finned fishes breathe with a two-stroke buccal pump.[27] The early vertebrate invaders possessed both gills and simple lung structures, and in their subsequent evolution they progressively lost their fish-like structures, such as gills, and began to breathe entirely with modified lungs.

Two other fish-like features that the vertebrate invaders eventually lost were lateral-line systems and shell-less, anamniote eggs. Lateral-line systems are a network of nerves located along the two sides of a fish that it uses to detect pressure waves, to "hear," in water. Such a system is useless in the thin air of the terrestrial realm. Instead, vertebrates had to evolve an entirely new method to detect pressure waves in air: tympanal ear systems. As we will see in chapter 7, no less than five different phylogenetic lineages of later tetrapod vertebrates independently and convergently evolved different tympanal ear systems to hear sound in air.

Last, most female fish lay shell-less, gelatinous eggs in water, and male fish spray the eggs with sperm.[28] Such a reproductive system functions in water, but it is useless on dry land. Many living amphibians still reproduce in a manner similar to fish and are thus restricted to reproducing in water. A key evolutionary innovation in the invasion of the terrestrial realm, an innovation that freed the vertebrates from dependency upon water to reproduce, was the evolution of the shelled, amniote egg. The evolution of that key innovation, like the evolution of tympanal ears, will be discussed in chapter 7.

Vertebrates are chordate animals, and the first chordates date all the way back to the beginning of the Cambrian, just as the arthropods do. Basal chordates like the sessile tunicates or sea squirts (urochordates; table 3.1), which have only a mobile swimming larval stage, and adult swimming lancelets (cephalochordates), which have neither eyes nor skulls, are known from the Early Cambrian, about 530 million years ago. The Early Cambrian lamprey-like species *Haikouichthys eraicunensis* from China indicates that the most primitive of the clade of vertebrate chordates, the petromyzontiforms (table 3.5), had also evolved by this time.[29] The jawless lampreys and other ancient armored fishes that possessed no jaws, now extinct, were once considered to form the clade "Agnatha" ("without jaws," the non-gnathostome vertebrates), but this grouping is now known to be paraphyletic (evolutionary grade Petromyzontiformes through Pituriaspida; table 3.5).

The first of the jawed vertebrates, the gnathostomes, evolved in the Late Ordovician, some 70 million years later.[30] There are four main types of jawed vertebrates, two of which are extinct (table 3.5). The most basal of these are the placoderms, peculiar fish that had massive bony armor in the skull and neck region of the body and an internal cartilaginous skeleton, like that of a shark (color plate 7). Some of these were fearsome predators, attaining lengths of seven meters (23 feet), but, surprisingly for predators, they had no teeth! Instead they had very sharp bone blades in their mouths that, similar to the sharp bone beaks of modern-day snapping turtles, functioned very efficiently in slicing up prey animals. These fishes, now extinct, are ancestors of the living sharks, the chondrichthyans, which also have an internal skeleton of cartilage but that possess a mouth full of teeth.

The acanthodians, now extinct, are known as spine-finned fishes because they possessed long bony spines in front of their fins and sometimes in a row along their bellies. They gave rise to the bony fishes, the osteichthyans, which have two major clades: the ray-finned fishes (actinopterygians) and the lobe-finned fishes (sarcopterygians). Both of these clades evolved in the Late Silurian, about 420 million years ago.[31] The actinopterygians are what we normally think of as "fish," in that the huge majority of fish species today are ray-finned fish. The characteristic ray fins of these fish can be examined by anyone who buys a whole fish in a local grocery store or fish market. For example, the translucent, plastic-like fin that we see in the front of the fish, near its gills, is supported by a series of very thin bony fin rays (lepidotrichia) that are attached to a parallel series of somewhat larger bones (radials) near the body of the fish. The parallel row of radials then articulates with the larger shoulder-girdle bones in the body of the fish.

The sarcopterygians are a bit different. The same fin in the front of the animal (the pectoral fin) has but a single large radial that articulates with the internal shoulder bones. This single large radial is the topmost bone, the bone closest to the body, shown in the pectoral lobe fin of the Devonian sarcopterygian fishes *Eusthenopteron foordi* and *Tiktaalik roseae* in figure 3.1. This sarcopterygian radial is later to become the

FIGURE 3.1 From lobed fins to limbs (from top to bottom): comparison of the bones present in the fin of the tristichopterid fish *Eusthenopteron foordi*, in the fin of the elpistostegalian tetrapodomorph *Tiktaalik roseae*, in the tetrapod of the aquatic tetrapod *Acanthostega gunnari*, in the forelimb of the terrestrial cynodont synapsid *Thrinaxodon liorhinus*, in the forelimb of the mammalian synapsid mountain lion *Felis concolor*, and in the arm of the human *Homo sapiens*. *Credit*: Illustration by Kalliopi Monoyios (modified from Shubin, 2009).

humerus in the arm in the tetrapod lineages, the large single bone that we can feel in our upper arm. Immediately below this radial bone, as seen in the front part of the pectoral lobe fin of *Eusthenopteron foordi* and *Tiktaalik roseae* (fig. 3.1), is the lobe-fin bone that will eventually become the radius in our lower arm—the bone that you can feel on the inside of your arm running from your elbow up to your wrist just below your thumb. Immediately behind that anterior-most bone in the sarcopterygian pectoral lobe fin is the bone that will become the ulna in our lower arms—the bone that you can feel on the outside of your lower arm, running from your elbow to your wrist just below your little finger.

The smaller, more distal radial bones in the sarcopterygian pectoral lobe fin, as seen in *Eusthenopteron foordi* and *Tiktaalik roseae* (fig. 3.1), will eventually be modified in our lineage to form the bones we can feel in our wrists (the carpals), the palms of our hands (the metacarpals), and our fingers (the digits). This evolutionary modification is produced by twisting the developmental axis of the bones in the lobe fin such that the budding bones in the distal part of the lobe fin are positioned opposite the more proximal radial bones that will form the ulna and radius.[32] The effect of this developmental twist can be seen in the forelimb of the Devonian tetrapod *Acanthostega gunnari* (fig. 3.1), in which eight digits are present (unlike the five that are present in our hands). Note that the fin is now lost—there are no thin, bony fin rays (lepidotrichia) radiating from the digits of *Acanthostega gunnari* (fig. 3.1). Still, the manus (hand) in the forelimb of *Acanthostega gunnari* is very different from our hands and was actually used as a swimming and crawling paddle by the animal.

This paddle-like forelimb in the aquatic tetrapods, like *Acanthostega gunnari*, was later modified to a forelimb that would bear the weight of the animal as it walked on dry land. At first the land-dwelling vertebrates walked and ran on the soles of their feet; that is, they were plantigrade animals. The plantigrade type of posture is illustrated by the ancient synapsid *Thrinaxodon liorhinus* (a member of our lineage; more on that later in chapter 7) in figure 3.1—note that the animal is walking on the metacarpals in its forefoot and that its wrist is at ground level. We, although highly derived synapsid vertebrates, still walk on the soles of our feet. Other highly derived, and much faster running, land dwellers actually walk and run on their toes, not the soles of their feet. This type of posture, known as digitigrade, is illustrated by the mountain lion in figure 3.1—note that the wrist and metacarpals of the lion's foot are elevated and that the animal is walking on its digits (toes). We humans have modified the forelimb still further in that we no longer walk on our forelimbs at all and walk (or run) only on our hind limbs (fig. 3.1).

The sarcopterygian fishes are not familiar to us today, but they greatly outnumbered the actinopterygians for most of the Devonian, in terms of species diversity. The sarcopterygian fishes began to die out in the Late Devonian, however, and today they are represented only by three genera of living lungfishes, one genus of coelacanth, and, of course, ourselves. Note that the sarcopterygian vertebrates first evolved in the oceans in the Late Silurian (the oldest known species at present is *Guiyu oneiros* from China[33]). At that time, the arthropods had already been on land for 41 million years! Making up for lost time, the sarcopterygian fishes rapidly spread out from the seas into freshwater rivers and on land in the Devonian. Fossils of three species of dipnoi, the lungfishes, are known from Early Devonian Lochkovian strata: *Youngolepis praecursor* and *Diabolepis speratus* in China and *Powichthys thorsteinssoni* in Canada. The first crossopterygian fishes are known from the Eifelian Age of the Middle Devonian: the species *Euporosteus eifeliensis* in Germany.[34] Species of the porolepiform crossopterygian genus *Holoptychius* are frequently found coexisting with the earliest tetrapodomorph species. The last group of sarcopterygian vertebrates, the tetrapodomorph fishes, is of major importance for it was this group that invaded the land. Phylogenetic analyses of the three groups of sarcopterygians, both morphological and molecular, have not yet revealed the definitive relationship between them, and for this reason they are shown as an unresolved trichotomy in table 3.5.[35]

At present, fossil evidence for the existence of six species of tetrapods—the first limbed vertebrates—is known from the Middle Devonian and from the Frasnian Age of the Late Devonian (table 3.6). The oldest fossil evidence comes from tetrapod trackways in Eifelian strata of the Middle Devonian in Poland, and the next oldest fossil evidence comes from tetrapod trackways in early Frasnian strata in Ireland (color plate 8). Thus our oldest evidence for the beginning of the vertebrate invasion of land comes from trace fossils, not fossils of the animals themselves, just as with the arthropod invaders. Fragmentary skeletal fossils of tetrapods are known for four species in the Frasnian Age: *Elginerpeton pancheni*, *Obruchevichthys gracilis*, *Sinostega pani*, and *Metaxygnathus denticulus* (table 3.6). However, as we shall see below, phylogenetic analyses of these tetrapod species indicate the existence of at least two other species lineages of tetrapod animals in the Frasnian—lineages that at present are "ghost" lineages, as no fossils have yet been found of these animals in Frasnian strata, even though they appear later in Famennian strata. The exact sequence of evolution of the first tetrapods within the

TABLE 3.6 Tetrapod invaders in the Middle Devonian and Frasnian Age of the Late Devonian.

Sarcopterygii (lobe-finned fishes + descendants; from table 3.5)
- I. Crossopterygii (coelacanths)
- II. Dipnoi (lungfishes)
- III. Tetrapodomorpha (tetrapod-like fishes + descendants)
  - A. Rhizodontia
  - B. Osteolepiformes
  - C. Tristichopteridae
  - D. Elpistostegalia
  - E. Tetrapoda (limbed vertebrates)
    1. Unknown early Eifelian tetrapod species in Zachelmie, Poland
    2. Unknown early Frasnian tetrapod species in Valentia Island, Ireland
    3. *Elginerpeton pancheni*
    4. *Obruchevichthys gracilis*
    5. *Sinostega pani*
    6. *Metaxygnathus denticulus*

Note: The numbered tetrapod groups are listed in the order in which they first appear in geologic time.

clade of the tetrapodomorphs remains controversial and is the subject of lively debate. The debate is particularly interesting in that we thought we definitely knew the answer to the question, "How did the tetrapods evolve?", only to have that theoretical answer overturned with new empirical fossil evidence.

At present, the most basal tetrapodomorph known is the species *Kenichthys campbelli* from the Early Devonian Emsian in China (table 3.7). It is known only from fossils of the skull bones, so we do not know what the rest of the animal looked like. The skulls do show, however, that these animals possessed an internal nostril (choana), a trait that is a key tetrapodomorph synapomorphy. The rhizodont tetrapodomorphs are more derived and appear in the fossil record in the Givetian Age of the Middle Devonian. Their phylogenetic relationships are not that well known, but ecologically they were very large predators that pursued our tetrapodomorph ancestors. The largest of these was the Early Carboniferous species *Rhizodus hibberti*, which was seven meters (23 feet) long and had a mouth full of rhizodont, knife-like teeth. Even though they are near-basal tetrapodomorph fishes, they did possess digit-like rays on their lobe fins, as seen in the Famennian species *Sauripterus taylori* in North America.

TABLE 3.7 Phylogenetic classification of the first tetrapod invaders.

Sarcopterygii (lobe-finned fishes + descendants)
– Crossopterygii (coelacanths)
– Dipnoi (lungfishes)
– **Tetrapodomorpha** (tetrapod-like fishes + descendants)
– – basal tetrapodomorphs
– – – *Kenichthys campbelli*
– – Rhizodontia
– – Osteolepidiformes
– – – Osteolepididae
– – – Megalichthyidae
– – – Eotetrapodiformes
– – – – Tristichopteridae
– – – – *Gogonasus andrewsae*
– – – – unnamed clade
– – – – – Elpistostegalia
– – – – – **Tetrapoda** (limbed vertebrates)
– – – – – – Family incertae sedis
– – – – – – – Unknown early Eifelian tetrapod species in Zachelmie, Poland
– – – – – – Family incertae sedis
– – – – – – – Unknown early Frasnian tetrapod species in Valentia Island, Ireland
– – – – – – Family Elginerpetontidae
– – – – – – – *Elginerpeton pancheni*
– – – – – – – *Obruchevichthys gracilis*
– – – – – – Family incertae sedis
– – – – – – – *Sinostega pani*
– – – – – – unnamed clade
– – – – – – – [Family incertae sedis]
– – – – – – – – [*Densignathus rowei*]
– – – – – – – unnamed clade
– – – – – – – – [Family incertae sedis]
– – – – – – – – – [*Ventastega curonica*]
– – – – – – – – unnamed clade
– – – – – – – – – Family incertae sedis
– – – – – – – – – – *Metaxygnathus denticulus*

Source: Phylogenetic data modified from Benton (2005), Lecointre and Le Guyader (2006), Long et al. (2006), Ahlberg et al. (2008), and Clack et al. (2012).

Note: Two "ghost" tetrapod lineages in the Frasnian Age are enclosed in brackets (e.g., [*Densignathus rowei*]). See text for discussion. Major clades of vertebrates are marked in bold-faced type.

The osteolepidids are more derived than the rhizodonts. The most basal species known at present is the osteolepiform *Osteolepis macrolepidotus* from the Eifelian Age of the Middle Devonian in Scotland. Curiously, the bones of the head of the animal, and its diamond-shaped scales, were covered with cosmine, a glossy layer composed of enamel and dentine. This trait is similar to that seen in the head of *Psarolepis romeri*, an ancient stem sarcopterygian found in the Pridolian Age of the Late Silurian in China.

The tristichopterids are more derived than the osteolepiforms and include the famous species *Eusthenopteron foordi*, long touted as the model for an animal intermediate in the fish–tetrapod transition (fig. 3.1). Well-preserved fossils of this species are found in the early Frasnian Escuminac Bay *Lagerstätte* of Canada. In addition to its lobed fins, it had a diphycercal, symmetrical tail that looked more like that of a coelacanth than the asymmetrical, heterocercal tail of other osteolepidids. Some tristichopterids, such as the Famennian species *Hyneria lindae*, were large predators and thus dangerous to the early tetrapods (color plate 5).

The excellently preserved fossils of the Australian tetrapodomorph fish *Gogonasus andrewsae* show traits that indicate it may be intermediate in evolution between the tristichopterids and the more derived elpistostegalians. It is early Frasnian in age, and will be considered in more detail later, as it is one player in the renewed debate concerning tetrapod origins.

The elpistostegalians are the most derived of the tetrapodomorph fishes (table 3.7). This group includes the famous late Givetian to early Frasnian species *Panderichthys rhombolepis* and *Livoniana multidentata* from Latvia, and *Elpistostege watsoni* and *Tiktaalik roseae* from Canada (fig. 3.1). Not long after the turn of the millennium, a couple of spectacular finds of these elpistostegalian fish fossils occurred, and it was thought that we had the definitive answer for the question, "How did tetrapods evolve?" Tetrapods were proposed to have evolved from the elpistostegalian tetrapodomorphs in the sequence: *Panderichthys rhombolepis* to *Elpistostege watsoni* to *Tiktaalik roseae* to *Livoniana multidentata* to *Elginerpeton pancheni*, which is the oldest known tetrapod species (table 3.6). The sequence of evolving traits in this proposed lineage made a nearly perfect evolutionary transition from fish to tetrapod. The elpistostegalian–tetrapod transition would then have occurred in the time interval from the early to middle Frasnian Age of the Late Devonian.

Unfortunately for the elpistostegalian–tetrapod scenario, in 2010 an equally spectacular find of tetrapod fossil trackways in Poland proved that tetrapods already existed on Earth in the early Eifelian Age, the very beginning of the Middle Devonian, and that they had probably evolved earlier in the Emsian Age of the Early Devonian! Anatomically, tetrapods possess an endoskeleton composed of hard bone, in contrast to the arthropod exoskeleton of fragile chitin. Thus tetrapods have a much higher probability of preservation in the fossil record than the arthropods; even so, the first evidence that we now have for the evolution of tetrapods comes from trace fossils rather than body fossils, just as for the arthropods. The discovery of the excellently preserved trackways of their footprints in present-day Poland prove that the tetrapods had evolved by the Eifelian of the Middle Devonian, some 398 million years ago.[36] Yet the oldest fossil bones of tetrapods yet discovered are from the species *Elginerpeton pancheni* in Scotland (table 3.6), which lived around 380 million years ago in the mid-Frasnian, some 18 million years later than the species that made the Polish fossil trackways.

Another tetrapod trackway has been found on Valentia Island in Ireland, dating to the Late Devonian Frasnian Age itself.[37] This trackway is radiometrically dated to 385 million years ago, at the very beginning of the Frasnian (see table 2.2 for the radiometric timescale), some 13 million years after the Eifelian trackways found in Poland.[38] The early Frasnian tetrapod species that made this trackway lived about five million years before the mid-Frasnian species *Elginerpeton pancheni*, which we shall consider shortly. Again, the identity of the actual species making the trackway has not yet been discovered. Given the close geographic proximity of the two fossil sites (Scotland and Ireland), could it be that the tracks in Ireland were made by an early individual of the Scottish species *Elginerpeton pancheni* itself, or a close relative?

As tetrapod bone is considered to have a high probability of preservation in the fossil record, it was proposed that the absence (thus far) of body fossils of the Middle Devonian and early Frasnian tetrapods that had to have existed somewhere on Earth (given the Eifelian trackways) was probably a function of small population sizes. The early tetrapods were rare. Alternatively, it also could be a function of the habitat of the first tetrapods. The Polish animals were marginal marine animals, and tidal-flat habitats are high-energy in nature—an unfavorable environment for the preservation of fossils. The Polish strata that preserve the tetrapod trackways are almost devoid of body fossils. The Warsaw

University paleontologist Grzegorz Niedźwiedzki and his colleagues, who have described the fossil trackways, suggest that the first appearance of tetrapods in the fossil record may reflect the timing of their migration into habitats with more favorable preservation probabilities than those inhabited by the original tetrapods, an ecological phenomenon rather than an evolutionary one.[39]

So, we once again have no definitive answer to the question, "How did tetrapods evolve?" The oldest known tetrapodomorph fish, *Kenichthys campbelli* from the late Emsian, is in the correct temporal position to have been the ancestor of the Polish tetrapods, but anatomically it is considered much too plesiomorphic (table 3.7). All of the more derived tetrapodomorph fishes evolved after the Emsian, in the Givetian and later. Rather than representing a transitional lineage from fish to tetrapod, the elpistostegalian tetrapodomorph fishes are now seen as a stable evolutionary group that coexisted with tetrapods for at least 10 million years. The tetrapods only appeared in the fossil record when they finally migrated into the same habitats where the elpistostegalians lived, perhaps to compete with the elpistostegalians for food.

The only other alternative is to propose that tetrapods evolved twice: once in the Emsian and again in the Frasnian! In that hypothetical scenario, the Polish trackways preserve evidence of an early tetrapod group that went extinct. Much later, in the Frasnian, a second group of tetrapods evolved from the elpistostegalian fishes. While this scenario might seem appealing, given the attractive elpistostegalian–tetrapod species sequence in the early to middle Frasnian, it still leaves the question of the ancestry of the Polish tetrapods unanswered. Who were the ancestors of these tetrapods?

Although attractive, the previously proposed Frasnian elpistostegalian–tetrapod transition lineage was not without its critics. The McGill University paleontologist Robert Carroll argued, "While *Tiktaalik* illustrates a plausible, intermediate way of life between marine fish and terrestrial amphibians, the absence of ossification of the vertebrae, their great number, and the divergent specialization of the forelimb suggest that this genus was not an immediate sister-taxon of any known tetrapods."[40]

The Australian tetrapodomorph fish *Gogonasus andrewsae* was mentioned briefly above in the lead-in to our consideration of the origin of tetrapods. This fish has an unusual mix of morphological features. For example, the Parisian paleontologists Philippe Janvier and Gaël Clément at the Muséum National d'Histoire Naturelle note that *Gogonasus andrewsae* had cosmine-covered rhombic scales and dermal bones like the basal osteolepidid fish *Osteolepis macrolepidotus,* "very different in aspect from the crocodile-like, vermiculate ornamentation of elpistostegalians and early tetrapods."[41] Thus one might at first conclude that *Gogonasus andrewsae* is a more basal tetrapodomorph fish, but the Australian National University paleontologist John Long and his colleagues, who described the new fish species, noted that a big surprise came in the discovery that its pectoral fin skeleton "shares several features with *Tiktaalik,* the most tetrapod-like fish" and suggested that "the basic tetrapod pectoral fin pattern, and the broad spiracular opening as a precursor to tetrapod middle ear architecture, might have originated further back than previously thought within the tetrapodomorph radiation."[42]

As noted before, the most basal of all the tetrapodomorphs discovered to date is *Kenichthys campbelli* (table 3.7), which is usually considered too plesiomorphic to have been a direct tetrapod ancestor. Long and colleagues point out, however, that this species "is known only from isolated skulls and cranial dermal bones, so we lack knowledge of its paired fins, endoskeletal girdles, gill arches, and overall body form to contribute to the understanding of early character development in the lineage."[43] Might the basic tetrapod pectoral fin pattern have originated at the very beginning of tetrapodomorph evolution?

In arguing that *Gogonasus andrewsae* is in fact more derived than the tristichopterid tetrapodomorphs (table 3.7), Long and colleagues conclude:

> Our new phylogeny replaces the tristichopterid *Eusthenopteron* as the typical fish model for the fish-tetrapod transition. It also raises the question of what environment the immediate stem group of the elpistostegalians inhabited. The marine environment inhabited by *Gogonasus* is in accord with the marginal marine environments of some elpistostegalians (*Panderichthys, Elpistostege, Tiktaalik*) and the tetrapod *Tulerpeton* [note: a Famennian tetrapod species that we will consider in detail in chapter 5]. Such observations support a model

> in which the first tetrapods, like their immediate piscine sister taxa, were capable of marine dispersal, thus explaining the widespread global distribution achieved shortly after their first appearance in the late Frasnian.[44]

Indeed, in 2010, the Polish tetrapod trackways revealed that the earliest now-known tetrapods inhabited marginal marine, tidal flat habitats.

Summarizing the debate in a *Nature* review article entitled "Muddy tetrapod origins," Philippe Janvier and Gaël Clément suggest: "Perhaps the earlier stages of the elpistostegalian and tetrapod lineages include elusive pre-Eifelian tetrapodomorph fishes that we may not be able to recognize as such, because the known elpistostegalians and early tetrapods are too 'derived'—that is, maybe they are too different from their common ancestor as we imagine that ancestor to have been."[45] They suggest that this ancestor may have been more osteolepidid-like, and the fact that *Gogonasus andrewsae* was covered with cosmine, like the basal osteolepidid fish *Osteolepis macrolepidotus*, lends some support to that proposal.[46] More recent phylogenetic analyses of the relationships between the lungfishes and *Kenichthys campbelli*, the most basal tetrapodomorph species yet known, suggest that the tetrapodomorph–lungfish lineages may have split in the Pridolian Age of the Late Silurian, and thus that the tetrapodomorphs may have evolved as early as the latest Silurian, not in the Emsian Age of the Early Devonian.[47]

If the common ancestor of the elpistostegalian fishes and the tetrapods lies deep in the past, back in the Early Devonian or even further back in the latest Silurian, how then do we explain the rather close morphological similarity of the late Givetian and Frasnian elpistostegalian fishes to mid-Frasnian tetrapod species? If there is no direct ancestor–descendant link between these separate phylogenetic lineages, the similar tetrapod-like features of the elpistostegalian fishes and the tetrapods must have evolved independently in these two separate lineages at the same time—a case of parallel evolution!

Parallel evolution is a common phenomenon in nature.[48] It has previously been proposed that some trends seen in tetrapodomorph evolution occurred in parallel in separate lineages. For example, the tetrapodomorph specialists Per Ahlberg of Uppsala University and

Zerina Johanson of the Australian Museum argue: "Tetrapod-like character complexes evolved three times in parallel within the Tetrapodomorpha. . . . Parallel evolution towards the morphology of a large predator, with reduced median fins and elaborate anterior dentition, occurred at about the same time in rhizodonts, tristichopterids, and elpistostegids+tetrapods. . . . The Tetrapoda thus arose out of one of several similar evolutionary 'experiments' with a large aquatic predator role."[49] The current decoupling of the ancestor–descendant "elpistostegids+tetrapods" lineage into two separate, independent lineages in the Frasnian brings the number of lineages evolving in the same direction at the same time to four, not three. The implications of parallel evolution of tetrapod-like morphologies in four separate lineages in the Givetian and Frasnian will be explored in more detail in chapter 8.

We have reached the point in this section of the chapter where tetrapods are finally found in the fossil record. Thus far in the book we have progressively considered finer and finer divisions of the geologic timescale. We started first with Eons, Eras, and Periods in our consideration of the evolution of life itself on Earth and the earliest forms of life on land (table 1.1). Then we used the finer time divisions of Epochs and Ages in our consideration of the invasion of land by plants (table 2.2). Now we need an even finer division of time—the Zone—to subdivide the Ages of the geologic timescale for our consideration of the invasion of land by vertebrates.

Two types of zones are used to finely divide time in the Devonian and Carboniferous: marine strata are dated using conodont zones, and terrestrial strata are dated using spore zones. Conodonts are tiny, marine, non-vertebrate craniate chordates (see table 3.1) that were abundant swimmers in ancient oceans, but that are now extinct. They looked somewhat like tiny eels or arrows—most were about the size of a toothpick—with elongated, cylindrical bodies and fins located on the dorsal and ventral surfaces of the posterior one-third or so of their bodies. Their soft bodies had an extremely low probability of preservation in the fossil record, but their very hard tiny teeth and dentary structures are numerous in marine sediments. Their importance lies in the fact they evolved rapidly, thus the evolutionary succession of their fossil species allows us to finely divide time in marine strata in the Devonian and Carboniferous.

TABLE 3.8 Finer time divisions (conodont zones) of the geologic timescale in the critical time interval of the Frasnian Age when the first tetrapod body fossils have been discovered.

Geologic Age	Conodont Zones
FRASNIAN	*linguiformis* Zone
	Late *rhenana* Zone
	Early *rhenana* Zone
	jamieae Zone
	Late *hassi* Zone
	Early *hassi* Zone
	punctata Zone
	transitans Zone
	Late *falsiovalis* Zone
	Early *falsiovalis* Zone, pars

Source: Modified from the 1990 international zonal timescale of Ziegler and Sandberg.

Terrestrial strata obviously have no fossils from marine animals, but they do contain fossils of the reproductive structures of plants, spores and pollen, which are very tough and have a high probability of preservation in the fossil record. Thus the evolutionary succession of spore and pollen fossil species allows us to finely divide time in terrestrial strata in the Devonian and Carboniferous. We shall consider spore zonations in further detail in chapter 5, where we shall use them to finely divide time in the Famennian Age, but at this point we now need to use the conodont zonation of the Frasnian Age, which is given in table 3.8. The Frasnian is divided into ten conodont zones, but it is not known exactly how much time each zone represents as some conodont species may have evolved faster than others. However, a rough time approximation may be made: the Frasnian Age was 10.8 million years long (table 2.2), thus one can roughly consider each of the ten zones of the Frasnian to be about one million years in duration. Some zones were probably longer, and some shorter, but one million years per zone gives a useful approximation of the passage of time in the Frasnian.

The early tetrapods slowly increased their numbers in the Middle Devonian, attaining ever larger population sizes and finally spreading away from the beaches and tidal flats in the Frasnian; as a consequence, the first bones and body fossils of tetrapod species begin to appear in the fossil record from the Middle Devonian. The strata containing fossils of the species *Elginerpeton pancheni* in Scotland (color plate 8) have been dated in the interval from the *jamieae* conodont to Early *rhenana* conodont zones,[50] thus are mid-Frasnian in age; this species is at present the oldest known tetrapod (table 3.6).

Fossils of another tetrapod species, *Obruchevichthys gracilis*, have been found to the east in Latvia and western Russia (color plate 8). The strata containing these fossils have been dated in the Early to Late *rhenana* conodont zones,[51] thus this species may be just slightly younger than *Elginerpeton pancheni* (table 3.6).

Further east, fossils of the tetrapod species *Sinostega pani* have been found in China (color plate 8). The strata containing this species are poorly dated at present, but the strata do contain fossils of galeaspid jawless fish (table 3.5), and all galeaspid fish are known to have gone extinct at the end of the Frasnian. Thus *Sinostega pani* is considered to be another Frasnian tetrapod, although the possibility exists that it may be older—perhaps even as old as the mid-Givetian in the Middle Devonian.[52] If the fossils are ever demonstrated to be Givetian in age, *Sinostega pani* would become the oldest known tetrapod species and would be closer in time to the unknown Eifelian species that made the trackways in Poland.

Last, and even further east, fossils of the species *Metaxygnathus denticulus* have been found in Australia (color plate 8). The strata containing this tetrapod species are also poorly dated at present,[53] but the best evidence thus far indicates that the species can probably be dated in the *rhenana* to *triangularis* conodont zonal interval,[54] thus it is of late Frasnian age and is the youngest known Frasnian species (table 3.6).

Just as we have seen for the arthropod invaders (table 3.3 versus 3.4), however, the fossil record of the vertebrate invaders (table 3.6) does not reveal the full complexity of that invasion (table 3.7). Phylogenetic analyses of the Frasnian species *Metaxygnathus denticulus* reveal that it is more derived than two tetrapod species known only from much younger strata in the Famennian Age. These two species, *Densignathus rowei* and *Ventastega curonica* (table 3.7), are more plesiomorphic than *Metaxygnathus denticulus* and thus represent lineages that must have evolved at some point earlier in time in the Frasnian, even though fossils of these species are only known from younger strata in the Famennian.[55] From these phylogenetic analyses, Ahlberg and colleagues conclude not only that the Famennian "*Ventastega* represents a lineage of Frasnian origin but that a substantial part of the Devonian tetrapod radiation occurred during the Frasnian."[56]

In summary, our knowledge of tetrapod evolution in the Frasnian is still fragmented. Phylogenetic analyses of the skeletal fossils of *Elginerpeton pancheni* indicate that it is the most basal of all of the known tetrapod species.[57] The skeletal material of *Obruchevichthys gracilis* is very similar to that of *Elginerpeton pancheni*, and the two species are considered to belong to the same family, the Elginerpetonidae (table 3.7).[58] *Sinostega pani* is known only from a partial jaw fragment, which yields little phylogenetic information other than it belongs to a tetrapod, and the phylogenetic relationships of the two unknown species that made the trackways in Poland and Ireland are, obviously, unknown. Thus the two trace-fossil species and *Sinostega pani* are shown in an unresolved quadrichotomy with the Elginerpetonidae in table 3.7. Phylogenetic analyses indicate that the ghost lineage of the species *Densignathus rowei* is more derived than that of the Elginerpetonidae, that the ghost lineage of the species *Ventastega curonica* is more derived than that of *Densignathus rowei*, and that *Metaxygnathus denticulus* is the most derived of the Frasnian tetrapods (table 3.7).

The First Vertebrate Invaders

What did these early tetrapods, our ancestors, look like? The Polish trackway reveals the presence of several individuals, ranging in length from 0.5 to 2.5 meters (a foot and a half to eight feet). Presumably these are all the same species, and thus the size range represents the presence of both juvenile and adult individuals. These animals clearly were walking with four limbs, and the adults had feet that were about 15 centimeters (six inches) wide (fig. 3.2). The trackway impressions of their footprints show the presence of digits on these feet, but it is not clear exactly how many digits the animals possessed on their feet. They clearly had at least five short, triangular-shaped toes on their hind feet, with two toes

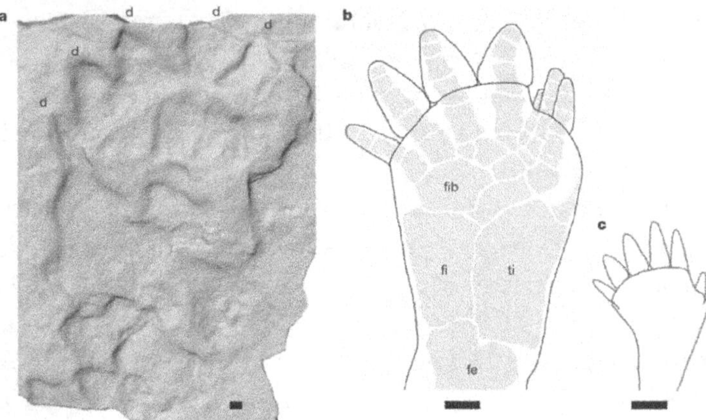

FIGURE 3.2 From left to right: (a) fossil footprint of the unknown Eifelian species of tetrapod in Zachelmie, Poland, where each digit (toe) impression is marked with a "d"; (b) reconstructed foot and internal bones of the Famennian tetrapod *Icthyostega stensioei*; and (c) outline of the foot of the smaller Famennian tetrapod *Acanthostega gunnari*. The foot of the unknown species of tetrapod (a) was about three times larger than the foot of *Icthyostega stensioei* (b), itself a large species in comparison with *Acanthostega gunnari* (c). (Length of scale bars in each figure are ten millimeters, or almost half an inch.) *Credit*: Macmillan Publishers Ltd: *Nature* (Niedźwiedzki et al., 2010), copyright © 2010. Reprinted with permission.

somewhat divergent from the other three.[59] They may have had smaller toes in addition to these five larger ones (as we will see in chapter 5, some Famennian tetrapods possessed as many as eight toes) that did not make a strong enough impression in the mud to be preserved. The footprints also reveal that the animals did not have claws on their digits.

The early Frasnian trackways in Ireland do not reveal as much detail about the animals that made them as the Polish ones. Several individuals with four limbs were clearly present, all heading in the same direction. The footprints are up to two centimeters (0.8 inches) deep and oval in shape, and they show no digit impressions, revealing that the animals were walking on soft mud and that their feet sank into the mud with each step. One trackway shows a shallow central furrow between the left and right footprints, revealing that the animal was pulling itself along in the soft sediment. From the spacing of the footprints, the average animal had a body length of 38 centimeters (15 inches), not including the length of the head or the tail (fig. 3.3). The total length of the animal is

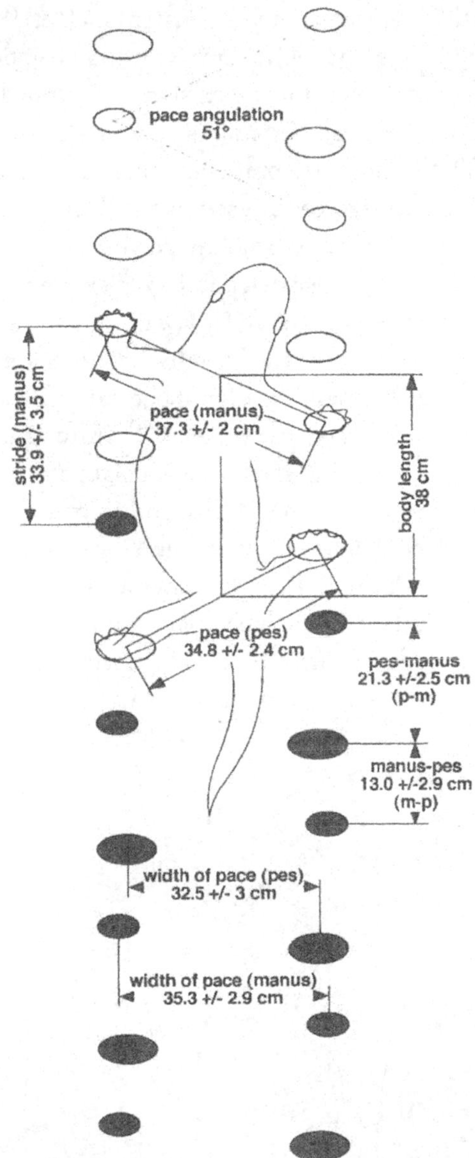

FIGURE 3.3 Reconstruction of the unknown early Frasnian species of tetrapod that produced the fossil trackways in Valentia Island, Ireland, based upon the width of the trackway and the spacing between the footprints ("manus" are the forefeet prints, and "pes" are the hind feet prints). *Credit*: From the *Journal of the Geological Society of London* (Stössel, 1995), copyright © 1995, reproduced with permission of The Geological Society Publishing House via Copyright Clearance Center.

thus estimated to have been about one meter (3.3 feet).[60] It is known that some living lungfish also make trackways with their lobe fins when they crawl; therefore, as evidence for the existence of limbed vertebrates, the Irish non-digitated trackways are not as definitive as the Polish digitated trackways.[61] Still, the large size of the animal and the fact that tetrapods are now known to have evolved as early as the Eifelian indicate that these trackways were probably made by a limbed vertebrate.

Body fossils of the Scottish tetrapod species *Elginerpeton pancheni* include more than 15 upper and lower jaw fragments, a postorbital skull fragment, a tibia and femur of the hind limb, one incomplete pelvic girdle, four incomplete shoulder girdles (pectoral girdles), a humerus of the forelimb, and one pre-sacral vertebra.[62] From these fragments we can reconstruct the animal as having had a skull about 400 millimeters (16 inches) long, a body with four limbs, and an overall length of at least 1.5 meters (five feet). The head of the animal was rather flat, with a triangular, pointed snout (fig. 3.4), and its jaws were filled with small, sharp teeth (fig. 3.5), including a pair of small coronoid fangs separated from the main tooth row.[63] Also of note here is the fact that *Elginerpeton pancheni*, our oldest yet discovered Frasnian tetrapod species known from

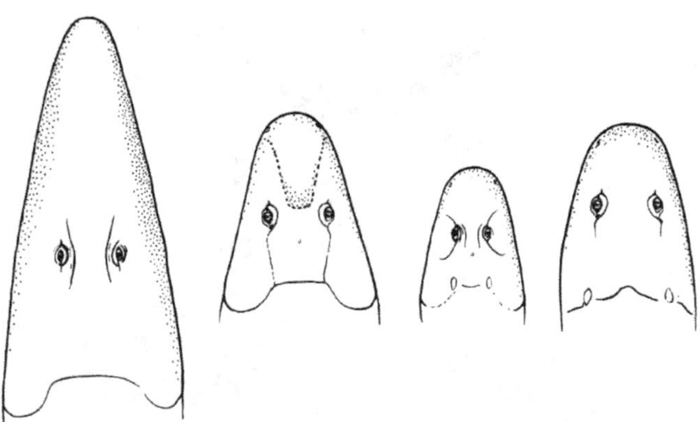

FIGURE 3.4 Reconstructions of the heads of the Frasnian species *Elginerpeton pancheni* and the Famennian species *Ventastega curonica*, *Acanthostega gunnari*, and *Ichthyostega stensioei*, arranged from least derived (left) to most derived (right). Note that although *Ventastega curonica* is a Famennian species, it is a member of a phylogenetic lineage that evolved in the Frasnian. *Credit*: Illustration by Kalliopi Monoyios.

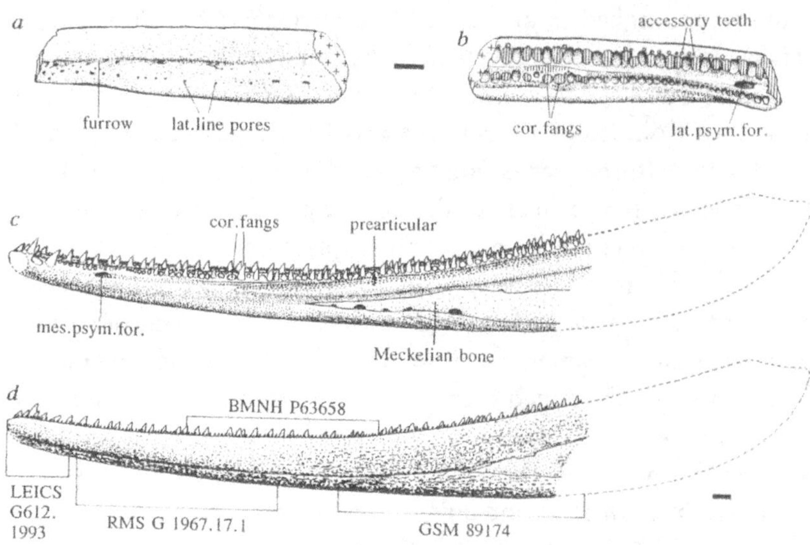

FIGURE 3.5 The fossil jaw of the Frasnian tetrapod *Elginerpeton pancheni*; (a) and (b) show lateral and dorsal views of two of the jaw fragments, and (c) and (d) show reconstructions of the mandible as seen from the outside of the jaw (c) and the inside of the jaw (d). Brackets and numbers in (d) indicate separate fossil fragments of the jaw preserved in different specimens. The animal possessed both coronoid fangs (abbreviated "cor. fangs") and a lateral-line system ("lat. line pores"), which are plesiomorphic traits. The scale bar is ten millimeters (almost half an inch) long. *Credit*: Macmillan Publishers Ltd: *Nature* (Ahlberg, 1995), copyright © 1995. Reprinted with permission.

skeletal fossils, was a much larger animal than the well-known Famennian tetrapod species *Acanthostega gunnari* and *Ichthyostega stensioei* (fig. 3.4), a fact that we shall examine in more detail in chapter 5 when we consider the aftermath effects of the Late Devonian mass extinction.

Body fossils of the Latvian species *Obruchevichthys gracilis* are only represented by fragments of the lower jaw. From these fragments, however, we can reconstruct skull and estimate the length of the skull to have been about 400 millimeters (16 inches), the same size as the skull of *Elginerpeton pancheni*. It possessed teeth like those of *Elginerpeton pancheni*, including the paired coronoid fangs, and the shape of the jaws suggests that the animal had a flat, triangular skull, also like *Elginerpeton pancheni*.[64] These morphological similarities between the two species, in addition to their similar ages in geological time and close geographic proximity, are argued to indicate a close phylogenetic relationship, so the two species are placed together in the Family Elginerpetondiae.[65]

Like *Obruchevichthys gracilis*, the Chinese species *Sinostega pani* is only known from fossils of its lower jaw. The single bone fragment is 81 millimeters (three inches) long and is the middle part of the lower jaw.[66] Little else is known about the anatomy of the species at present; its chief significance lies in its geographic location.

The Frasnian ghost lineage species *Densignathus rowei* is known only from two fragments of its left lower jaw and one fragment of its right lower jaw, found in Famennian strata in Pennsylvania, North America. However, the jaw contained coronoid fangs, which is a plesiomorphic, fish-like trait. Philadelphia Academy of Natural Sciences paleontologist Ted Daeschler notes that the "condition of the coronoids is very similar to that seen in *Ventastega* and differs from the reduced coronoid dentition seen in *Ichthyostega* and *Acanthostega*,"[67] two famous Famennian genera of tetrapods (fig. 3.4) that we shall examine in detail in chapter 5. Daeschler also notes that the ventral margin of the jaw is twisted anteriorly, and that the coronoid region of the jaw has a square-shaped cross-section, traits also found in *Ventastega curonica*. The fossil of the left lower jaw is 173 millimeters (6.8 inches) in length, and the total animal may have been approximately one meter (3.3 feet) in length.

Body fossils of the Frasnian ghost lineage species *Ventastega curonica* include almost the entire skull, most of the shoulder girdle, part of the pelvis, one rib, and numerous tail fin rays (caudal lepidotrichia), and these fossils were found in Famennian strata in Latvia, like the known Frasnian species *Obruchevichthys gracilis*. The lower jaw of the animal was 210 millimeters (8.3 inches) in length and contained numerous small, sharp teeth along the jaw margin and two coronoid fangs in the middle part of the jaw. The head of the animal had a snout that was not as triangular as that of *Elginerpeton pancheni*, and it had a trough-like midline depression, also unlike *Elginerpeton pancheni*. The head had large eyes that were situated dorsally (fig. 3.4); thus the head of *Ventastega curonica* looked more like that of the Frasnian elpistostegalian fish *Tiktaalik roseae* (fig. 3.1), whereas the post-cranial body of the animal was more like the more derived Famennian tetrapod species *Acanthostega gunnari* (which we will consider in detail in chapter 5),

FIGURE 3.6 Reconstruction of the tetrapod *Ventastega curonica* from Latvia. Note the large tail fin used in swimming. Much of the central skeleton is missing, so the limbs are modeled after the somewhat more derived species *Acanthostega gunnari*. It is likely that the limbs in *Ventastega curonica* were shorter and even more paddle-like than those shown here. The living animal was about 1.2 meters (four feet) long. *Credit*: Illustration by Kalliopi Monoyios.

and the tail of the animal clearly possessed a fin like *Acanthostega gunnari* (fig. 3.6) as evidenced by the fossil tail fin rays. At present, nothing is known about the morphology of the limbs of *Ventastega curonica* or what type of digits it may have possessed. The animal was about 1.2 meters (3.9 feet) in length.[68]

Last, the Australian species *Metaxygnathus denticulus* is also known only from fossils of its lower jaw. However, the jaw is complete, unlike the mid-jaw fragment of *Sinostega pani*. The fossil bone is the right lower jaw, is 120 millimeters (five inches) long, and contains numerous small, sharp teeth (fig. 3.7). Three of the teeth are somewhat larger in the middle of the jaw and may represent small coronoid fangs on the three coronoid bones of the jaw.[69] If the fossil jaws of *Sinostega pani* and *Metaxygnathus denticulatus* are from adult animals, then these two species had jaws that were only 30 percent of the length of the jaws of *Elginerpeton pancheni* and *Obruchevichthys gracilis*. Presumably, the total length of the animals was similarly scaled; *Sinostega pani* and *Metaxygnathus denticulatus* were probably about 0.5 meters (a foot and a half) in length.

Although the morphological data that the fossils of *Sinostega pani* and *Metaxygnathus denticulus* can give us are sparse, the geographic

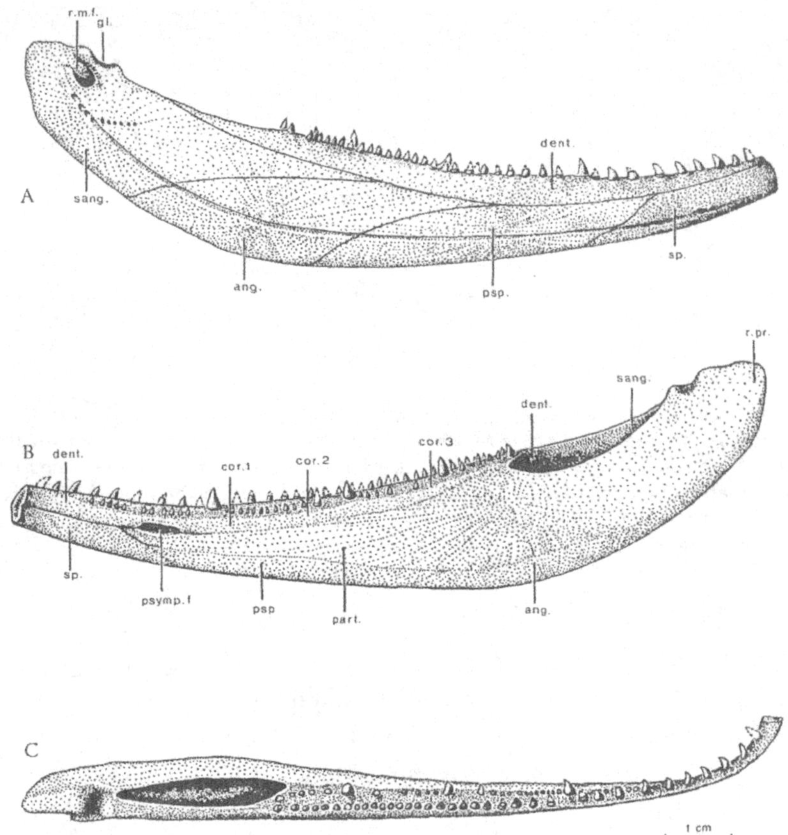

FIGURE 3.7 The fossil jaw of the Frasnian tetrapod *Metaxygnathus denticulus*. (a) The side of the mandible as viewed from the outside of the jaw; (b) the side of the mandible as seen from the inside of the jaw; (c) the upper surface of the mandible with tooth lines. *Credit*: From "A primitive amphibian from the Late Devonian of New South Wales" by K. S. W. Campbell and M. W. Bell, *Alcheringa* 1:369–381 (1977), reprinted by permission of the publisher (Taylor & Francis Ltd, http://www.tandf.co.uk/journals).

data they give us are of major significance: Frasnian tetrapod species had a worldwide geographic distribution, from Europe to China to Australia. To have achieved a worldwide range in the Frasnian, they must have evolved earlier than the Frasnian. As we now know, tetrapods were present on Earth in the early Eifelian, thus the first tetrapods probably evolved in the Emsian Age of the Early Devonian.

Ecology and Biogeography of the First Vertebrate Invaders

The oldest fossil evidence (Eifelian to mid-Frasnian) for the existence of tetrapods on Earth comes from strata found in Poland, Ireland, Scotland, Latvia, and Russia (color plate 8), as discussed in the previous section of the chapter. All of these present-day countries were located on a single large continent in the Devonian, variously called Laurussia, Euramerica, or the Old Red Sandstone continent. The "Lau" in the term Laurussia refers to the Laurentian shield of Canada, located on the western side of the Laurussian continental landmass, and the term "russia" refers to present-day Russia, located on the eastern margin of the continent. The Laurussian continent itself was formed by the collision and fusion of the older continents Laurentia, Baltica, and Avalonia, previously discussed in chapter 2 (color plate 4).

In the Late Devonian, the Laurussian continent straddled the equator, which bisected present-day Greenland (color plate 8). The western Canadian and northern European region of the continent extended up to 30° north of the equator, and most of the region of the present-day United States of America extended down to 30° south of the equator. To the northeast, the once separate continental region of Kazakhstan was joined to Laurussia along the belt of the Ural mountains, and to Kazakhstan was joined the once separate continental region of Siberia along the belt of the Central Asian mountains. This peninsular region extended up to 60° north of the equator.

Two large mountain chains stretched across Laurussia: the Appalachian-Caledonian mountains and the Variscian mountains. The Appalachian-Caledonian mountain chain extended along the eastern margin of North America and northwards through Nova Scotia in Canada, across northern Scotland, and along the eastern margin of Greenland and western margin of Norway (all joined together, unlike the modern world, in which North America and Greenland are on one side of the Atlantic Ocean, and Scotland and Norway on the other). The Variscide mountain chain started in the east, in present-day Germany and Poland, extended westward through France and southern England, then turned south along the western margin of Portugal and continued on into northern Africa (present-day Morocco).

To the southeast of Laurussia was located another large continent, Gondwana, which was composed of the present-day continental regions of South America, Africa, India, Antarctica, and Australia all combined. A shallow epicontinental seaway separated the two landmasses. In the Silurian, Laurussia and Gondwana had been separated by a deep-water seaway, the Iapetus Ocean, but by the Late Devonian the tectonic plates holding the two continents had moved them closer, resulting in the collision of Laurussia and Gondwana and the creation of the Variscide mountain chain, further uplifting of the Appalachian end of the Appalachian-Caledonian mountain chain. To the west of Laurussia was the large Panthalassa Ocean, and to the east was the smaller embayment of the Paleotethys Ocean. Two large islands existed in the Paleotethys Ocean: present-day northern and southern China, then separate landmasses.

Laurussia was thus a tropical landmass and, similar to climatic conditions today, was arid and desert-like in its interior, whereas its seaward coasts experienced monsoonal storms. As discussed in chapter 2, land plants were only well developed in the coastal regions of the continent. As a result, any rains that occurred on the interior of the continent, or in the mountainous highlands, would trigger flash floods as there existed little if any plant cover to impede erosion. Sediments from arid and highland regions were highly oxidized and reddish in color, and the deposition of these sediments in thick alluvial fans and plains in both North America and Europe is the reason Laurussia was initially called the "Old Red Sandstone continent" by European geologists.

The first tetrapods evolved in marine tidal flat environments and lagoons around the margins of tropical Laurussia,[70] as evidenced by the Eifelian trackways in Poland. These intertidal environments provided our ancestors with a generally dependable food twice a day: marine animals stranded on the tidal flats. The early tetrapods had only to clumsily crawl over the tidal flats to consume the helpless animals left behind when the tide retreated. These environments also allowed the first tetrapods to gradually evolve abilities to explore the new terrestrial realm without yet being totally dependent upon the land for a source of food.

In the Frasnian, tetrapods and closely related elpistostegalian tetrapodomorph fish began to actively spread up the rivers and eventually into lakes far from the ocean. Their food source now was both fish in the rivers and arthropods out on the riverbanks and up on the land. Still, some tetrapod species remained in the marginal marine habitats of their ancestors. For example, *Obruchevichthys gracilis* lived in shallow, near-shore marine environments in tropical Laurussia (present-day Latvia), and the Frasnian ghost lineage species *Ventastega curonica* (table 3.7) still lived in these habitats in the later Famennian.

The remaining three fossil species of Frasnian tetrapods known to us, the Frasnian ghost lineage species *Densignathus rowei*, and the unknown species that made the trackways in Ireland, lived in non-marine environments. The Irish field site was bounded by the Caledonide mountains to the north, from which flash floods shed sediments into a broad alluvial plain into the Munster basin in the south. The alluvial plain was not channelized, thus the tetrapods crossing it were not exclusively river-dwelling animals.[71]

Elginerpeton pancheni (figs. 3.4, 3.5) lived in rivers feeding into the Orcadian Lake, a large freshwater lake in the Caledonide mountains that covered all of northeastern Scotland. With a length of 1.5 meters (five feet) and with jaws 400 millimeters (16 inches) long, *Elginerpeton pancheni* was the second largest vertebrate predator in aquatic ecosystems in Scotland. The largest predator was the porolepiform crossopterygian fish *Holoptychius giganteus*, which was a bit over 1.5 meters (5 feet) long and had deep, blunt-snouted skulls and jaws, 470 millimeters (18.5 inches) long, with large fangs.[72] Thus, while the first species of tetrapod known to us was already a large predator, it in turn had to deal with an even larger fish predator. Escaping predators by leaving the water and living on land is one possible ecological stimulus for the vertebrate invasion of land itself.

Of the other freshwater vertebrate fauna found with *Elginerpeton pancheni*, the antiarch placoderm fish (color plate 9) *Bothriolepis paradoxa* is the most abundant, but additional fishes present included the jawless heterostracan pteraspidomorph species *Psammosteus falcatus* and *Traquairosteus pustulatus*, the acanthodian *Cosmacanthus malcolmsoni*, the lungfish *Conchodus ostreiformis*, and the porolepiform crossopterygian *Duffichthys mirabilis*[73] (for the phylogenetic relationships of these various fish groups, see table 3.5). *Bothriolepis paradoxa* was a bottom-dwelling fish, specializing in eating the organic detritus of plants and animals that lay on the lake floor. The jawless *Psammosteus falcatus, Traquairosteus pustulatus,* and the spine-finned fish *Cosmacanthus malcolmsoni* ate small prey, chiefly molluscs and aquatic arthropods. *Duffichthys mirabilis* and *Conchodus ostreiformis* were smaller predatory fish. The large predators were our ancestor, *Elginerpeton pancheni,* and the fearsome *Holoptychius giganteus.*

The two remaining fossil Frasnian tetrapod species known to us ventured away from the equator into Earth's higher latitude regions. In the Famennian, the Frasnian ghost lineage species *Densignathus rowei* (table 3.7) also lived in floodplain habitats located 25° south of the equator in southern Laurussia, but we cannot prove that species of this lineage also lived there in the Frasnian, as no Frasnian species in the lineage have as yet been found (thus it is not shown in color plate 8).

Fossils of the Frasnian species *Sinostega pani* show that it lived in non-marine habitats in present-day North China, which was a large, isolated island in the Late Devonian, located about 15° north of the equator (color plate 8). It is found with the freshwater antiarch placoderm fishes of the genera *Remigolepis* and *Sinolepis,* an unnamed lobe-finned fish, and the lycopod land plants *Leptophloeum rhombicum* and *Sublepidodendron mirabile.*[74] Our early tetrapod ancestors seem to have preferred habitats that included bottom-dwelling, placoderm detritivores, such as *Bothriolepis paradoxa* in Scotland. In China, however, the coexisting placoderm fishes were usually species of the genera *Remigolepis* and *Sinolepis* instead of *Bothriolepis.* The highly successful species of *Bothriolepis* (color plate 9) first evolved in the Emsian in the islands of China, and then they spread throughout the world and were present in Laurussia in the Frasnian. Species of *Remigolepis* and *Sinolepis* also evolved in the islands of North and South China, but later, in the Frasnian. *Remigolepis* is particularly interesting in that species of this genus spread to Gondwana in the Frasnian, then migrated west across Gondwana and north into Laurussia in the Famennian, where they then are found living with later Famennian tetrapods. Species of *Sinolepis,* however, remained endemic to the east and never migrated to Laurussia.[75]

In the opposite latitudinal direction from *Sinostega pani, Metaxygnathus denticulatus* lived in river environments on the eastern margin of the great southern supercontinent Gondwana (present-day Australia), located about 30° south of the equator (color plate 8). It is found together with fossils of the freshwater antiarch placoderm fishes *Bothriolepis* sp., *Remigolepis* sp., and *Phyllolepis* sp., and the lungfish *Soederberghia* sp.[76] Note that species of both the older genus of detritivorous placoderm fish, *Bothriolepis,* and species of the newly evolved genus *Remigolepis,* are found living with *Metaxygnathus denticulatus* in the Frasnian in Gondwana. Species of the third coexisting placoderm genus *Phyllolepis* appear to have evolved in eastern Gondwana itself, not in the islands of China, and they also migrated west to Laurussia in the Famennian.[77] The fossils of all these animals are found in an overbank deposit adjacent to a river, and the river deposits themselves also contain fragments of the lycopod land plant *Leptophloeum* sp.

In our world, environments located 30° away from the equator are no longer tropical, but rather fall in the temperate climatic zone. However, the Frasnian world was hotter than ours; evidence for this assessment will be discussed in the next chapter. In the ancient Late Devonian world, the equator-to-pole temperature gradient was not as steep as today, so the habitats of *Metaxygnathus denticulus,* 30° south of the equator, were probably still quite hot, unlike the modern world.

In summary, Frasnian tetrapods had a worldwide distribution (color plate 8). Tetrapods evolved in the tropical, marginal marine environments of Laurussia, probably in the Emsian Age of the Early Devonian, judging by the evidence of the Eifelian trackways found in Poland. By the Frasnian Age of the Late Devonian, the tetrapods had spread out around the world. They no doubt achieved this widespread geographic dispersal by traveling along the marine coastlines of the continents and islands of the Middle Devonian world.[78] In the Frasnian, the tetrapods began to venture up the rivers and away from the coastlines of the various continents they inhabited, as evidenced by the fossil finds of the Frasnian species *Elginerpeton pancheni* (figs. 3.4, 3.5), *Sinostega pani,* and *Metaxygnathus denticulus.* Still, all of the Frasnian tetrapod species known to us are considered to be "aquatic tetrapods." That is, they possessed limbs and feet, but they used those limbs and feet mostly as paddles for swimming in open water and crawling through tangled water vegetation, algae, and lagoonal plants in marginal marine settings and terrestrial plants in freshwater environments. They could venture out on land when they needed to, to escape a predator in the water or to snap up arthropod prey on the riverbanks, but they were not fully terrestrial animals.

Prelude to Disaster

Life was good. The arthropods had firmly established themselves on land, and the vertebrates were well on the way. The Frasnian tetrapod species were still aquatic tetrapods, however, and they had a few more evolutionary modifications to make before they could become lords of the land. They never made it. Instead, they were caught up in the environmental catastrophe that triggered the end-Frasnian extinction.

PLATE 1 Peculiar Ediacaran animals of the Neoproterozoic Era. The vertical, pleated, frond-like structures are not algae, but rather the enigmatic animal *Charniodiscus*. Several individuals of another odd pleated mat-like Ediacaran animal, *Dickinsonia*, are shown gliding or resting on the sea floor. The *Tribrachidium*, an Ediacaran with very unusual tri-radial symmetry, is shown on the sea floor in the center of the bottom edge of the figure. Also shown swimming are more familiar, primitive, non-bilaterial animals: the jellyfishes. *Credit*: Photography courtesy of the Smithsonian Institution.

PLATE 2 Banded-iron strata exposed in Fortescue Falls, Karijini National Park, Australia. The scale of the outcrop exposure is given by the people in the center of the photograph. *Credit*: Photograph courtesy of Dr. Robert Kopp, Department of Earth and Planetary Sciences, Rutgers University.

PLATE 3 Banded-iron strata exposed in Mount Tom Price Mine, Pilbara region, Australia. *Credit*: Photograph courtesy of Dr. Robert Kopp, Department of Earth and Planetary Sciences, Rutgers University.

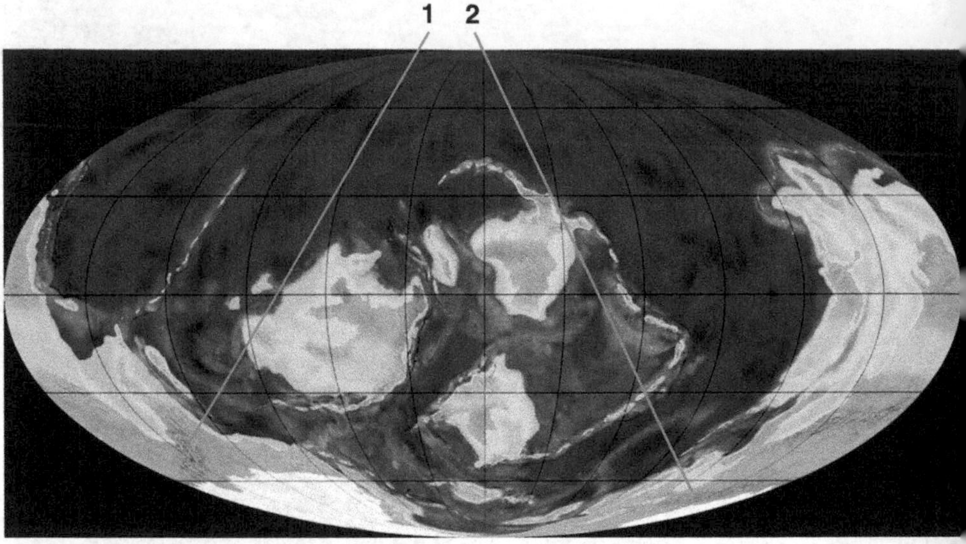

PLATE 4 Paleogeography of the Middle Ordovician world, showing the locations of the fossil finds of the earliest land plants in Argentina (1) and Oman (2) on the southern supercontinent of Gondwana. *Credit*: Paleogeographic map copyright © Ron Blakey, Colorado Plateau Geosystems, Inc. Used with permission.

PLATE 5 A river scene in the Famennian world. The four-limbed, fish-like animal standing on the river bank on the right, and swimming in the river on the right, bottom center, and left, is the tetrapod *Hynerpeton bassetti*. The large predatory fish swimming in the center is the tristichopterid tetrapodomorph *Hyneria lindae*, and the tall trees on both river banks are the non-spermatophyte lignophyte *Archaeopteris hibernica*. The strange, pole-like, furry-looking trees on the left riverbank, about half as tall as the *Archaeopteris* trees, are unusual lycophyte trees that are relatives of our modern little club mosses. *Credit*: art by Kazuhiko Sano/National Geographic Society Stock. Used with permission.

PLATE 6 A young onychophoran velvet worm (*Peripatus*) from the rainforests of Panama. *Credit*: Photograph courtesy of Dr. Lynn Kimsey, Bohart Museum of Entomology, University of California at Davis.

PLATE 7 The Late Devonian armored predator *Dunkleosteus terrelli*. The living fish was seven meters (23 feet) long. *Credit*: Illustration by Dmitry Bogdanov. Used with permission.

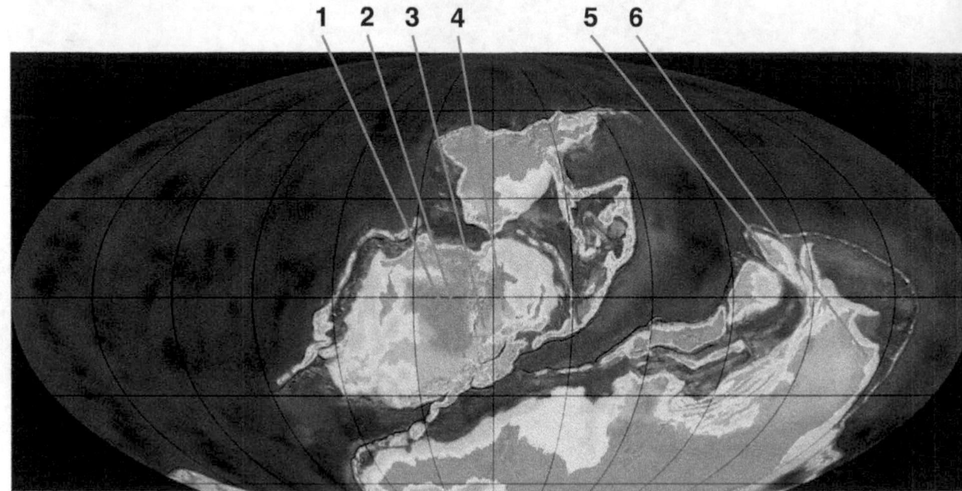

PLATE 8 Paleogeography of the Frasnian world showing the locations of the fossil finds of the trackways in Ireland (1), *Elginerpeton pancheni* (2), the trackways in Poland (3), *Obruchevichthys gracilis* (4), *Sinostega pani* (5), and *Metaxygnathus denticulatus* (6). Note the worldwide distribution of tetrapods in the Frasnian, as opposed to the Famennian (color plate 11). *Credit*: Paleogeographic map copyright © Ron Blakey, Colorado Plateau Geosystems, Inc. Used with permission.

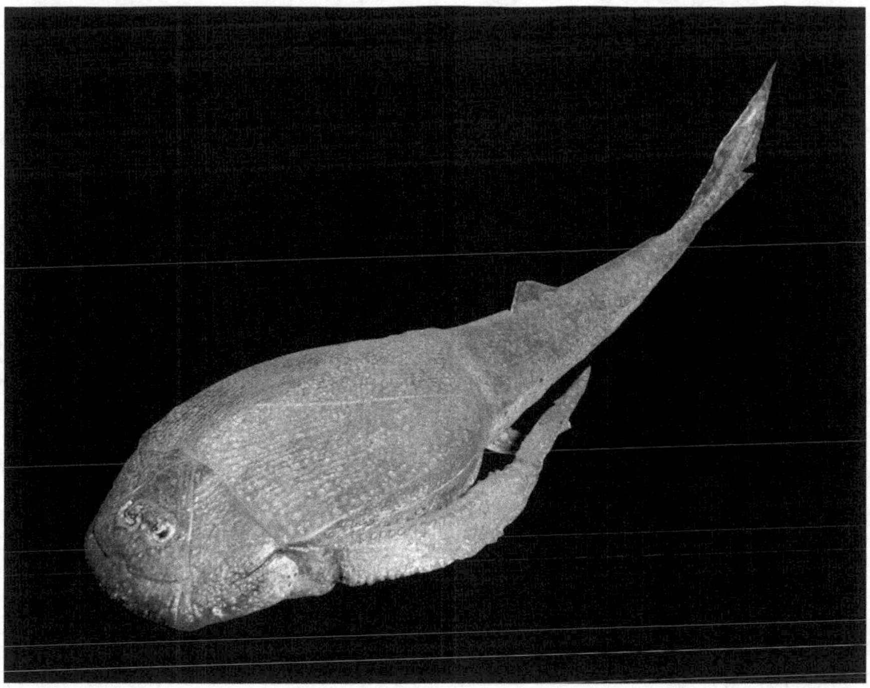

PLATE 9 Reconstruction of the armored fish *Bothriolepis canadensis*, which lived in the same aquatic habitats and coexisted with our tetrapod ancestors during the span of the Late Devonian. The living animal was about 32 centimeters (a little over a foot) long. *Credit*: Illustration © Citron / CC-BY-SA-3.0 (http://creativecommons.org/licenses/by-sa/3.0/). Used with permission.

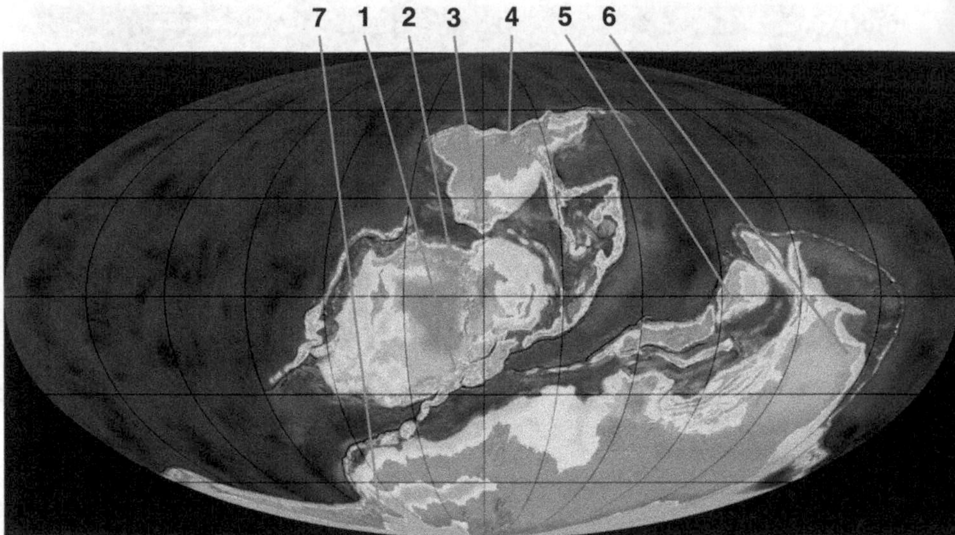

PLATE 10 Paleogeography of the Frasnian world showing the positions of active tectonic regions in the late Frasnian. The continent Laurussia was bounded on the southeast and east by the Appalachian and Variscide mountains (1) and on the northwest and north by the Antler and Ellesmeride mountains (2). The Asian continent was bounded on the south by the Ural mountains (3), and the Central Asian mountains (4) were located in the continental interior. Mountain belts occurred on the southern edge of both of the subcontinental islands of North China and South China (5), and the supercontinent Gondwana was bounded on the east by the Lachlan mountains (6) and on the west by the Bolivianide mountains (7). *Credit*: Paleogeographic map copyright © Ron Blakey, Colorado Plateau Geosystems, Inc. Used with permission.

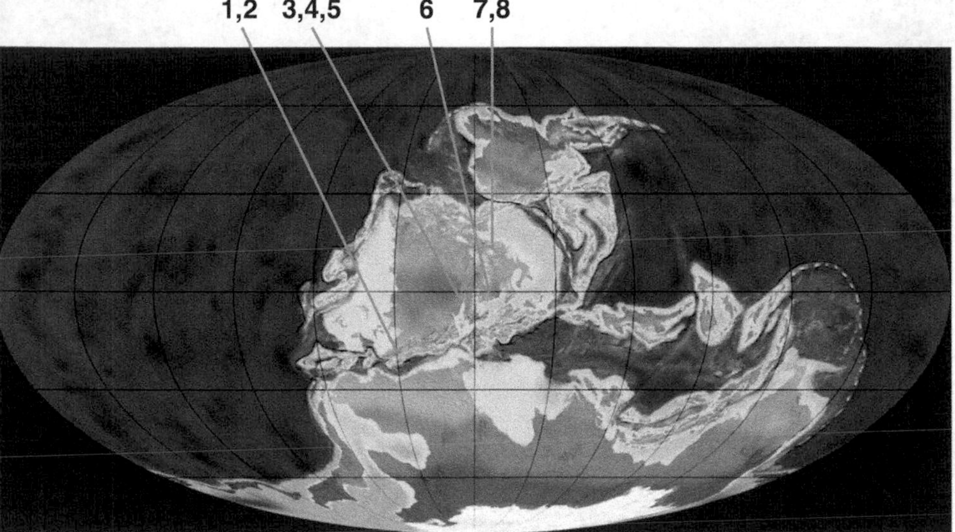

PLATE 11 Paleogeography of the Famennian world showing the locations of the fossil finds of *Hynerpeton bassetti* (1), *Densignathus rowei* (2), *Acanthostega gunnari* (3), *Ymeria denticulata* (4), species of *Ichthyostega* (5), *Ventastega curonica* (6), *Jakubsonia livnensis* (7), and *Tulerpeton curtum* (8). Note the restricted geographic distribution of tetrapods in the Famennian, as opposed to the Frasnian (color plate 8). *Credit*: Paleogeographic map copyright © Ron Blakey, Colorado Plateau Geosystems, Inc. Used with permission.

PLATE 12 Sunset on Mars. Note the blue tint in the pink sky, as opposed to an Earth sunset with a red tint in a blue sky. The view is to the west in Gusev Crater on Mars; the bottom part of the sun is just below the rim of the crater. *Credit*: Photograph courtesy of NASA.

CHAPTER 4

The First Catastrophe and Retreat

The End-Frasnian Catastrophe

The Late Devonian biodiversity crisis struck the Earth.[1] In the waning years of the Frasnian, the vertebrates and arthropods on land began to die in greater numbers. Reproduction rates in the living fell, and fewer and fewer young vertebrates and arthropods were born into terrestrial habitats. Population sizes inexorably declined, generation after generation, as land animals produced fewer matings and fewer offspring. With the deaths of their last members, vertebrate and arthropod species began to vanish from the land. By the end of the Frasnian, all of the known Frasnian tetrapod species had perished. Gone.

They were not alone. In the rivers and estuaries, the highly derived elpistostegalian tetrapodomorph fishes, close relatives to the tetrapods, also perished. The tetrapodomorph genetic lineage that had produced the parallel evolution of so many tetrapod-like traits in *Panderichthys rhombolepis*, *Elpistostege watsoni*, *Tiktaalik roseae* (fig. 3.1), and *Livoniana multidentata* was terminated. It would evolve no further.

The elpistostegalian fishes were not alone. Three entire clades of the great armored fishes, the Placodermi (color plates 7, 9), died: the Acanthothoraci, the Rhenanida, and the Petalichthyida (table 4.1). The armored fishes lost half of their phylogenetic lineages in the end-Frasnian

TABLE 4.1 The effect of the end-Frasnian extinction on the fishes.

Vertebrata (vertebrate animals)
– Petromyzontiformes (lampreys; living jawless fishes)
– unnamed clade
– – Pteraspidomorphi
– – unnamed clade
– – – Anaspida
– – – unnamed clade
– – – – Thelodonti **†Frasnian**
– – – – unnamed clade
– – – – – unnamed clade
– – – – – – Osteostraci
– – – – – – Galeaspida **†Frasnian**
– – – – – – Pituriaspida †Eifelian
– – – – – Gnathostomata (jawed vertebrates)
– – – – – – Placodermi (armored fishes)
– – – – – – – Acanthothoraci **†Frasnian**
– – – – – – – unnamed clade
– – – – – – – – unnamed clade
– – – – – – – – – Rhenanida **†Frasnian**
– – – – – – – – – Antiarchi
– – – – – – – – unnamed clade
– – – – – – – – – Arthrodira
– – – – – – – – – unnamed clade
– – – – – – – – – – Petalichthyida **†Frasnian**
– – – – – – – – – – Ptyctodontida
– – – – – – unnamed clade
– – – – – – – Chondrichthyes (cartilaginous fishes)
– – – – – – – unnamed clade
– – – – – – – – Acanthodii
– – – – – – – – Osteichthyes (bony fishes + descendants)
– – – – – – – – – Actinopterygii (ray-finned fishes)
– – – – – – – – – **Sarcopterygii** (lobe-finned fishes + descendants)

Source: Modified from the phylogenetic classification of Benton (2005) for extinct fishes, and the phylogenetic classification of Lecointre and Le Guyader (2006) for living fishes.

Major clades of vertebrate land invaders are marked in bold-faced type.

† extinct taxa. Where a period is given, taxa went extinct in that specific geologic period. Fish groups that perished in the end-Frasnian extinction are marked **†Frasnian.**

extinction; they were dealt a severe blow. The ancient jawless fishes suffered just as severely: they lost two entire clades (table 4.1), the Thelodonti and the Galeaspida, cutting their phylogenetic lineages in half as well.[2]

Out in the oceans, the impact was far worse. The great reefs were dying. The Devonian reefs were the largest reefs ever to exist in Earth history. They covered 5,000,000 square kilometers (3,105,000 square miles) of shallow marine seafloor, ten times the areal extent of modern reef ecosystems. By the end of the Frasnian, 4,999,000 square kilometers (3,104,400 square miles) of reefs had died.[3] The areal extent of reefs on Earth shrank by a factor of 5000 in the end-Frasnian extinction. The dominant framework builders of the great Devonian reefs, the stromatoporoid sponges and tabulate corals, never recovered from the diversity losses they suffered—the end-Frasnian extinction had precipitated a permanent change in the evolution of reef ecosystems on Earth.

Unlike our modern world, where the molluscs are the dominant shellfish in the oceans, in the Paleozoic that ecological role was held by the dominant brachiopods. The Devonian had been a "Golden Age" for the highly diverse brachiopod shellfish; the strata are full of fossil seashells of numerous brachiopod species. By the end of the Frasnian, three-quarters of all of the brachiopod species had died. Death spread across the seafloors of the Earth, taking not only brachiopods but other bottom-dwelling species as well. The tournayellid foraminifera lost 70 percent of their species, the phyllocarid arthropods lost 68 percent of their species, and on and on.[4]

Not even the tiny animals floating up in the water columns of the Earth's oceans were spared. The little tentaculitoids, a major element of the world's zooplankton, perished in huge numbers. Species of only two families managed to survive into the Famennian, but then they too died out.[5] The genetic lineage of a major type of Paleozoic zooplankton was terminated. Gone.

The actual severity of the end-Frasnian extinction may never be fully known to us. After all, it occurred over 374.5 million years ago—separated from us in the present by a huge chasm of time. As has been stressed thus far in the book, the fossil record is imperfect in precisely recording first and last occurrences of species, and in capturing rare species and species whose anatomy gave them a very low probability of preservation. Yet, all in all, we know that at least 70 percent of all species alive in the oceans—and perhaps as much as 82 percent of all species alive in the oceans—died in the end-Frasnian extinction.[6]

It is also known that the end-Frasnian extinction was a pulsed event, which may lead some to question why it is called an "event" if it was not confined to a single instant in time. In analogy, consider the historical event of World War II. That war was not confined to the horrific pulse that occurred near its end, when the flash of a nuclear explosion took place over Hiroshima, but consisted of a whole series of horrific pulses distributed over four years. Yet we still consider World War II to have been an "event" in human history.

In the intensely analyzed Steinbruch Schmidt stratigraphic section in Germany, the catastrophic end-Frasnian event took place in five pulses: pulse 1 of the extinction took place in the earliest part of the Late *rhenana* conodont zone (see table 4.2 for the zonal timescale; the Famennian timescale will be discussed in the next section of the chapter). The great reefs began to die in this pulse, as did many marine animals that lived on the sea bottoms. Then three severe pulses of extinction occurred in quick succession, within a single conodont zone: pulse 2 took place in the middle part of the *linguiformis* conodont zone (table 4.2), pulse 3 in the late part of *linguiformis* conodont zone, and pulse 4 in the very latest *linguiformis* conodont zone, the end of the Frasnian Age. These pulses killed organisms in every ecological zone of the oceans, not just the sea bottoms. Fishes and swimming cephalopods, as well as the tiny floating zooplankton that lived in the water column, had not been severely affected by pulse 1 of the extinction. They died in massive numbers in these following three pulses. Last, the repercussions of the extinction pulses continued into the Famennian: pulse 5 occurred in the earliest part of the Early *triangularis* conodont zone (table 4.2), the dawn of the Famennian Age, and eliminated a few poor species groups that had managed to survive the end of the Frasnian in a final coupe de grâce.[7] Elsewhere in the world, it may not be possible to distinguish between each of these five pulses, but the twin pulses of the Lower and Upper Kellwasser Horizons usually are recognizable. The end-Frasnian extinction "event" was not a single-event phenomenon, just like World War II was not.

Curiously, it can be shown that the maximum pulse in land plant extinction took place before the maximum pulses of marine extinction that occurred in the *linguiformis* conodont zone at the end of the Frasnian. In northwestern Laurussia (present-day western Canada),

TABLE 4.2 Conodont zonal timescale for the Frasnian and the Famennian.

Geologic Age	Conodont Zones	
FAMENNIAN	Late *praesulcata* Zone	
	Middle *praesulcata* Zone	
	Early *praesulcata* Zone	
	Late *expansa* Zone	
	Middle *expansa* Zone	
	Early *expansa* Zone	
	Late *postera* Zone	
	Early *postera* Zone	
	Late *trachytera* Zone	
	Early *trachytera* Zone	
	Latest *marginifera* Zone	
	Late *marginifera* Zone	
	Early *marginifera* Zone	Famennian Gap
	Late *rhomboidea* Zone	Famennian Gap
	Early *rhomboidea* Zone	Famennian Gap
	Latest *crepida* Zone	Famennian Gap
	Late *crepida* Zone	Famennian Gap
	Middle *crepida* Zone	Famennian Gap
	Early *crepida* Zone	Famennian Gap
	Late *triangularis* Zone	Famennian Gap
	Middle *triangularis* Zone	Famennian Gap
	Early *triangularis* Zone	Famennian Gap
FRASNIAN	*linguiformis* Zone	
	Late *rhenana* Zone	
	Early *rhenana* Zone	
	jamieae Zone	
	Late *hassi* Zone	
	Early *hassi* Zone	
	punctata Zone	
	transitans Zone	
	Late *falsiovalis* Zone	
	Early *falsiovalis* Zone, pars	

Source: Modified from the 1990 international zonal timescale of Ziegler and Sandberg (1990).

Note: The Famennian Gap spans the time from the Early *triangularis* to Late *marginifera* conodont zones in the Famennian Age of the Late Devonian; see text for discussion.

31 of 71 species of land plant spores (44 percent) went extinct in the Late *rhenana* conodont zone[8]—only three of the surviving 40 species (8 percent) went extinct in the following *linguiformis* conodont zone.[9] In southeastern Laurussia (present-day New York State), the diversity

of land-plant macrofossil genera dropped from 24 in the mid-Frasnian *punctata* conodont zone[10] to 13 in the Late *rhenana* conodont zone,[11] a 46 percent reduction in standing diversity. Land plants maintained the same minimum level of diversity, 13 macrofossil genera, across the Frasnian/Famennian boundary and into the middle Famennian before finally beginning to recover.

Thus the land plants suffered their worst extinction in the Late *rhenana* conodont zone, at a time in which the marine extinctions were just beginning. In the oceans, pulse 1 was merely a harbinger of what was to come, yet on land the terrestrial plants do not appear to have suffered further diversity losses at the same time that huge diversity losses were occurring out in the marine realm in the *linguiformis* conodont zone. On land all was quiet—too quiet, as we shall see in the next section.

The Famennian Gap

At the dawn of the Famennian Age, 374.5 million years ago, the land was strangely silent. Out in the oceans, waves broke over endless tracks of dead reefs around the world. The tides rose and fell, but the sandy beaches were empty of brachiopod seashells. On land, the valiant attempt of the vertebrates to invade the terrestrial realm had failed—none of the Frasnian tetrapod groups known from the fossil record survived the end-Frasnian extinction (table 4.3).

Obviously, things were not quite that dire. Some few tetrapods managed to survive somewhere on Earth, else we would not be here today; the tetrapods were, however, very, very close to total extinction. None of the known fossil tetrapod groups survived the end of the Frasnian Age, but phylogenetic analyses indicate that the two ghost lineages of tetrapods survived as their respective species are known from the Famennian fossil record. Also, we do not know the identity of the tetrapod groups that produced the Eifelian trackways in Poland or the early Frasnian trackways in Ireland. We do not know if these groups survived or perished in the end-Frasnian extinction, thus I have listed these two tetrapod lineages as "†Frasnian?" in table 4.3. Perhaps one of these lineages also managed to survive.

TABLE 4.3 The effect of the end-Frasnian extinction on the tetrapods and tetrapodomorph fishes.

Sarcopterygii (lobe-finned fishes + descendants)
– Crossopterygii (living coelacanths)
– Dipnoi (lungfishes)
– **Tetrapodomorpha** (tetrapod-like fishes + descendants)
– – basal tetrapodomorphs
– – – *Kenichthys campbelli*
– – Rhizodontia
– – Osteolepidiformes
– – – Osteolepididae
– – – Megalichthyidae
– – – Eotetrapodiformes
– – – – Tristichopteridae
– – – – *Gogonasus andrewsae* **†Frasnian**
– – – – unnamed clade
– – – – – Elpistostegalia **†Frasnian**
– – – – – **Tetrapoda** (limbed vertebrates)
– – – – – – Family incertae sedis **†Frasnian?**
– – – – – – – Unknown early Eifelian tetrapod species in Zachelmie, Poland
– – – – – – Family incertae sedis **†Frasnian?**
– – – – – – – Unknown early Frasnian tetrapod species in Valentia Island, Ireland
– – – – – – Family Elginerpetontidae **†Frasnian**
– – – – – – – *Elginerpeton pancheni* **†Frasnian**
– – – – – – – *Obruchevichthys gracilis* **†Frasnian**
– – – – – – Family incertae sedis **†Frasnian**
– – – – – – – *Sinostega pani* **†Frasnian**
– – – – – – unnamed clade
– – – – – – – [Family incertae sedis]
– – – – – – – – [*Densignathus rowei*]
– – – – – – – unnamed clade
– – – – – – – – [Family incertae sedis]
– – – – – – – – – [*Ventastega curonica*]
– – – – – – – – unnamed clade **†Frasnian**
– – – – – – – – – Family incertae sedis **†Frasnian**
– – – – – – – – – – *Metaxygnathus denticulus* **†Frasnian**

Source: Phylogenetic data modified from Benton (2005), Lecointre and Le Guyader (2006), Ahlberg et al. (2008), and Clack et al. (2012).

Note: Two "ghost" tetrapod lineages in the Frasnian Age are enclosed in brackets (e.g., [*Densignathus rowei*]). See text for discussion. Major clades of vertebrate land invaders are marked in bold-faced type.

† extinct taxa. Tetrapods and related groups that perished in the end-Frasnian extinction are marked **†Frasnian.**

The previously immensely successful and seemingly unstoppable arthropod invaders were in full retreat at the dawn of the Famennian. Their population sizes dropped to such low levels that they disappeared from the fossil record, a disappearance that is lamented by Cambridge University paleontologist Jennifer Clack: "Little is known of terrestrial arthropods from the period between the Frasnian and the middle of the Carboniferous. Their evolution has to be inferred by comparing what is known of the Mid Devonian forms with those from the Late Carboniferous. This huge gap in the invertebrate fossil record is unfortunate because the radiation of terrestrial invertebrate faunas undoubtedly influenced that of vertebrates."[12] All previously known vertebrate invaders were also gone (table 4.3), thus there were precious few vertebrate survivors on land to be influenced by any other organisms in the early Famennian.

Only very late in the Famennian do tetrapod species reappear in the fossil record. I refer to this gap in the tetrapod fossil record as the "Famennian Gap."[13] The conodont zonation for the Famennian Age is given in table 4.2, in which the Famennian Gap spans the interval of time from the Early *triangularis* conodont zone to the Late *marginifera* conodont zone.[14] The Famennian Age is divided into 22 conodont zones in total, and the Famennian Gap spans 10 of those zones, or roughly one-half of the Famennian Age. Dividing the 15.3-million-year duration of the Famennian Age (table 2.2) by the 22 conodont zones it contains results in an estimated duration of 0.7 million years per zone. Thus the Famennian Gap, or 10 conodont zones, can be estimated to have spanned about seven million years of the Famennian Age.

The Famennian Gap is also recognizable in the fossil record of the land plants and in the marine realm as well. For example, this interval of time has been formally designated as the "Early and Middle Famennian vegetation crisis" in land plants[15] and will be discussed in detail later in this chapter. In the marine realm, the stromatoporoid sponges and tabulate corals of the great Devonian reefs never recovered their previous diversity after the end-Frasnian extinction, but the rugose corals did. Even here, however, University of Liège paleontologist Eddy Poty notes that "the Rugosa remained almost totally absent . . . in the long time interval between the end-Frasnian crisis and the early part of the late

Famennian. . . . They first reappear in the Upper [= Late] *marginifera* Zone"[16]; that is, the rugose corals did not recover until after the Famennian Gap.

In this critical seven-million-year interval, something was seriously wrong with the terrestrial environments of the Earth. Whatever it was, it only began to abate in the late Famennian Age, and as a result tetrapod species once again began to recover sufficient population sizes and reappeared in the fossil record.

What Went Wrong?

What are the empirical observations—the facts—about environmental conditions on Earth during the late Frasnian and the seven million years of the Famennian Gap? It is only after we have considered all of the available empirical evidence concerning climatic conditions in this critical interval of time that we can begin to attempt to understand what went wrong on Earth at the end of the Frasnian.

FACT 1: *Major mountain-building events, driven both by the tectonic mechanisms of continental crustal block collisions and by oceanic subduction with terrane accretion, began in the Frasnian and continued through the Famennian.* In the early to mid-Frasnian, in the short geologic time period of only four million years, major mountain-belt deformation and uplift spread across the Laurussian continent driven by the incipient collision of two major continental crustal blocks. Metamorphic deformation of the Appalachian mountain belt, the southern end of the Appalachian-Caledonian mountain chain that stretched across Laurussia, has been dated to 384 million years ago in the earliest Frasnian (see table 2.2 for the radiometric ages of the geologic timescale). Then the northern African mountain belt of the Variscide mountain chain began to deform 381 million years ago, and this deformation and uplift spread northward and westward into the European mountian belt of the Variscides 380 million years ago (color plate 10). These crustal deformations and mountainous uplifts were driven by the collision of the southeastern margin of the Laurussian crustal block with the northwestern margin of the Gondwanan crustal block and the gradual closure of the Iapetus Ocean between them.[17]

On the northeastern margin of Laurussia, metamorphic deformation of the southern Ural mountain belt has been dated to 382 million years ago, also in the early Frasnian (color plate 10). This crustal deformation and uplift was driven by the collision of the Kazakhstan crustal block with Laurussia. Then, further westward, the Central Asian mountain belt (color plate 10) formed with the collision of the Siberian crustal block with the eastern margin of the Kazakhstan crustal block, leading to the formation of the Laurasian supercontinent.[18]

In addition to crustal deformation and mountainous uplift triggered by these continental block collisions with Laurussia, metamorphic deformation of the northwestern margin of Laurussia also has been dated to the late Frasnian. This deformation led to the uplift of the Antler mountain belt 379 million years ago in the present-day western margin of the United States and Canada, extending eastward into the Ellesmerian-Svallbardian mountian belt 378 million years along the present-day northern and arctic margin of Canada (color plate 10). This deformation was not triggered by continental block collisions, but rather by oceanic subduction along the northeastern margin of Laurussia and by the accretion of island-arc terranes with the mainland. Similar Late Devonian deformation and mountainous uplift driven by oceanic subduction occurred in Gondwana (color plate 10), in the Bolivianide mountain belt on the western margin of Gondwana (in present-day South America) and in the Lachlan mountain fold-belt on the eastern margin of Gondwana (in present-day eastern Australia, and in Antarctica).[19]

In summary, radiometric dating of tectonically deformed strata around the world prove that the Late Devonian was a period of intense tectonic activity. Additional empirical evidence of major mountain uplift and greatly increased silicate weathering is seen in the deposition of thick clastic wedges of Late Devonian age around the world, such as the Catskill Red Sandstone in North America and the Old Red Sandstone in Europe. One last source of empirical data on mountain-building activity during the Late Devonian is given in the ratio of the magnitudes of the isotopes strontium-87 and strontium-86 in sea water. Strontium-87 is radiogenic and is primarily found in continental rocks; strontium-86 is nonradiogenic and is primarily found in seafloor rocks. Higher ratios of strontium-87 to strontium-86 indicate higher rates of

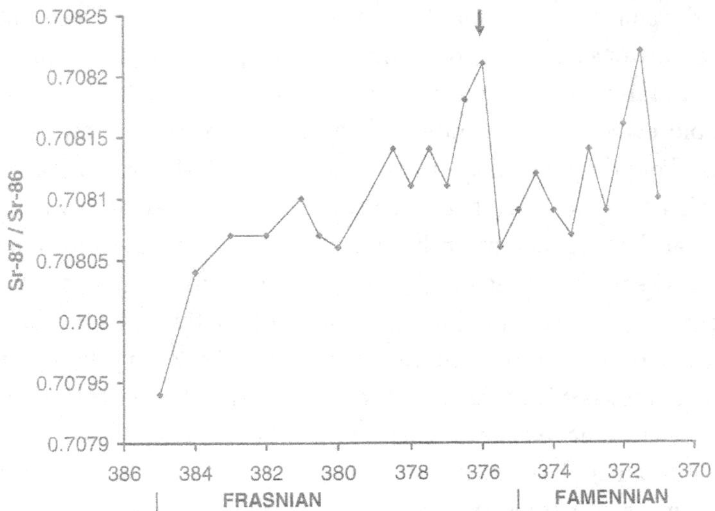

FIGURE 4.1 Variation in average ratio magnitudes of isotopes of strontium (Sr) in the Frasnian and early Famennian. Note the elevated spike (indicated by the arrow) in the isotopic record near the end of the Frasnian, indicating that an intensification of continental chemical weathering occurred at this critical time, and a second pulse of intense continental weathering in the Famennian Gap. Data from Averbuch et al. (2005).

chemical weathering and erosion of crustal rocks on the continents, with the delivery of more strontium-87 to the oceans; lower ratios indicate the reverse (or increased seafloor hydrothermal activity).[20] The average of measured strontium-87 to strontium-86 ratios increased during the Frasnian and spiked in the late Frasnian,[21] indicating an intensification of continental chemical weathering at this critical time (fig. 4.1). A second spike in average strontium-87 to strontium-86 ratios occurred in the Famennian Gap, indicating a second pulse of intense continental weathering during this interval of time. Tentative early analyses of sulfur isotopes also show a negative spike in the ratio of sulfur-34 to sulfur-32 at the end of the Frasnian, a spike that also is compatible with an increase in continental chemical weathering.[22]

There remains a third major tectonic mechanism that might also have been active in the Late Devonian, in addition to continental crustal block collisions and oceanic subduction with terrane accretion: mantle plume volcanism. In mantle plume volcanism, a super-plume of magma rises from deep in the Earth's mantle and, when it intersects continental crust at the Earth's surface, erupts almost unimaginable volumes of basaltic lava and injects a huge amount of ash and gasses into the atmosphere. Possible evidence for mantle plume volcanism in the Late Devonian comes from the Viluy Traps in eastern Siberia.[23] The original size of the Viluy Traps is unknown because they are highly eroded. Estimates suggest that the areal extent of these lava flows may have been as much as 6,000,000 square kilometers (3,726,000 square miles), but this estimate remains to be proved. Radiometric dating of the traps is not precise, but it does place the eruptions in the interval of 377 to 350 million years ago; that is, somewhere in time from the middle Frasnian to the late Tournaisian, which is indeed the critical time interval of both the end-Frasnian and end-Famennian extinctions. Smaller igneous provinces, some with kimberlite magmatism that proves a mantle source for the volcanism, are also known in other areas of the ancient Laurussian continent in the same interval of time from 376 to 350 million years ago.[24]

FACT 2: *The concentration of oxygen in the Earth's atmosphere was very low in the Frasnian but began to increase after the Famennian Gap.* The empirical data that support this conclusion come from the distribution of charcoal deposits in the sedimentary rock record. Charcoal is produced by wildfires, and wildfires will not occur unless specific levels of oxygen are present in the atmosphere. Despite extensive stratigraphic searches, the Frasnian Age is known for the rarity of charcoal deposits in its strata. The absence of wildfires in the Frasnian is quite striking, given the spread of Earth's first forests during this same time interval (chapter 2). Geologists Andrew Scott, of the University of London, and Ian Glasspool, of the Field Museum of Natural History, comment that:

> *Archaeopteris* [color plate 5], the first large woody tree, evolved in the Late Devonian and spread rapidly. By the Mid-Late Frasnian, monospecific archaeopterid forests dominated lowland areas and coastal settings over a vast geographic range. Despite this extensive biomass, charcoal occurrences are rare with only isolated fragments of charred *Callixylon* (archaeopterid) wood reported from this interval and small amounts of inertodetrinite (microscopic charcoal fragments) preserved in early Late Devonian Canadian coals.[25]

In fact, the entire interval of the Eifelian through the Frasnian has been termed the "charcoal gap," as there is very little evidence of wildfires anywhere during this period of time.[26]

In contrast, the Famennian Age is well known for numerous charcoal deposits in its strata, deposits that prove that wildfires were common. The oldest Famennian charcoal deposits yet discovered come from strata in Pennsylvania dated to the Middle *expansa* conodont zone,[27] six conodont zones after the end of the Famennian Gap (table 4.3). Frequent wildfires occurred in widely separated regions of the Earth in the time span represented by the last five conodont zones in the Famennian.[28]

An oxygen level of at least 13 percent has to be present in the atmosphere in order for wildfires to ignite.[29] Thus hard empirical data exist to prove that the atmosphere of the Earth contained 13 percent oxygen, or more, in the last 3.5 million years or so of the Famennian.[30] The absence of charcoal in strata in the Famennian Gap, and in the mid-to late Frasnian, can be used to argue that oxygen levels present in the Earth's atmosphere were below 13 percent during this span of time. Woody plant material was abundant in the Frasnian; if wildfires occurred, then charcoal should be present in Frasnian strata.

In addition to empirical data, another type of evidence exists that supports the previous conclusion concerning oxygen levels in the Earth's atmosphere during the Late Devonian: the "empirical model," a model whose predictions not only depend upon the mathematics of the model's assumptions but also upon input from actual empirical data. Two empirical models, Rock-Abundance and Geocarbsulf, have been constructed by the geochemist Robert Berner of Yale University and his colleagues in an attempt to predict atmospheric oxygen concentrations throughout the span of the Phanerozoic on Earth.[31] Both models predict that the concentration of oxygen in the Earth's atmosphere increased from the Frasnian into the Famennian and continued to increase throughout the Carboniferous into the Permian (fig. 4.2). However, the models yield contradictory predictions for the period of time from the Silurian through the Frasnian. The older Rock-Abundance model predicts a slight decrease in the amount of oxygen in the Earth's atmosphere during this interval of time. The Geocarbsulf model predicts an increase in oxygen in the atmosphere from the Silurian into the Early Devonian,

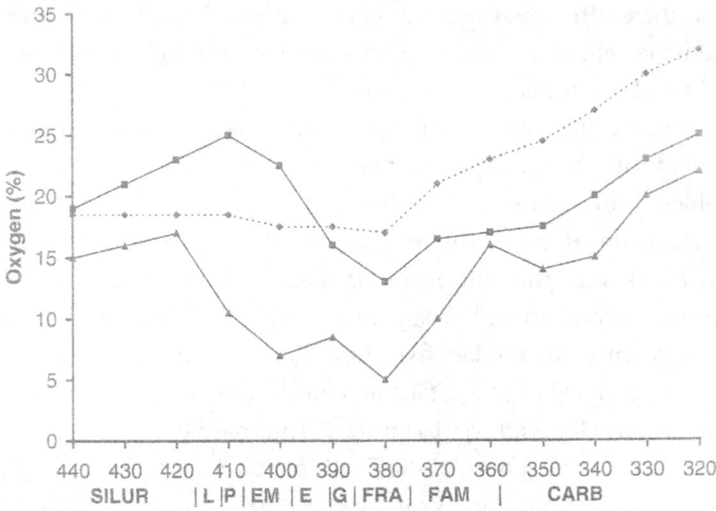

FIGURE 4.2 Modeled fluctuations in atmospheric oxygen content from the Silurian through the Early Carboniferous. The diamond data points and dotted line show the predictions of the Rock-Abundance model; the square data points and solid line show the predictions of the Geocarbsulf model; and the triangular data points and solid line show the predictions of the C_{org}–P-ratio model; see text for discussion. Geologic timescale abbreviations: SILUR, Silurian; L, Lochkovian; P, Pragian; EM, Emsian; E, Eifelian; G, Givetian; FRA, Frasnian; FAM, Famennian; CARB, Carboniferous. Data from Scott and Glasspool (2006) and Algeo and Ingall (2007).

and then predicts a sharp decline in oxygen concentrations through the Middle Devonian into the Frasnian (fig. 4.2). Both models predict minimum oxygen concentrations in the atmosphere in the mid-Frasnian (380 million years ago): 17 percent in the Rock-Abundance model and 13 percent in the Geocarbsulf model, and higher oxygen concentrations in the Famennian: 21 percent to 23 percent in the Rock-Abundance model and 16 percent to 17 percent in the Geocarbsulf model.

However, Scott and Glasspool urge some caution in accepting either model's predictions: "Predictions of the degree to which O_2 fluctuated [in the Late Devonian] are based on data-driven models. However, the complexity of the feedback mechanisms that govern O_2 levels results in a large degree of uncertainty in these models, as is evident in their frequent refinement. Additional data sets are invaluable to these refinements."[32] They then note that their charcoal data set better agrees with the results of the Geocarbsulf model than the Rock-Abundance model in the interval of time from the Silurian through the Frasnian. Occurrences of charcoal are known from strata in the Late Silurian and Early Devonian, and this supports the model prediction that oxygen concentrations actually increased in the Earth's atmosphere during this period of time (fig. 4.2). Lower oxygen concentrations in the Middle Devonian and Frasnian atmosphere fit with the "charcoal gap"[33] in the strata of this time period. However, given the frequency of charcoal occurrences in late Famennian strata, they suggest that oxygen concentrations in the atmosphere in the late Famennian may have been higher than the 17 percent predicted by the Geocarbsulf model, and more in accord with the 23 percent concentration predicted by the Rock-Abundance model. The discrepancy between the charcoal data set and the Geocarbsulf models becomes even more marked in the Carboniferous: "Collectively, these data suggest levels of O_2 modeled for this interval rising from 17% to 23.5% [in the Geocarbsulf model] are inappropriate and instead favor prior, higher levels modeled at ≈23–31.5% [in the Rock-Abundance model], values further supported by the occurrence of very large arthropods at this time."[34] The appearance of gigantic arthropods on Earth in the Carboniferous will be considered in detail in chapter 7.

To add to the uncertainty about the oxygen content of the Late Devonian atmosphere, yet another empirical geochemical model has been proposed by sedimentologist Thomas Algeo at the University of Cincinnati and his colleague Ellery Ingall at the Georgia Institute of Technology, that is based upon the ratio of organic carbon to phosphorus in ancient sedimentary rock.[35] The C_{org}–P-ratio model predicts even lower oxygen concentrations in the atmosphere than either the Rock-Abundance or Geocarbsulf models (fig. 4.2). The shape of the C_{org}–P-ratio model is more in accord with the Geocarbsulf model in that is shows a higher concentration of oxygen in the atmosphere in the Silurian, a decrease in atmospheric oxygen in the Middle Devonian, and an oxygen minimum at 380 million years ago in the mid-Frasnian. However, the C_{org}–P-ratio model predicts an atmospheric oxygen level of only 5 percent in the mid-Frasnian minimum (fig. 4.2), a value less than half of the 13 percent value predicted by the Geocarbsulf model. The C_{org}–P-ratio model also predicts much lower oxygen values in the

Carboniferous than the Rock-Abundance model, with oxygen levels rising above present atmospheric levels (21 percent) only in the very latest Early Carboniferous (320 million years ago; fig. 4.2). As such, these values are subject to the same criticisms leveled against the Geocarbsulf model predictions for the Carboniferous by Scott and Glasspool (2006), as discussed above.

In summary, we have two lines of evidence that oxygen was in short supply in the atmosphere of the Frasnian world, the charcoal distribution data and the empirically based geochemical models. This fact leads immediately to the question of why oxygen was in short supply. One obvious way to address that question is to consider the organisms that are producing oxygen in the first place: the plants. As outlined at the beginning of the chapter, it is known that land plants suffered a major loss in diversity in the Late *rhenana* conodont zone in the late Frasnian. The number of known macrofossil genera dropped to 13, a minimum diversity level that persisted unchanged through the Famennian Gap.

Taking a look at longer timescales, the University of Texas paleobotanists Anne Raymond and Cheryl Metz have shown that land plants were losing diversity through the entire Middle Devonian into the Frasnian,[36] though not as severely as the diversity loss that occurred in the late Frasnian. Land plant macrofossil diversity was at its maximum in the Early Devonian Emsian, with 38 genera known. Diversity dropped to 31 genera in the Middle Devonian Eifelian, dropped further to 24 genera in the middle Frasnian, and then precipitously dropped to a low of 13 genera in the late Frasnian. In summary, the decline in land plant diversity seen in the fossil record roughly parallels the decline in oxygen levels in the atmosphere predicted by the Geocarbsulf and C_{org}–P-ratio models in the Middle Devonian and Frasnian, suggesting a causal relationship between land plant diversity and atmospheric oxygen levels.

FACT 3: *Climatic instability and seasonal stress increased during the Frasnian.* Beginning in the middle Frasnian, species of the large lignophyte tree *Archaeopteris* (color plate 5) began to increase rapidly in abundance such that by the late Frasnian "monospecific archaeopterid forests dominated lowland areas and coastal settings over a vast geographic range," as noted by paleobotanists Scott and Glasspool (2006) in the previous section (Fact 2). At the same time, the previously diverse aneurophytalean lignophytes declined precipitously in abundance, as did herbaceous lycophytes. A climatic trigger would be one possible explanation for this marked change in the dominance structure of land plant species, and cooling due to the early onset of glaciation in Gondwana has been suggested.[37]

In the late Frasnian, the *Archaeopteris* trees in North America also began to show growth rings and became deciduous, indicating some type of seasonal climatic instability.[38] However, the shedding of deciduous branches with laminar leaves and the onset of slowed growth, producing growth rings, could equally have been produced by the onset of dry seasons following wet ones, or by cold seasons following warm ones. Whatever the climatic trigger was, late Frasnian *Archaeopteris* trees experienced seasonal stress and climatic instability that was not present in the early and middle Frasnian.

FACT 4: *Massive amounts of marine-derived organic carbon were deposited in the shallow seas of the Earth during the Frasnian and the Famennian Gap.* Evidence for this conclusion comes from the worldwide distribution of black shale deposits in Frasnian and Famennian Gap strata. These strata were deposited in shallow seaways, generally less than 200 meters (656 feet) in water depth, that covered wide areas of the continental landmasses in the Late Devonian. The amount of organic carbon present in the Late Devonian black shale deposits is massive, and these shales are an important source of oil and natural gas in many parts of the world in the present day.[39] When we consider the extensive, hydrocarbon-rich deposits that exist in many Late Devonian strata, two questions immediately come to mind: How did all this carbon accumulate? Where did it come from?

To answer the first question, there was apparently insufficient oxygen present in the bottom waters of the shallow seas of the Earth for bacteria to oxidize the carbon and to release it as carbon dioxide into the seawater in the normal process of aerobic respiration. The center-basin, deeper-water regions of the shallow seaways were depleted of oxygen, and it is in these areas that the majority of the black shale deposits are found. In aqueous environments, just how oxygen-depleted is "depleted of oxygen"? In chemical redox terminology, bottom waters are described as oxic if they contain more than 1.0 milliliter of oxygen (O_2) per liter of water, dysoxic if they contain between 1.0 and 0.1 milliliters of oxygen per liter of water, and anoxic if they contain less than 0.1 milliliter of oxygen per liter of water down to no oxygen content at all.

The deeper water regions of the Frasnian shallow seas were often anoxic. The depletion of oxygen even in the shallow marine waters of the Frasnian is not surprising, in that the concentration of oxygen in the Earth's atmosphere itself was also very low in this same interval of time, as discussed above (the section on Fact 2). Also, there exist other mechanisms to deplete the oxygen in marine basins in addition to that provided by the existence of low partial pressures of oxygen in the atmosphere. Stagnation, in which basin waters become vertically stratified and no longer mix, will also lead to anoxia in bottom waters as aerobic bacteria use up all the oxygen in respiration and no new sources of oxygen are mixed in from surface waters. Eutrophication, overproductivity of algae and phytoplankton in the surface waters, will also lead to anoxia in bottom waters. The rain of excess organic material from the algal blooms in the surface waters triggers an equal excess of bacterial oxidation of the material in the bottom waters, eventually using up all the available oxygen in aerobic respiration.

However, evidence does exist that anoxic waters were not confined to the deeper-water basin centers in the Frasnian shallow seas. The presence of the chemical compound isorenieratane and its diagenetic products in some Frasnian black shales demonstrates that anoxic waters extended up all the way into the photic-zone surface waters. Isorenieratane is produced by green-sulfur bacteria of the family Chlorobiaceae, a group of bacteria that are anaerobic photosynthesizers. Because the green-sulfur bacteria are photosynthetic, they require light; as anaerobes, however, they also require the presence of hydrogen sulfide and the absence of oxygen. Thus the discovery of isorenieratane biomarkers in some Frasnian black shales indicates that even the photic-zone surface waters were anoxic during the deposition of these shales, as green-sulfur bacteria were flourishing in the shallow-water photic zone.[40]

One last black-shale oxygen fact is the most surprising of all: it can be proved that bottom waters in many black-shale depositional areas were not constantly devoid of oxygen during the Late Devonian, and that in some cases deeper waters actually contained some oxygen at the very same time as the shallow waters were totally anoxic. Fossils of bottom-dwelling foraminifera are frequently found in Frasnian and Famennian black shales in both the Appalachian and Illinois basins in North America. Related modern foraminifera can tolerate short periods without oxygen but generally live in marine environments with an oxygen concentration of around 0.05 milliliters of oxygen per liter of water, which is technically still anoxic but not entirely devoid of oxygen.[41] In Europe, some Famennian black shales have been shown to contain geochemical signatures of both photic-zone anoxia in the water column simultaneous with the presence of oxygen in the bottom waters. The latter is indicated by the presence of large pyrite framboids and low ratios of uranium to thorium in the shale, which are taken as evidence that the sediment was deposited in oxic to dysoxic waters, yet the additional presence of isorenieratane biomarkers in the shale indicate that the photic-zone waters above the sea floor were anoxic.[42]

This last piece of evidence disproves the theory that the Late Devonian organic-rich sediments accumulated in stagnant, vertically stratified water masses in which the bottom waters were perpetually devoid of oxygen. Vertical mixing of the water masses and renewal of the bottom waters had to have occurred, at least occasionally, to explain the presence of some oxygen in these waters. Thus anoxia in Late Devonian shallow seawaters was not the product of stagnation but rather of eutrophication—the excess photosynthetic production of hydrocarbons was itself the trigger for oxygen depletion in marine waters that led to the accumulation of unoxidized hydrocarbons in the sediments on the seafloor.

Additional evidence supporting eutrophication comes from geochemical analyses that indicate that the black shale deposition episodes were confined to the shallow seas near the coasts of the continents, and that they were not oceanic phenomena. There appears to have been no change in the global pattern of sulfur cycling in the oceans during the Late Devonian, indicating no change in oceanic redox conditions.[43] The entire oceans of the Late Devonian world were not depleted of oxygen, just the epicontinental seaways and the shallow shelf regions facing the open oceans. It is in these shallow-water settings that eutrophication typically occurs, triggered either by nutrient runoff from land or nutrient upwelling from submarine currents rising along the continental shelf margins.

This conclusion leads immediately to another question: What triggered the excess photosynthetic productivity that produced eutrophication in Frasnian shallow seas? This question in turn leads us back to the second question posed at the beginning of this section of the chapter: Where did all the carbon come from? To answer the second question, the overwhelming majority of the carbon deposited in the black shales of the Frasnian and the Famennian Gap is organic and marine in origin, produced by phytoplankton and marine algae. Terrestrially derived organic material, chiefly spores and tracheids, generally constituted less than 20 percent of the total kerogen content of mid-Frasnian black shales in Europe[44] and less than 10 percent in the lower Huron Black Shale of North America, dated to the Famennian Gap.[45] This is in sharp contrast to strata deposited after the Famennian Gap, in which the terrestrially derived organic content reached as high as 60 percent of the total kerogen, as will be discussed in chapter 6.

Another line of evidence on the source of the carbon deposited in the Frasnian black shales comes from the analysis of carbon isotopes. As discussed in chapter 1, when we considered the geochemical evidence for the earliest life on Earth, photoautotrophic organisms prefer the lighter isotope of carbon in photosynthesis; that is, plants preferentially extract carbon-12 from the atmosphere to synthesize hydrocarbons, and leave the heavier isotope carbon-13 behind. The ratio of carbon-12 to carbon-13 can be used to track the degree of productivity of photosynthetic organisms in the stratigraphic record: enrichment of carbon-13 in sea water indicates periods of high productivity and preferential removal of carbon-12 from the environment by photoautotrophs, and the reverse indicates periods of low productivity. In addition, the oxidation of organic hydrocarbons by bacteria in aerobic respiration released the carbon-12 back into the environment, thus the enrichment of carbon-13 in sea water is evidence that these hydrocarbons were not oxidized but rather were buried in the sediments on the sea bottoms. This enrichment of carbon-13 in the oceans is termed a "heavy carbon anomaly," in that the heavier isotope of carbon is anomalously concentrated.

Geochemical analyses that my colleagues and I conducted in 1985 detected two large heavy carbon anomalies in the Lower and Upper Kellwasser Horizons in Germany.[46] As summarized at the beginning of the chapter, the pulsed extinctions that occurred in the marine realm during the latest Frasnian took place at the same time as the deposition of these black shales and bituminous limestones, which was the reason we were interested in analyzing the carbon isotope ratios present in these strata. Subsequent geochemical analyses by scientists in many nations have demonstrated that these heavy carbon anomalies have been found all around the world; that is, they are evidence that major productivity episodes by phytoplankton and marine algae occurred simultaneously in the shallow seas of the Earth in the latest Frasnian, and that massive burial of carbon-12 fixed in unoxidized organic hydrocarbons occurred in the marine sediments.

The conclusion of this section of the chapter still leaves us with an unanswered question: What triggered the excess photosynthetic productivity that produced eutrophication in Frasnian shallow seas? It can be shown that this eutrophication was episodic, not continuous, as will be demonstrated in the next section of the chapter.

FACT 5: *The Frasnian was punctuated by episodes of extensive and apparently abrupt increases in the geographic extent of black-shale depositional regions in the shallow seas of the Earth, the largest of which occurred in the latest Frasnian and which correlates with marine extinction events. The exact causes of these black-shale-expansion episodes remain unknown.* In the well-studied stratigraphic sections of New York State in North America, nine distinct episodes of black-shale expansion occurred in the Frasnian.[47] In each episode, the depositional region of black shales spread from the deeper-water regions of the Appalachian basin in the west of New York into the shallower-water regions in the east. The youngest two of these episodes are temporally correlated with the Lower and Upper Kellwasser Horizons in Europe, thus these episodes of black-shale expansion occurred in widely separated, shallow basins of the Earth at approximately the same time.

What could trigger episodes of black-shale expansion to occur simultaneously in many geographically isolated shallow seas? One clear way to trigger a simultaneous worldwide change in the marine realm would be to change sea level. Because black shales are usually deeper-water facies, an episode of black-shale expansion would represent an expansion of the geographic extent of deeper-water regions; that is, such an episode would indicate a rise in sea level.

Alternatively, it has been proposed that sea level and water depth remained the same and that episodes of black-shale expansion indicate that formerly oxygenated bottom waters in the Earth's shallow seas were abruptly replaced by oxygen-depleted shallow bottom waters. One mechanism by which shallow water regions can abruptly switch from oxygen rich to oxygen poor is familiar to many farmers with fresh water ponds on their property that receive too much nutrient runoff from surrounding fertilized fields: eutrophication of the ponds. Pond algae experience a bloom in their productivity, given the excess nutrients in the ponds that field runoff produces, and create a green scum that entirely covers the surface of a pond in a eutrophication event. Aerobic bacterial decay of the excess algal organic matter rapidly depletes oxygen levels in the pond's waters and will eventually kill all the fish in the pond as they soon have no oxygen to breathe, much to the farmer's dismay. Similarly, episodes of black-shale expansion in Frasnian shallow seas have been proposed to indicate eutrophication episodes, not rises in sea level. The evidence of heavy-carbon anomalies presented in the previous section of the chapter (Fact 4 above) indicates that major eutrophication episodes did occur on a worldwide basis in the latest Frasnian. This leads us back to the unanswered question posed in previous section: What triggered the excess photosynthetic productivity that produced eutrophication in Frasnian shallow seas?

In an analogous consideration of the eutrophication of a farmer's pond due to excessive nutrient runoff from the surrounding fertilized farmland, it has been proposed that eutrophication in Frasnian shallow seas was triggered by excess nutrient runoff from the Frasnian landmasses. Evidence for increased nutrient production on land comes from the consideration of one particular activity of the newly evolved forests of the Frasnian: the weathering of terrestrial silicate rocks. Prior to the Frasnian, most land plants were restricted to the wetlands in the lowlands and had shallow, unbranched or sparsely branched root systems. In the Earth's oldest known forest in the Middle Devonian Givetian, the famous Gilboa fossil forest of New York State, the large cladoxylopsid trees *Wattieza (Eospermatopteris) erianus* (fig. 2.9) were over eight meters (26 feet) tall and had trunks that were a half-meter to a meter (1.6 to 3.3 feet) in diameter, yet these trees were anchored only by a broad, bulbous base that had numerous short roots.[48]

In contrast, the giant lignophyte *Archaeopteris* (color plate 5) trees of the Frasnian towered 30 meters (100 feet) into the air and had deep, branched root systems that penetrated more than a meter (3.3 feet) into the ground. These trees produced deep soil horizons both by chemical and mechanical weathering of the rocks into which they were rooted. Chemical weathering of silicate rocks would increase the available supply of nutrients, particularly phosphorus and silica, that could be transported by rainfall and rivers from the land into the shallow seas. The evidence that the newly evolved *Archaeopteris* forests could extensively weather silicate rocks has led to the proposal that they greatly increased the nutrient runoff from the land, which in turn triggered eutrophic blooms of marine phytoplankton and algae in the seas, which in turn led to the depletion of oxygen in the shallow seas and the accumulation of unoxidized organic hydrocarbons in black shales.[49]

Evidence also exists that the Frasnian was a time of extensive mountain-building activity around the Earth (Fact 1 above). Uplift of silicate rocks in towering mountains would also expose these rocks to extensive erosion and weathering, which would deliver increased amounts of the nutrients phosphorus and silica into the oceans. Thus we have empirical evidence of two new sources of increased nutrient production acting simultaneously on land in the Frasnian: weathering of silicate rocks in newly evolved *Archaeopteris* forests and in newly uplifted mountain chains.

Some models of global biochemical cycles[50] predict that a 40 percent increase in the runoff of phosphorus into the Earth's oceans would be sufficient to trigger eutrophication, and the resultant carbon-12 fixing and burial would be sufficient to produce the two heavy-carbon anomalies seen in the late Frasnian strata (discussed in Fact 4 above). Unfortunately, other geochemical and climatic modeling of eutrophication in Frasnian seas suggests a more complicated trigger than just increased nutrient runoff from land. For example, the "eutrophication pump" model of the Northwestern University geochemist Adam Murphy and colleagues[51] proposes that the shallow seas of the Frasnian were thermally stratified on a seasonal basis: that bottom waters were anoxic and unrenewed during climatically stable warmer seasons, and dysoxic to oxic during mixing episodes in stormier, cooler seasons. This seasonally linked stratification-versus-mixing cycle in bottom waters became more pronounced throughout the Frasnian with the observed enhancement of seasonality during the same time interval (Fact 3 above). The cycle is also proposed to be linked to variation in sea level, with anoxia in bottom waters during times of rising sea level and decreased input of sediment from the land, and dysoxic to oxic conditions in bottom waters in times of falling sea level and increased sediment influx from land.

Deposition of the Kellwasser black shales in the eutrophication pump model is proposed to be ultimately triggered by the progressive increase in the magnitude of seasonal climatic variations that occurred in the late Frasnian, an observed increase that is in turn proposed to be linked to global cooling in the same interval. If true, the Kellwasser episodes of black-shale expansion should correlate with drops in global temperature, and we shall see in the next section of the chapter that some geochemical temperature data indeed support that conclusion, though we also shall see that the data are disputed.

The eutrophication pump model has been criticized on several grounds, particularly with regard to the connection of episodes of black-shale expansion with global cooling events, events that traditionally have been assumed to occur during warm climatic periods and sea level highstand.[52] Murphy and colleagues agree that episodes of black-shale expansion should also correlate with times of sea level highstand, times of sediment starvation, and hence greater concentrations of organic matter in bottom waters in their model, but they defend the linkage of episodes of black-shale expansion to cooling events by noting that a similar linkage has been proposed for the black-shale expansion and marine extinction event that occurred at the end of the Famennian,[53] a topic that we shall consider in more detail in chapter 6.

Also, it can be shown that the style and intensity of marine anoxia was not the same during the Lower and Upper Kellwasser Horizons, hence the eutrophication pump model would have to be modified somewhat to take this observation into account. The Lower Kellwasser Horizon in Germany is associated with anoxia in shallow bottom waters, while the deeper basinal waters often remained oxic. In the Upper Kellwasser Horizon, the deep bottom waters were anoxic, while the shallow bottom waters were often only dysoxic.[54] While both phases of marine anoxia are driven by eutrophication, the intensity of anoxia associated with the Lower Kellwasser Horizon was less and was more viable over time; for example, starting in the earlier part of the Late *rhenana* conodont zone in Germany but becoming more developed in the later part of the Late *rhenana* condont zone in Poland.[55] In contrast, anoxia associated with the Upper Kellwasser Horizon is synchronous around the world, occurring in the latest part of the *linguiformis* conodont zone, and may be a function of both eutrophication and maximum sea-level highstand occurring simultaneously.[56]

An entirely different global warming model for the eutrophication of Frasnian seas has been proposed by the University of Oregon paleopedologist[57] Gregory Retallack and colleagues,[58] one that totally contradicts the eutrophication pump model of Murphy and colleagues. Evidence for two unusually hot and humid climatic periods on land in the Frasnian of North America comes from the analysis of deep-calcic fossil soil horizons. Deep-calcic paleosols have horizons that contain carbonate nodules that are over 650 millimeters (26 inches) deep, and they indicate warm and humid climates with higher precipitation and longer soil formation than that seen in other fossil soil horizons.

Retallack and colleagues have proposed that the observed deep-calcic-soil horizons A and B that exist in North America correlate with the Lower and Upper Kellwasser Horizons in Europe. If true, this would mean that the Kellwasser Horizons were deposited during two periods of global warming, not global cooling. Global warming episodes would also result in the melting of any polar ice on Earth, which would lead to rises in sea level that could explain the globally simultaneous nature of the Lower and Upper Kellwasser black-shale expansion episodes. In addition, increased rainfall would lead to increased weathering of silicate rocks on land, both by the spread of *Archaeopteris* trees (color plate 5) and by increased mountain erosion, which would contribute to the eutrophication of shallow seas and could likewise explain the Lower and Upper Kellwasser black-shale-expansion episodes. Thus this model invokes both sea-level rises and terrestrial-nutrient-induced eutrophication pulses as triggers for the globally-simultaneous Kellwasser black-shale expansion episodes in the late Frasnian. On the other hand, the critical assumption of this model is the proposed pedostratigraphic correlation of the observed deep-calcic soil horizons A and B in North America with the Lower and Upper Kellwasser Horizons in Europe. No zonal evidence, either spore or conodont, is offered to support this temporal correlation. Instead, a lithostratigraphic argument is offered, based on a similar scaling in the thickness of sedimentary units in the two areas of the Earth.[59]

We have now reached the sad state of affairs where we have facts, empirical observations, for which we have no unequivocal causal explanations. This situation will get worse in the following sections of the chapter. As summarized at the beginning of the chapter, the first pulse of globally simultaneous extinction in the marine realm in the late Frasnian correlates with the onset of the Lower Kellwasser black-shale expansion. The three severest pulses of extinction occurred during the onset and depositional time span represented by the Upper Kellwasser black-shale expansion. Those are the facts—but exactly what triggered the episodes of black-shale expansion? For that question we have only models that make contradictory predictions. Why did the marine extinctions occur during the late-Frasnian Kellwasser black-shale expansion but not with the several episodes of black-shale expansion that occurred earlier in the Frasnian? We shall consider this last question in detail in the Kill Mechanism section at the end of this chapter.

FACT 6: *Data concerning global temperature levels during the Famennian Gap are contradictory.* The majority of the paleobiological data supports the hypothesis that the Earth was cold during the Famennian Gap and began to warm again after the Famennian Gap. On land, a marked latitudinal contraction takes place in equatorial global floras in the latest Frasnian; that is, equatorial fossil spore assemblages reached their maximum latitudinal width in the Frasnian—floral evidence that the Frasnian world was indeed hot—but that latitudinal width contracts sharply at the end of the Frasnian.[60] Plants that had geographic ranges up into the subpolar regions of the Earth in the Frasnian are now found only in the tropical regions in the Famennian Gap, and it is difficult to see how this pattern of latitudinal contraction in the geographic dispersal of land plants could be produced by anything other than global cooling. In addition, the University of Liège paleobotanist Maurice Streel and his colleagues also note that a severe loss of spore diversity, and hence of the diversity of land plants producing the spores, occurred in the latest Frasnian and continued in the Famennian Gap, and they attribute this to global cooling: "The diversity loss of miospores, observed in the Late Frasnian, is obvious also in the Early and Middle Famennian but becomes more and more severe from the paleo-tropical to the paleo-subpolar regions. A strong climatic gradient from a warm equator to a cool pole was apparently operating during the Early and Middle Famennian. Global cooling, except probably in the equatorial belt, forced land plants to migrate from high latitudes to lower latitudes"[61]; they explicitly refer to this interval of time as "the Early and Middle Famennian vegetation crisis." The Early and Middle Famennian Sub-Ages of Streel and colleagues span the time interval from the Early *triangularis* through the Late *marginifera* conodont zones,[62] which almost exactly coincide with the time span of the Famennian Gap (table 4.3). Streel and colleagues further note that: "From the start of the Late Famennian, miospore diversity increases again. . . . The climatic gradient was probably less marked because the same miospore zones are recorded again (like in the Frasnian) from paleo-tropical to paleo-subpolar regions."[63] The Late Famennian Sub-Age of Streel and colleagues begins with the Latest *marginifera* conodont zone, one conodont zone after the end of the Famennian Gap (table 4.3),[64] thus the paleobotanical evidence indicates a warming trend on land following the Famennian Gap.

In the seas, similar latitudinal contractions in the equatorial geographic range after the end-Frasnian extinction are seen in stromatoporoids, brachiopods, foraminifera, and trilobites. These marine animals had geographic ranges that extended well up into higher latitudes of the Earth in the Frasnian, but in the Famennian Gap they are found only in low-latitude regions, close to the equator.[65] As with the land plants, it is difficult to see how this pattern of latitudinal contraction in the geographic dispersal of tropical marine animals could be produced by anything other than global cooling. Last, siliceous sponges adapted to cold water actually flourished and experienced abundant blooms on a global scale in the late Frasnian when just about everything else was dying, blooms that persisted through the Famennian Gap; this is taken as additional evidence of a shift to colder temperatures in the Earth's oceans in this period of time.[66]

Two last lines of empirical geochemical evidence can also be used to argue for global cooling in the late Frasnian: the two heavy-carbon anomalies that occurred worldwide at the times of the Lower and Upper Kellwasser Horizons in Europe (see Fact 4 above), and the greatly increased rates of chemical weathering of silicate rocks that occurred on the continents in the late Frasnian, as evidenced by the spike in ratios of strontium-87 to strontium-86 isotopes that occurred at this time (see Fact 1 above). Massive amounts of carbon-12 extracted from the atmosphere by photosynthesis must have been buried in sediments to produce the heavy-carbon anomalies. Likewise, the increased rate of weathering of silicate rocks in newly-uplifted mountain chains around the Earth extracted carbon dioxide from the atmosphere, as calcium released in the weathering of silicates combines with carbon dioxide to form carbonates.[67] Both of these processes acting in concert would have removed a significant amount of carbon dioxide from the Earth's atmosphere in the late Frasnian.

The removal of significant amounts of carbon dioxide from the Earth's atmosphere in the late Frasnian should have triggered a reverse-greenhouse effect; that is, it should have led to global cooling. There are two important questions concerning this effect: How much cooling, and for how long? These two questions immediately lead to another: Does any way exist to determine temperature from the fossil record? This is in fact possible with the use of oxygen isotopes.

Oxygen has two primary isotopes: oxygen-16 and oxygen-18 (another isotope of oxygen, oxygen-17, exists but is very rare). Three different mechanisms can lead to the differential removal of the lighter isotope, oxygen-16, from seawater: biomineralization, freezing, and evaporation. In biomineralization, marine organisms preferentially use oxygen-16 in the construction of the minerals that make up their skeletons. In freezing, water molecules that contain oxygen-16 are preferentially frozen out in producing ice crystals. In evaporation, water molecules that contain the lighter isotope of carbon also preferentially evaporate into the atmosphere first. Both evaporation and freezing of seawater also causes a rise in the level of salinity in the seawater, and thus the ratio of oxygen-16 to oxygen-18 can also be used to track salinity changes in oceans over time. Even increased rainfall on land and the resultant increased volume of freshwater runoff into the oceans will have an effect of the ratio of oxygen-16 to oxygen-18 in seawater.

However, the degree of fractionation of the isotopes of oxygen in biomineralization is sensitive to temperature, and ratios of oxygen-18 to oxygen-16 found in the skeletons of marine organisms potentially can be used to reconstruct past temperatures. The worst problem with oxygen isotope ratios is their notorious susceptibility to diagenetic alteration—the analyst must ensure that the skeletal minerals used in measuring these ratios was not chemically altered in any way since the death of the animal. Otherwise the ratio value found is meaningless and so is the temperature determination based upon it.

Brachiopod seashells are frequently used to obtain oxygen isotope ratio data and to determine paleotemperatures in the Carboniferous. Now we encounter a major mystery concerning the Devonian: oxygen isotope data obtained from brachiopod seashells in this interval of geologic time consistently yield temperature predictions that are clearly wrong. These isotopic data predict temperatures that ranged from 36°C to 54°C (97°F to 129°F) in the oceans during the entire Devonian, spiking as high as 60°C (140°F) at the Frasnian/Famennian boundary.[68] The problem is that the lethal thermal limit for most marine invertebrate species is around 38°C (100°F); according these data, nothing should have been alive in the Earth's oceans for just about the entire Devonian, yet marine life clearly flourished for most of the Devonian. What is wrong? It is suspected that the calcite minerals in Devonian brachiopod seashells have been diagenetically altered, thus yielding oxygen isotope ratios that are not representative of the true ratios of oxygen-18 to oxygen-16 that were present in the seawater of Devonian oceans. Why this diagenetic alteration took place in the Devonian, but not in the Carboniferous, remains a mystery. The alternative possibility is that the oxygen-isotope composition of seawater in the Devonian was significantly different from that of the Carboniferous and from that of the modern world.[69]

More recently, carbonate-fluor-apatite minerals in the dentary skeletal structures of conodonts have been used by University of Erlangen geochemist Michael Joachimski and his colleagues as an alternative to oxygen isotope ratios in the Devonian. These analyses have shown that two positive shifts in the oxygen isotope ratios occurred at the same time as the two heavy-carbon anomalies (discussed in Fact 4 above) during the deposition of the Lower and Upper Kellwasser strata in Germany. Interpretation of the magnitude of these shifts indicates a drop in tropical sea-surface water temperatures of 5°C to 7°C (11°F to 15°F) in these two critical time intervals: in the Early *rhenana* conodont zone sea-surface temperatures ranged from 31°C to 33°C (88°F to 91°F) in tropical Laurussia (present-day Germany), but then dropped sharply to 26°C (79°F) during the deposition of the Lower Kellwasser Horizon in the Late *rhenana* conodont zone. Likewise, sea-surface temperatures increased again to 34°C (93°F) in the *linguiformis* conodont zone, only to sharply drop to 27°C (81°F) during the deposition of the Upper Kellwasser Horizon.[70]

The sharp temperature drops predicted by the conodont oxygen isotope data are in accord with the paleobiological data that also predict global cooling in the late Frasnian. Twin drops of 5°C to 7°C (11°F to 15°F) in tropical sea-surface temperatures in the latest Frasnian would be comparable to those reconstructed for the Pleistocene glaciations, where sea-surface temperatures dropped by 4°C to 8°C (8.5°F to 17°F), but in low-latitude regions rather than in the tropics,[71] and the Pleistocene glaciations are known to have triggered extinctions both in the sea and on land (as will be discussed in the Kill Mechanism section of this chapter). Moreover, twin temperature drops at the Lower and Upper Kellwasser Horizons are also in accord with the eutrophication pump model of Murphy and colleagues for the Kellwasser Horizons, but not with the proposed pedostratigraphic correlation of deep-calcic soil formation with these horizons, which predicts twin warming episodes instead (discussed in Fact 5 above).

Because the conodont oxygen isotope data yield temperature predictions that are biologically reasonable and in agreement with the paleobiological data, Joachimski and colleagues extended the analysis to the entire Devonian, and data from other regions of the Earth were included with additional data from Europe.[72] The twin sharp drops in sea-surface temperatures at the Lower and Upper Kellwasser Horizons were detected in strata outside of Europe, although with varying magnitudes—for example, sea-surface temperature only dropped by 4°C in China as opposed to 7°C in Europe at the end of the Frasnian.[73]

The global results for sea-surface temperature values in the Frasnian and Famennian interval are shown in figure 4.3.[74] The twin sea-surface temperature drops during the Lower and Upper Kellwasser Horizons were too closely spaced in time to plot on the expanded timescale of the entire Devonian in the original study by Joachimski and colleagues, but figure 4.3 presents the average data for only the Late Devonian and only the single datum for a 4°C (8.5°F) drop in average sea-surface temperature at the Upper Kellwasser Horizon at the very end of the Frasnian (fig. 4.3).

The reconstructed sea-surface temperature curve shows a low of 23°C (73°F) in the early Frasnian, steadily rising to a high of 31°C (88°F) in

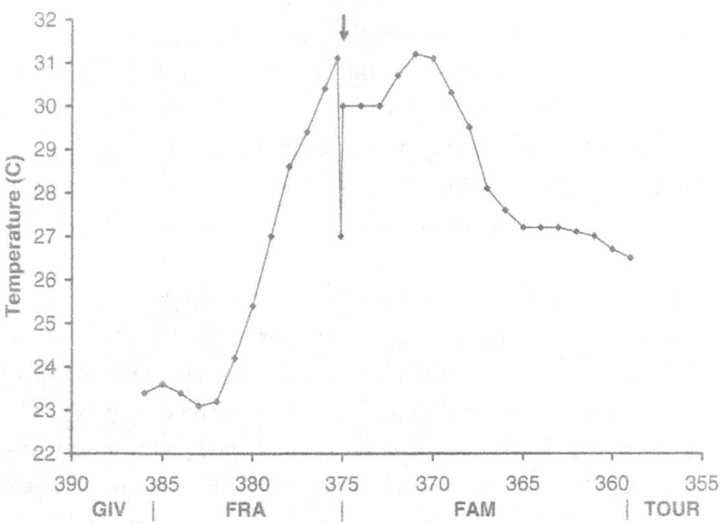

FIGURE 4.3 Average sea-surface temperatures in the Frasnian and Famennian, based on oxygen isotope analyses of conodont dentary skeletal structures. Note the sharp drop in sea-surface temperatures (indicated by the arrow) that occurred at the end of the Frasnian; see text for discussion. Geologic timescale abbreviations: GIV, Givetian; FRA, Frasnian; FAM, Famennian; TOUR, Tournaisian. Data from Joachimski et al. (2009).

the late Frasnian and then dropping sharply to 27°C (81°F) at the very end of the Frasnian. Sea-surface temperatures are then reconstructed to remain at the slightly cooler 30°C (86°F) level for only three million years in the early Famennian Gap (375–373 million years ago). Thereafter, temperatures began to rise again and returned to 31°C (88°F), the same level as the late Frasnian, after five million years into the Famennian Gap (371 million years ago). In the latest Famennian Gap (369 million years ago), temperatures are reconstructed as falling steadily to a low of 26.5°C (80°F) by the end of the Famennian.[75]

We have now reached the confusing state of affairs where different data sets yield contradictory predictions for Famennian temperatures. The paleobiological data indicate an interval of seven million years of cold climatic conditions in the Famennian Gap, whereas the conodont oxygen isotope curve reconstructs a much shorter interval of three million years, with only a very small 1°C (2.1°F) decrease from the hot late Frasnian high temperature of 31°C (88°F). The remaining four million years of the Famennian Gap are reconstructed as very hot, eventually as hot as the 31°C maximum reached in the late Frasnian itself (fig. 4.3). The paleobiological data indicate climatic warming after a cold Famennian Gap, whereas the conodont oxygen isotope curve shows the exact opposite: temperatures are reconstructed to fall after the seven million years of the Famennian Gap, and to continue to fall though the remainder of the Famennian. We do know in fact that the latest Famennian was indeed cold, as continental ice sheets formed on Earth at this time (this will be discussed in chapter 6).

Are the geochemical data wrong? Are the oxygen isotope ratios in conodont skeletal apatite giving a false signal, just as those in brachiopod skeletal calcite are known to have done? The ratio of oxygen-16 to oxygen-18 can also be affected by changes in ocean water salinity and the rate of evaporation of seawater, as well as the amount of precipitation and freshwater runoff from land. Do the conodont oxygen isotope ratio changes reflect long-term climatic changes affecting the degree of salinity and evaporation in the Devonian oceans and not temperature?[76]

Are the paleobiological data wrong? Are the patterns of latitudinal contraction into the tropics that are seen to have occurred in both plant and animal distributions in the late Frasnian—latitudinal contractions taken as evidence of late Frasnian cooling—instead mere artifacts of incorrect paleogeographical reconstructions of ancient continental positions? Are the blooms of siliceous-secreting sponges that occur in the late Frasnian oceans triggered by increased volcanic introduction of the limiting nutrient silicon dioxide into oceanic waters, and not by the onset of colder temperatures?[77] Are the paleobotanical data, the "Early and Middle Famennian vegetation crisis" data that indicate a cold early and middle Famennian (as discussed above), in fact incorrectly interpreted?

A final possibility: are the calibrations of our geologic timescales wrong? The paleobiological data indicate an interval of seven million years of cold conditions during the Famennian Gap (375–369 million years ago), followed by a warming interval of four million years (368–365 million years ago), then followed by a cooling interval of five million years (364–359 million years ago), and eventually resulting in the formation of glaciers on Earth, to be discussed in chapter 6. The geochemical data (fig. 4.3) indicate a slightly cooler interval of three million years in the early Famennian (375–373 million years ago), a much hotter interval of four million years in the middle Famennian (372–369 million years ago), and falling temperatures throughout the remainder of the Famennian.

The Famennian Gap spans ten conodont zones (table 4.2), and this span of time is estimated to have been seven million years (as discussed in the Famennian Gap section of this chapter). Is this time duration estimate wrong? Did these ten conodont zones in fact represent only three million years, as in figure 4.3? Alternatively, the data used to construct the temperature curve in figure 4.3 come from geographically distant regions around the Earth, and it is often difficult to accurately correlate time relationships on global scales—thus are the time duration estimates in figure 4.3 wrong? Is the cooler interval of three million years shown in figure 4.3 actually seven million years long instead? In either case, figure 4.3 would indicate a temperature in the Famennian Gap—whatever its actual time duration—that is far from cold at 30°C (86°F). The absolute magnitudes of the reconstructed sea-surface temperatures in figure 4.3 are dependent upon assumptions concerning the sea-water chemical composition of the Devonian oceans, compositions that are in fact unknown,[78] but still the *relative magnitudes* of the plotted values will remain the same even if different assumptions are made regarding the chemical composition of Devonian sea water. That is, the sea-surface temperature during the first three million years of the Famennian will always plot as *only slightly cooler* than the hot late Frasnian, regardless of any changes in the estimate of what the absolute magnitude of that temperature might be.

In summary, did the Famennian start out very cold, and then warm up? Or did it start out very hot, and then cool down? The two climatic scenarios contradict each other perfectly, and at present we cannot prove either one beyond the shadow of a reasonable doubt.

FACT 7: *Data indicating a major drop in sea level at the Frasnian/Famennian boundary are disputed.* In 1985, the Oregon State University paleontologist Jess Johnson and his colleagues published a proposed Devonian sea-level curve,[79] based upon stratigraphic analyses conducted in multiple areas of North America and Europe (fig. 4.4). The similarity of the analytic results in these widely separated areas on the ancient Laurussian continent led them to propose that the argued sea-level fluctuations probably affected the entire world, thus were indeed global fluctuations in sea level and were not simply due to local uplift or subsidence of the coastline that was confined to the Laurussian continent. Numerous small sedimentary cycles in late Frasnian strata in particular are proposed to have been caused by global eustatic sea-level changes, and to be evidence for the waxing and waning of alpine glaciers in Gondwana.[80] Subsequent stratigraphic analyses conducted in other areas of the world tend to confirm the major features of the proposed sea-level curve, thus corroborating the global nature of the sea-level fluctuations.

Johnson and colleagues argued that global sea level was at its lowest stand in the Early Devonian Pragian. Sea level then began to rise and, punctuated by a few smaller drops in sea level, continued to rise until it reached a global highstand at the end of the Frasnian (fig. 4.4). Sea level then began to fall and, punctuated by a few smaller sea-level rises, continued to drop throughout the Famennian. So far so good; the proposed overall pattern of sea-level rise and fall in the Devonian is generally not disputed, but rather corroborated, by subsequent workers.

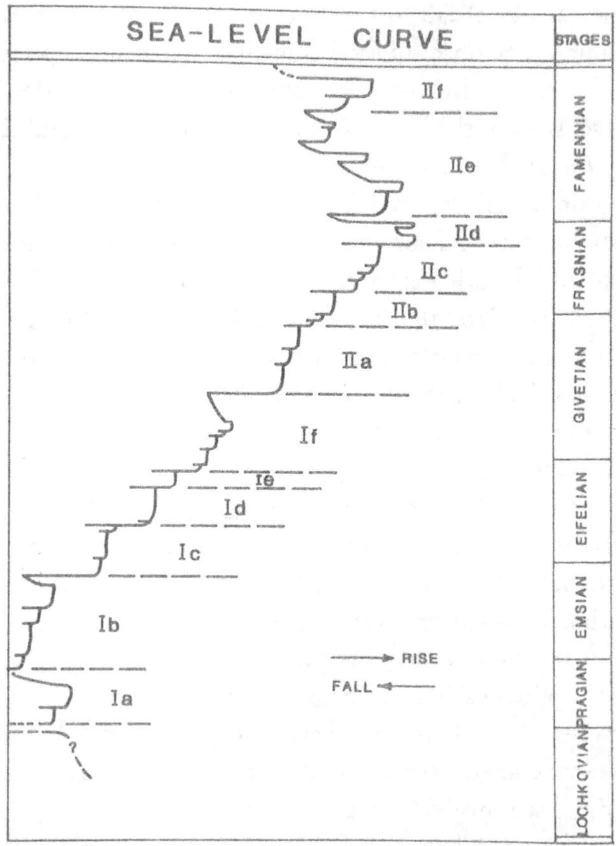

FIGURE 4.4 The Devonian sea-level curve of Johnson et al. (1985). Note the maximum sea level highstand is reconstructed to have occurred in the late Frasnian (eustatic cycle IId), followed by a sharp fall in sea level. *Credit*: From the *Geological Society of America Bulletin* (Johnson et al., 1985), copyright © 1985. Reprinted with permission.

However, the proposed finer-scale sea-level rises and falls have been, and continue to be, disputed.

I will now focus the discussion on Johnson and colleagues' proposed sea-level fluctuations for the critical time interval of the late Frasnian and early Famennian. This interval of time was proposed to correspond to their eustatic cycle "IId" (fig. 4.4): it consisted of a sea-level rise in the Early *rhenana* conodont zone, a small sea-level fall in the Early *linguiformis* conodont zone, another sea-level rise in the Middle *linguiformis* zone, and a major sea-level fall in the Early *triangularis* conodont zone of the Famennian. The onset of the black-shale expansion episodes of the Lower and Upper Kellwasser Horizons were taken as evidence for the two sea-level rises in the Early *rhenana* conodont zone and the middle part of the *linguiformis* conodont zone.[81]

The two proposed sea-level falls in the eustatic IId cycle have been more controversial. Johnson and colleagues considered the proposed sea-level fall in the early part of the *linguiformis* conodont zone to have been small in scale, and the sea-level fall in the Early *triangularis* conodont zone to have been major. Of importance to note is the fact that the first sea-level fall was proposed to occur *before* the Upper Kellwasser Horizon, and the second sea-level fall was proposed to occur *after* the Upper Kellwasser Horizon.

Evidence for a major drop in sea level in the latest Frasnian–earliest Famennian interval does exist in western Canada, southern China, and western Australia. In these three regions of the world, preexisting submarine carbonate platforms were exposed to dry air after that sea-level fall, as the marine waters retreated from the land. The subsequent subaerial erosion of the carbonates produced extensive paleokarst horizons with characteristic scalloped surfaces, dissolution pits, and neptunean dykes. The evidence for sea-level fall is clear, but unequivocally dating the time of the sea-level fall or falls is not.

In the Elephant Hill section at Guilin in China, strata dated to the Lower *triangularis* conodont zone sit directly on the paleokarst horizon, which is dated in the *linguiformis* conodont zone. Thus it was proposed that the major sea-level fall that occurred in this area took place after the Upper Kellwasser Horizon and is evidence for the second IId sea-level fall described by Johnson and colleagues. Yet it is also clear that the strata corresponding to the Upper Kellwasser Horizon in other Chinese sections are missing in the Elephant Hill section.[82] Arguments for a major fall in sea level in the Famennian Early *triangularis* conodont zone propose that these strata were originally present but were eroded due to the sea-level fall, and thus the sea-level fall took place *after* the time of the Upper Kellwasser Horizon. Arguments against a major fall in sea level in the Famennian Late *triangularis* conodont zone maintain that the sea-level fall took place *before* the Upper Kellwasser Horizon, because the marine *linguiformis* conodont zone strata corresponding to this time interval are missing in the section and were probably never deposited due to the low sea-level stand in this area at that time.[83]

Similar problems of interpretation exist with the Canadian paleokarst strata. In the Abitibi River section, strata dating to the Lower *triangularis* conodont zone sit directly on the paleokarst horizon, which is dated to the early part of the *linguiformis* conodont zone. Strata corresponding to the Upper Kellwasser Horizon are missing, hence the same problem exists here as in China. The situation is more challenging in western Canadian sections, where strata dating to the Middle *triangularis* conodont zone sit directly on strata dated to the Late *rhenana* conodont zone, and the entire *linguiformis* conodont zone and Early *triangularis* conodont zones are missing.[84] Interpretation of the age of the paleokarst horizons in Australia are similarly contentious.[85]

A second line of argumentation for a major drop in sea-level after the Upper Kellwasser Horizon is based upon the proliferation of species of the conodont genus *Icriodus* in many strata dated to the Early *triangularis* conodont zone. In the Frasnian world, *Icriodus* species were apparently adapted to shallow water. The sudden abundance of *Icriodus* species in the earliest Famennian strata around the world has been argued to be evidence of a sudden shift to shallow water conditions, hence a global sea-level fall. The counterargument points out that *Icriodus* species were some of the few conodonts to survive the end-Frasnian extinction, and their post-extinction habitat range may have been quite different from that in the Frasnian. That is, sea-level may have remained exactly the same while the *Icriodus* species survivors in the Famennian opportunistically expanded their habitat ranges into deeper water regions that were off limits to them in the Frasnian due to the presence of competitors adapted to deep water.[86]

In summary, did a major sea-level drop occur in the Earth's oceans before the critical Upper Kellwasser Horizon? Or did a major sea-level drop occur immediately after the critical Upper Kellwasser Horizon? At present, we cannot prove either possibility beyond the shadow of a reasonable doubt.

FACT 8: *The loss of biodiversity in both marine animal and land plant groups in the late Frasnian was triggered by two separate phenomena acting in concert: an increase in extinction rates but also a decrease in speciation rates.* I first commented on this peculiarity in the end-Frasnian biodiversity crisis in the marine realm in a paper that I presented to a meeting of the Paleontological Society in Providence, Rhode Island, back in 1984, and published a follow-up analysis in the journal *Paleobiology* in 1988.[87] The accuracy of that analysis has since been corroborated by other scientists, using the massive data set for fossil genera compiled by Jack Sepkoski.[88] The Virginia Tech paleobotanist Steve Scheckler noticed the same phenomenon in Frasnian land plants in 1986, and it too was subsequently corroborated by other scientists using larger data bases.[89]

The catastrophic loss of biodiversity at the end-Frasnian would not have been nearly as severe if it had been produced by extinction alone. Some animal groups, such as the brachiopod shellfish, actually flourished during time intervals characterized by elevated extinction rates because their speciation rates were higher, per time interval, than corresponding extinction rates. That is, the brachiopods were adding new species into the ecosystem faster than the ecosystem was losing species by extinction. That pattern abruptly ceased in the late Frasnian: speciation rates dropped precipitously. In essence, the ecosystem suddenly collapsed; large numbers of species were being lost from the ecosystem by extinction without being offset by any new species additions into the ecosystem. Marine diversity plummeted.

The Ohio University paleontologist Alycia Stigall points out that there are two types of speciation that can be discerned in the fossil record: vicariant and dispersive. In vicariant speciation, an ancestral species population is passively split (often by tectonic events such as the uplift of a mountain chain or the rifting apart of two continents) into two large subpopulations that then genetically diverge through time until they become separate species. In dispersive speciation, a small subpopulation of the ancestral species population actively migrates away to a new geographic region and then genetically diverges in isolation until it becomes a new species. Stigall notes that studies of modern and fossil species groups frequently "document speciation by vicariance in much higher frequencies than speciation by dispersal," but her research shows that in the late Frasnian "speciation by vicariance is almost entirely absent during the crisis interval."[90]

In viewing the end-Frasnian biodiversity crisis from an ecological perspective, one must ask more than the question: What triggered the elevated extinction rates? One must also ask: What was the inhibiting factor that caused the cessation of new species originations?

The Kill Mechanism

Now let us consider the theories that seek to link the empirical observations listed in the previous section of the chapter with a hypothetical kill mechanism (or mechanisms) for the end-Frasnian extinction. To distill the results of many years of debate from the very onset of this discussion, current research is focused on two principal kill mechanisms: hypoxia and/or hypothermia (table 4.4). That is, suffocating to death and/or freezing to death.[91]

Definite proof that either one of these hypothetical kill mechanisms was the trigger for the end-Frasnian extinction—or even proving that they both worked in concert as the kill mechanism—remains elusive.

TABLE 4.4 Summary of the kill mechanism and speciation suppression models that have been proposed for the end-Frasnian extinctions.

- I. Hypoxic Kill Mechanism
  - A. Marine eutrophication model
  - B. Atmospheric oxygen minimum model
  - C. Atmospheric hydrogen sulfide model
- II. Hypothermic Kill Mechanism
  - A. Glaciation model
  - B. Impact winter model
  - C. Volcanic winter model
  - D. Biological weathering model
  - E. Chemical weathering model
  - F. Equatorial convergence model
- III. Combination Kill Mechanism
  - A. Hypothermia followed by hypoxia model
- IV. Speciation Suppression Mechanism
  - A. Marine habitat impoverishment model
  - B. Marine invasive species model
  - C. Terrestrial floral monocultures model

Note: These kill mechanism and speciation suppression models are discussed sequentially in the text.

That is, to use a courtroom analogy, it remains to be proved beyond a shadow of a reasonable doubt that either, or both, of these hypothetical kill mechanisms is guilty for the loss of so many animal and plant species on Earth that occurred 374.5 million years ago.

Let us consider the argument for a hypoxic kill mechanism first. We have seen in the previous section that the oxygen content of the Earth's atmosphere was below 13 percent during the Frasnian (Fact 2), and that eutrophication produced widespread regions of anoxic water in the Earth's shallow seas during this same period of time (Facts 4 and 5). Oxygen was indeed in short supply in many areas of the Earth during the Frasnian.

In 1985, Jess Johnson and his colleagues, in their study of sea-level changes in the Devonian discussed above, were the first to propose that the spread of vast areas of marine anoxic waters during a sea-level rise at the end of the Frasnian triggered the extinction of marine animals by hypoxia.[92] The hypoxic kill mechanism hypothesis for marine organisms does have an important fact in its favor: the worldwide distribution of black shales and other geological evidence for marine anoxia at the Upper Kellwasser Horizon in the latest Frasnian, which is also the time horizon of the major pulses of extinction.

On the other hand, the hypoxic kill mechanism hypothesis has a significant problem: how to explain the suffocation all of the organisms that are known to have perished at the end of the Frasnian. The spread of anoxic bottom waters will indeed suffocate bottom-dwelling animals that either cannot flee because they are sessile or are mobile but cannot flee quickly enough to escape the deadly waters. This phenomenon can clearly be seen in the "dead zones" that occur in modern oceans, dead zones that sometimes are triggered by eutrophication produced by nutrient runoff from the fertilized farmlands of humans living in coastal areas. Thus the extinctions that are observed to have occurred in Frasnian benthic marine organisms could be explained by hypoxia.

Suffocation of animals that live up in the water column, the actively swimming nekton and the passively floating plankton, is more problematic. In general, the upper 50 to 100 meters (164 to 328 feet) of the water column is aerated directly by oxygen diffusion from the atmosphere, assisted by agitation of the water by wind in producing waves, and by tidal motion. However, oxygen concentrations in the atmosphere were low in the Frasnian (Fact 2), and the presence of isorenieratane in some Frasnian black shales proves that anoxia did indeed extend all the way up into surface waters (Fact 4), thus hypoxia could potentially kill both nekton and plankton during these periods of time.

One problem still remains in explaining Frasnian marine extinctions with hypoxia: the latitudinal contractions in equatorial geographic range after the extinction that are observed to have occurred in stromatoporoids, brachiopods, foraminifera, and trilobites (Fact 6). Those latitudinal contractions are compatible with the global cooling of marine waters, where warm-adapted marine species retreat from higher latitudes to the safety of the warm tropics. It is difficult to explain this observed pattern of retreat from high latitudes with hypoxia. For example, why should warm-adapted stromatoporoid sponges have smothered in higher-latitude temperate and polar sea bottoms but not in low-latitude tropical ones? If anything, one might predict that the hot tropical waters would be more oxygen-depleted than cooler, temperate zone waters, and that the tropical stromatoporoids should have perished as well, but they did not.

More serious problems arise for hypoxia as a killing mechanism in the transition from marine extinctions to terrestrial extinctions—why should land-dwelling organisms be affected by oxygen depletion in the sea? Land-dwelling organisms breathe air, thus to kill them with hypoxia the atmosphere itself has to become oxygen-depleted. Indeed, evidence exists for low oxygen levels in the atmosphere—but not just during the time of the extinctions at the end of the Frasnian. We have previously considered the "charcoal gap" that spanned the entire interval of the Eifelian through the Frasnian (Fact 2)[93] as evidence that oxygen concentrations in the atmosphere were at or below 13 percent for some 23 million years, not just at the end of the Frasnian Age. The fact that the oldest yet discovered charcoal from the Famennian occurs in the late Famennian, the Middle *expansa* conodont zone (table 4.3), indicates that the "charcoal gap" actually spanned the interval from the Eifelian through the Famennian Gap, some 30 million years.

However, rather than suffocating, terrestrial tetrapods and elpistostegalians made major evolutionary advances during this same interval of time. The tetrapod specialist Jennifer Clack notes that the evolution of air breathing occurred independently in the tetrapodomorphs and the lungfishes in this same interval of time:

> The dipnoans present an interesting parallel case to that of the tetrapodomorphs, in their acquisition of air-breathing adaptations from the Middle Devonian onwards. . . . Potentially, then, the increased size of the spiracular notch among osteolepids, and the onset of air-breathing specialization of dipnoans, which both occurred in the Eifelian/Givetian, may correlate with the gradual decline of oxygen levels through the Middle Devonian, and initiated trends that continued in their respective lineages as oxygen levels continued to decline throughout the Late Devonian.[94]

Actually, as we have seen in this chapter, oxygen concentrations in the atmosphere did not continue to decline throughout the Late Devonian, but rather began to increase after the Famennian Gap. Clack further notes that:

> Though other sarcopterygian and possibly even actinopterygian groups may have exploited air-borne oxygen to a degree, it seems to have been the tetrapodomorphs that were able, for some reason, to adapt to its use more extensively. One possible reason is that in shallow waters, their robust endochondral fin skeletons allowed them to support their heads out of water to gulp air, and the parallel elaboration of the spiracular mechanism with the increase in size and complexity of the pectoral limb is compelling. A second factor may be connected with the evolution of the choana, or internal nostril, which is a tetrapodomorph synapomorphy. . . . Appearance of the choana may at least have provided an "exaptation" for narial breathing in later tetrapods.[95]

The tetrapodomorphs that Clack refers to as having increased sizes in their pectoral limbs, or forelimbs, which allowed them to elevate themselves out of water, are advanced elpistostegalian fish such as *Tiktaalik roseae* (fig. 3.1). It is now known that the elpistostegalian tetrapodomorphs are not the ancestors of the tetrapods, but rather represent an independent coexisting lineage with the tetrapods (as discussed in chapter 3). The tetrapods, our ancestors, were already present on Earth in the early Eifelian, as witnessed by their fossil trackways preserved in present-day Poland. Thus three, not two, separate vertebrate lineages evolved air-breathing adaptations from the Eifelian through the Frasnian in parallel: the lungfishes, the elpistostegalian fishes, and the tetrapods. The low oxygen concentration present in the atmosphere during the "charcoal gap" was thus a *stimulus* for the evolution of air breathing, an evolutionary *advancement* in the invasion of land, not the reverse.

To argue in favor of hypoxia as a killing mechanism on land, one could propose that oxygen levels in the Earth's atmosphere reached a critical absolute minimum during the very latest Frasnian and the Famennian Gap, a minimum oxygen concentration so low that it caused serious difficulty for air-breathing animals. Animals would have had to retreat from the highlands, where the air had become unbreatheable, and to concentrate in the lowlands near sea level. Crowding and competition for food, rather than actual suffocation, might then trigger increased extinction rates in terrestrial animals. It is interesting that both the Rock-Abundance and Geocarbsulf models (Fact 2) predict an oxygen minimum that occurred in the Frasnian: atmospheric oxygen at 17 percent in the Rock-Abundance model and at 13 percent in the Geocarbsulf model (fig. 4.2). As there is no fossil charcoal in the late Frasnian stratigraphic record, oxygen levels in the atmosphere actually had to be *lower* than the 13 percent minimum predicted by the Geocarbsulf model. As outlined at the beginning of the chapter, it is known that land plants—the organisms that produce oxygen—suffered a major loss in diversity in the Late *rhenana* conodont zone in the late Frasnian, and that this minimum diversity persisted through the Famennian Gap. Thus hypoxic conditions even on land, as well as in the oceans, is a distinct possibility for the late Frasnian–Famennian Gap interval.

We have now arrived at the most serious problem with hypoxia as a killing mechanism on land: the late Frasnian extinction in land plant species, and the "Early and Middle Famennian vegetation crisis" that we considered in Fact 6. In addition, latitudinal contraction in the geographic range of land plant distributions also occurred in the late

Frasnian and Famennian Gap, where land plant species retreated from high latitude regions into the tropics of the Earth (Fact 6). None of these observations are compatible with hypoxia as a killing mechanism, whereas they are compatible with hypothermia. To land plants, oxygen is a waste product. Unlike terrestrial arthropods or vertebrates, plants do not need oxygen—they need carbon dioxide. In the fossil record, we see repeatedly that land plants flourished during times of low oxygen content and high carbon dioxide content in the atmosphere. Therefore, land plants should have also flourished in the late Frasnian and during the "charcoal gap" of the Famennian Gap. They did not.

A proposed modification of the hypoxic kill mechanism hypothesis invokes a possible side effect of marine eutrophication as an additional kill mechanism to explain the observed pattern of extinction in land plants in the late Frasnian: in times of extreme oceanic anoxia, anaerobic photosynthetic bacteria in surface waters could metabolize sulfur to produce hydrogen sulfide, which could then diffuse into the atmosphere of the Earth. Pennsylvania State University geochemist Lee Kump and his colleagues[96] constructed a model for atmospheric chemistry transport during times of surface-water anoxia in the seas and reached some deadly conclusions: lethal amounts of hydrogen sulfide could be released from the oceans and accumulate in the low-altitude level of the atmosphere on land, and could diffuse further upwards into the stratosphere to destroy the protective ozone layer of the Earth's atmosphere. On land, atmospheric concentrations of hydrogen sulfide could reach 100 parts per million. Land plants are progressively poisoned as hydrogen sulfide levels in the atmosphere exceed three parts per million, and animals are poisoned at levels exceeding five parts per million. Both land plants and animals would be asphyxiated long before the predicted levels of 100 parts per million would be reached. Plants could not escape death even in regions of the Earth where hydrogen sulfide possibly did not reach lethal levels: the simultaneous destruction of the high-altitude ozone layer of the Earth's atmosphere would lead to radiation poisoning of the terrestrial flora below.

The hydrogen sulfide kill mechanism for land plant extinction, induced by hypoxic oceans, was proposed by Kump and colleagues in 2005, and it lasted only three years before being disproved. Rather than a simple one-dimensional altitudinal model for atmospheric chemistry transport, as used by Kump and colleagues, Cambridge University geochemist Michael Harfoot and his colleagues[97] used a more complex and realistic two-dimensional model that included both altitudinal and latitudinal variations in the atmosphere. Their modeling showed that hydroxyl radicals produced in the Earth's atmosphere in equatorial regions would quickly oxidize hydrogen sulfide in the atmosphere and prevent damage to the high-altitude ozone layer. The maximum concentration of hydrogen sulfide in the atmosphere at low-altitude levels on land would only reached 1.2 parts per million, as opposed to the lethal 100 parts per million predicted by the one-dimentional model of Kump and colleagues. In summary, Harfoot and colleagues state: "We conclude that massive H_2S escape from anoxic oceans . . . is unlikely to seriously damage Earth's ozone shield or induce mass poisoning of the terrestrial biota during the end-Permian. Our conclusion is also applicable to other times in the Phanerozoic when extinctions and biotic turnover coincide with extreme oceanic anoxia,"[98] that is, in the late Frasnian.

In summary, the proposed hypoxic kill mechanism for the end-Frasnian extinction could clearly account for marine extinctions, both in sea bottom and surface water environments, and it may possibly account for animal extinctions on land as well, given the low oxygen concentrations present in the Earth's atmosphere in the Frasnian. However, hypoxia cannot account for the observed extinction that occurred in land plants. Hypoxia as a killing mechanism also cannot account for the observed latitudinal contractions in equatorial geographic range that occurred in animals in the oceans and in plants on the land after the end-Frasnian extinction.

Now let us consider the argument for a hypothermic kill mechanism. We have examined the paleobiological and geochemical evidence for global cooling at the end of the Frasnian (Facts 3 and 6). We have seen, however, that the paleobiological and geochemical evidence is contradictory regarding global temperature levels on Earth during the Famennian Gap (Fact 6).

The hypothermic kill mechanism hypothesis has no difficulty in explaining simultaneous extinctions both in the sea and on land, unlike the hypoxic kill mechanism hypothesis. In the geologically recent

Pleistocene glaciations, we see that both marine and terrestrial animals experienced marked increases in extinction rate. Hypothermia killed land plants as well as animals and triggered the retreat of land plant species from the high latitude regions of the Earth, which were either covered in ice and were devoid of plants or saw only the development of tundra vegetation adapted to the cold. These Pleistocene empirical observations are seen in the late-Frasnian world as well.

In 1977, Laurentian University paleontologist Paul Copper was the first to propose global cooling and hypothermia as a kill mechanism for the end-Frasnian extinction.[99] The evidence offered for this hypothesis was paleobiological (included in Fact 6 above), not physical, as no geological evidence exists for glaciation on Earth at the end of the Frasnian. It is possible that glaciation did occur at this time, but that it was a very short-lived cold pulse or series of pulses and that it did not produce extensively distributed tillites or other characteristic sedimentary deposits linked with more extensive glacial episodes in the fossil record. If so, however, even a short-term glacial event should have produced a drop in sea level. As we have seen above (Fact 7), evidence for and against a major sea-level drop occurring at the Frasnian–Famennian boundary continues to be disputed, hence this line of evidence for a glacial pulse at the Frasnian–Famennian boundary remains unproved. The alternative scenario (Fact 7) describes a major sea-level drop, and hence perhaps a short-lived glacial pulse, as occurring in the early part of the *linguiformis* conodont zone before the critical Upper Kellwasser Horizon. Yet the geochemical data suggest global temperature drops that coincide with the Kellwasser Horizons, not to the interval of time between them (Fact 6).

Thus the hypothermic kill mechanism hypothesis has a major problem: how to explain how the Earth suddenly became cold at the end of the Frasnian, if it did in fact do so. An unusual mechanism for cooling the Earth very rapidly indeed was proposed only three years after Copper's original global cooling hypothesis: impact the planet with an asteroid. In the "impact winter" model, the collision of a large asteroid or comet with the Earth would inject a vast amount of vaporized bolide and target rock material into the atmosphere, produce a globe-spanning dust and gas cloud that would block sunlight from reaching the Earth's surface.[100] Cutting off the radiative heat of sunlight would trigger lethally cold temperatures around the globe, even at the equator.

In 1980, the research team led by physicist Louis Alvarez and his geologist son, Walter, proposed that the Earth had been impacted by a large asteroid at the end of the Cretaceous, and that the climatic effects resulting from the impact winter triggered the collapse of the dinosaur ecosystem, an enormously successful terrestrial ecosystem that had persisted for some 150 million years.[101] At the time, this proposal was highly controversial and triggered protests from many geologists who considered it to be a return to the catastrophist (and often creationist) geological models of the 1800s. The main piece of evidence for such an impact that the Alvarez team had discovered was an anomalous concentration of the element iridium in clays dated precisely to the time horizon of the end-Cretaceous mass extinction. Iridium is depleted in the Earth's crust but is enriched in some asteroids, thus an obvious source for an anomalous concentration of the element in sediments on Earth's surface would be for it to come from outside the Earth—from an asteroid. Further research since 1980 has corroborated the Alvarez hypothesis and has added additional geological evidence for asteroid impact at the end-Cretaceous, including the impact crater itself in Chicxulub, Mexico.

In 1981, I was on research leave from Rutgers, working at the Field Museum of Natural History in Chicago. Over coffee and at lunch I spent some time discussing the asteroid impact hypothesis with the museum's meteorite expert, Ed Olsen, and an obvious question eventually arose from our discussions: Could there be an anomalous concentration of iridium in strata at the end-Frasnian as well? That is, could the end-Frasnian crisis have been triggered by an asteroid impact? I knew of well-dated stratigraphic sections that contained the Frasnian–Famennian boundary in western New York, so, in the cold spring of 1981, Ed and I set out from Chicago to sample those strata.[102] In the fall of 1981, we outlined our project to test the asteroid–impact hypothesis for the Late Devonian crisis at the now famous Snowbird Conference in Snowbird, Utah, which had been called to debate (hotly!) the asteroid impact hypothesis for the end-Cretaceous crisis. At the conference we met nuclear chemist Carl Orth from the Los Alamos National Laboratory, who expressed interest in our project and joined as the analyst. In the summer of 1982, I was working at the University of Tübingen, Germany, and traveled to Brussels, Belgium, to meet with Paul Sartenaer at the Belgian Royal Institute of Natural Sciences. The Ardennes Mountains of southern Belgium contain well-dated stratigraphic sections of the Frasnian–Famennian boundary, and Paul directed me to the best one to sample, thus adding European data to our project as well.

By 1982, I was beginning to have doubts that a single asteroid impact could explain the pattern of biodiversity loss that was beginning to emerge from analyses of biostratigraphic data from the Frasnian. Rather than a single pulse of extinction, such as is seen in the end-Cretaceous, the end-Frasnian extinctions appeared to have occurred in a series of pulses (as discussed at the beginning of the chapter). So, rather than a single impact, I proposed that the end-Frasnian extinctions could have been triggered by a series of impacts, the multiple impacts hypothesis.[103] The Frasnian does appear to have been a time of an increased flux of extraterrestrial impacts, based on the number of known impact craters on Earth of that age.[104] In particular, the Siljan impact crater in Sweden has been radiometrically dated to be 377.2 million years old,[105] exactly the same as the radiometrically determined age of an ash bed in the Late *rhenana* conodont zone in the Frasnian at 377.2 ± 1.7 million years old.[106] The original diameter of the Siljan impact crater is estimated to have been between 65 kilometers and 85 kilometers (40 miles to 53 miles). This is smaller than the 180-kilometer diameter (112-mile diameter) of the Chicxulub impact crater that is associated with the end-Cretaceous extinction, but could the Siljan impact have triggered the first pulse of the end-Frasnian extinction in the Late *rhenana* conodont zone? The age of the Frasnian–Famennian boundary is currently estimated to be 374.5 ± 2.6 million years old,[107] thus the Siljan impact is within the range of uncertainty of the age of the Frasnian–Famennian boundary itself and the most severe pulses of extinction.

Back in the United States, the analytic results began to emerge from Carl's lab at Los Alamos: there were no anomalous iridium concentrations in any of our stratigraphic samples at the Frasnian–Famennian boundary. We had sampled the Frasnian–Famennian boundary in four sections, on two continents, and found nothing unusual. The time interval that we had crossed included what is now known to be pulse 2 through pulse 5 of the biodiversity crisis, and we had found no evidence for asteroid impact in this interval. We wrote up our discouraging results and sent them off to the journal *Nature*.[108]

Pulse 1 of the biodiversity crisis, which occurred in the Late *rhenana* conodont zone is known to be exactly the same age as the Siljan impact crater in Sweden, remained unexamined. In 1984, at a meeting of the Paleontological Society in the United States, I discussed the possibility of sampling other stratigraphic sections with Willi Ziegler, director of the Senckenberg Research Institute in Frankfurt, Germany. He stressed the importance of the Steinbruch Schmidt section in Germany, which he had spent years working with and which now had conodont zonal control at the centimeter level. In the summer of 1984, while I was at another research stay at the University of Tübingen, I visited Willi in Frankfurt and we drove up to the Steinbruch Schmidt section in the Rhenish Slate Mountains. The complete Steinbruch Schmidt section covers the entire time interval from the Late *rhenana* condont zone in the Frasnian to the Late *triangularis* conodont zone in the Famennian, hence it included all five of the extinction pulses. Again, the stratigraphic samples were shipped off to Carl at Los Alamos, and again, he found no anomalous iridium concentrations that would support the hypothesis of an asteroid impact.[109]

The absence of an iridium anomaly does not disprove the impact of an extraterrestrial body, it simply does not provide any evidence for an impact. The impacting body may have been a comet or a highly fractionated asteroid, both of which contain small amounts of iridium. The number of smaller impact craters that have been dated to the Frasnian remains suggestive, but at present the extensive and multifaceted lines of geological evidence that have been amassed to prove the end-Cretaceous impact–extinction linkage have not been proved for the end-Frasnian, even after 30 years of searching by scientists around the world.

Interestingly, an alternative Earth-based catastrophic model has been proposed to produce the exact same climatic effects as the impact winter model: the "volcanic winter" model.[110] In the volcanic winter model, the same globe-spanning dust and gas cloud is proposed to result, but is produced by injecting vast amounts of ash, dust, and gas into the Earth's atmosphere during a mantle-plume volcanic event. We now have the dilemma of two proposed catastrophic triggers for global cooling: the extraterrestrial impact winter and the Earth-based volcanic winter models, which produce the same climatic effect, thus making the differentiation of the two potential causes of extinction problematical in the fossil record.[111] In the geological record, however, some evidence for catastrophic mantle-plume volcanism during the end-Frasnian extinction does exist, but both the magnitude and timing of that volcanism remains at present unproved (Fact 1).

Both the impact winter and volcanic winter models are catastrophic models for global cooling in that they are triggered by geologically rare, and geologically rapid, events. Two alternative models for global cooling invoke the effects of common processes on Earth that happen to have unusual timing or intensity: the biological weathering and chemical weathering models. The biological weathering model was proposed by University of Cincinnati sedimentologist Thomas Algeo and his colleagues.[112] This model predicts global cooling triggered by atmospheric carbon dioxide depletion, which in turn was triggered by the development of geographically widespread lignophyte trees and other vascular plants on land. Algeo and colleagues stress the potential chemical and mechanical weathering of silicate rocks that could have been produced by the spread of *Archaeopteris* trees in the Frasnian. These trees have already been discussed (see Fact 5) as a potential source for increased nutrient runoff from the land into the Frasnian shallow seas, as they were quite large (30 meters high) and had major root systems that penetrated more than a meter into the ground.

In the biological weathering model, three different mechanisms are proposed to have acted in concert to remove significant amounts of carbon dioxide from the atmosphere. First, photosynthetic productivity by the vascular plants that covered vast geographic areas of coastal and lowland areas of the Earth by the late Frasnian removed carbon from the atmosphere to form organic hydrocarbons.[113] The increased burial rate of organic carbon on land began the process of depleting carbon dioxide in the atmosphere.

Second, mechanical and chemical weathering of silicate rocks by *Archaeopteris* root systems to produce deep soil horizons also removed carbon dioxide from the air by forming carbonates.[114] Plant roots not only fracture rocks mechanically, exposing more rock surface area to contact with carbon dioxide and water and thus potentially to greater weathering, they also introduce organic acids that directly weather the rock chemically.

Third, they argued that enhanced soil formation by vascular plants "resulted in elevated fluxes of soil solutes (especially biolimiting nutrients) as a consequence of (1) enhanced mineral leaching, (2) fixation of nitrogen by symbiotic root microbes, and (3) shedding of plant-derived detrital carbon compounds. . . . Elevated river-borne nutrient fluxes may have promoted eutrophication of semi-restricted epicontinental seas and stimulated algal blooms."[115] Eutrophic blooms of marine phytoplankton and algae in the seas would in turn lead to the depletion of oxygen in the shallow seas and the accumulation of unoxidized organic hydrocarbons in black shales, rather than returning the carbon to the atmosphere as carbon dioxide in the normal process of aerobic respiration by bacteria. The resultant carbon dioxide downdraw from the atmosphere is proposed to have led to global cooling in a reverse greenhouse effect.

The model for attributing the end-Frasnian extinction to the spread of vascular plants on land drew immediate criticism. University of Birmingham paleontologist Tony Hallam and his colleague, University of Leeds paleontologist Paul Wignall, comment that "perhaps the only question arising from the Algeo model lies in the degree to which chemical weathering increased in the Late Devonian. The *Archaeopteris* forests [color plate 5] were restricted to floodplain environments, whereas more upland areas may not have been colonised until later in the Famennian, with the appearance of seed plants. Increased chemical weathering may not therefore have become significant until the very end of the Devonian,"[116] that is, during the end-Famennian extinction event, not the end-Frasnian event.

The pulsed nature and timing of the end-Frasnian extinction also is not explained by the biological weathering model. In the model predictions, a progressive spread of land plants should have led to a gradual decline in carbon dioxide in the atmosphere and a gradual cooling of the Earth through the middle and late Frasnian. Instead, global temperatures seem to have dropped sharply in pulses only at the Lower and Upper Kellwasser Horizons in the latest Frasnian. If the contested conodont oxygen-isotope curve for temperature fluctuations in the Late Devonian is correct (see discussion in Fact 6 above), global sea-surface temperatures actually increased steadily from a low of 23°C (73°F) in the early Frasnian to a high of 31°C (88°F) in the late Frasnian, the exact opposite of what is predicted by the biological weathering model.

A different model for global cooling triggered by atmospheric carbon dioxide depletion has been proposed by University of Lille sedimentologist Olivier Averbuch and his colleagues.[117] They stress the potential effect of the nonbiological chemical weathering of vast amounts of exposed silicate rocks, thrust up into the atmosphere in the extensive mountain chains that were created by tectonic forces in the Late Devonian (Fact 1). In particular, we have already examined the strontium isotope data that show a prominent spike in the late Frasnian, evidence that an intensification of the chemical weathering of silicate rocks on the continents occurred at this critical time, and a second pulse of intense continental weathering in the Famennian Gap (fig. 4.1).

In the chemical weathering model, two different mechanisms are proposed to have removed significant amounts of carbon dioxide from the atmosphere, but they are also somewhat different from the mechanisms proposed in the biological weathering model. First, subaerial chemical weathering of silicates removes carbon dioxide from the air by forming carbonates.[118] Biological removal of carbon dioxide is seen to be a secondary process: it is estimated that during the Phanerozoic about 80 percent of the carbon that was removed from the atmosphere was deposited as carbonates and only about 20 percent was deposited as organic matter.[119] Thus, as a second mechanism, chemical weathering of silicates on land would release limiting nutrients, particularly phosphorus and silica, which would be transported down the rivers and into the shallow seas. The input of these additional nutrients into the marine ecosystem would trigger marine algal and phytoplankton blooms, as well as the proliferation of silica-secreting marine organisms. Algal and phytoplankton overproductivity would produce eutrophication, and eutrophication would deplete the oxygen in the shallow seas. Widespread anoxia in the shallow seas would in turn lead to massive amounts of organic carbon burial; that is, in the absence of oxygen in seawater that could be used to oxidize it and return it to the atmosphere as carbon dioxide in aerobic respiration by bacteria, the organic carbon is instead buried in sediments on the seafloor and hence removed from the carbon cycle. The two heavy-carbon anomalies at the Lower and Upper Kellwasser stratigraphic horizons are evidence that massive removals of carbon-12 from oceanic waters did occur in the late Frasnian (as discussed in Fact 4), and the removal of that carbon in the seas must have likewise removed a significant amount of carbon dioxide present in the atmosphere of the Earth.[120] The effect of this second mechanism in the chemical weathering model is the same as the first: global cooling by a reverse greenhouse effect.

On the other hand, the chemical-weathering model offers an alternative explanation for the observed worldwide bloom in siliceous-secreting species that occurred in the late Frasnian and in the Famennian Gap. This bloom is usually taken as paleobiological evidence for global cooling, particularly for the siliceous sponges (Fact 6), but instead this could have been produced by the increased influx of the limiting nutrient silica into late Frasnian shallow seas.[121] Alternatively, the proliferation of the siliceous sponges could have been the result of both triggers acting in concert—global cooling and increased silica availability—in that the two factors are not mutually exclusive.

The chemical weathering model removes the difficulties noted above with the biological weathering model: first, mountain-building episodes are typically pulsed in geologic time, not continuous or gradual. In addition, the strontium-isotope data (fig. 4.1) give empirical support to the interpretation of a particularly strong pulse of tectonically induced uplift and subsequent subaerial erosion at the end of the Frasnian. Second, radiometric dating of tectonically deformed strata around the world proves that the Late Devonian was a period of intense tectonic activity, beginning in the early to mid-Frasnian (Fact 1). In contrast, the evolution of spermatophyte lignophytes, the seed plants, did not occur until the late Famennian (chapter 2). It is only these seed plants that had the capability to colonize, and hence weather, the silicate highlands and mountains far from the riverbanks and coastlines where the non-spermatophyte lignophyte *Archaeopteris* forests (color plate 5) were confined.

More recent in geologic time, the global cooling trend that resulted in the major glaciations of the Pleistocene has also been proposed to have been triggered by the tectonic upthrust of the extensive Alpine–Himalayan mountain chains and the exposure of towering expanses of silicate rock to chemical weathering.[122] However, the Cenozoic cooling trend began in the Early Eocene, about 50 million years ago, before the onset of chemical weathering of uplifted mountainous regions in the latest Middle Eocene, about 38 million years ago. Rutgers University geologist Dennis Kent and his colleague Giovanni Muttoni of the University of Milan have suggested a modification of the mountain-chain weathering scenario to explain the early phases of the Cenozoic cooling trend.[123] Instead of mountains, they attribute the carbon dioxide downdraw and resultant cooling to the chemical weathering of the Deccan and Rajmahal Traps in India as the Indian subcontinent crossed the equator of the Earth, riding the tectonic plate that would eventually collide India with Asia and produce the Himalayan mountain chain. The Indian subcontinent split off from the supercontinent Gondwana in the Southern Hemisphere in the Jurassic. By the Early Eocene, 50 million years ago, the Deccan and Rajmahal Traps of India straddled the equator and were located entirely in the equatorial humid zone. The combination of high precipitation and high temperature intensifies the process of chemical weathering, and it is estimated that half of the original 1,000,000-square-kilometer (621,000-square-mile) expanse of the Deccan Traps were eroded away at this time, consuming an enormous amount of atmospheric carbon dioxide, and is proposed to have initiated the Cenozoic global cooling trend. This addition of the climatic effects of the continental positioning of major silicate rock exposures in the equatorial humid zone of the Earth to the chemical weathering model has been named the "equatorial convergence theory."[124]

It is important to note that the Late Devonian was not only a period of intense tectonic activity (Fact 1), but also that the majority of the extensive mountain chains uplifted during this period of time were located in the equatorial region.[125] Invoking the equatorial convergence theory, carbon dioxide depletion of the atmosphere should have been even more intense during this geologic time period than in times when the major mountain chains of the Earth were located outside of the equatorial region. Thus both the uplift of major mountain chains and the positioning of those exposed silicate rock areas in the equatorial zone of the Earth during the late Frasnian are empirical data supporting the hypothesis that global cooling should have taken place at this time in Earth history due to carbon dioxide depletion in the atmosphere, triggered by the chemical weathering of silicate rocks.

There exists one last possibility to consider for the end-Frasnian extinction: perhaps the kill mechanism was not just hypoxia or hypothermia acting alone, but the effect of both occurring together that triggered the catastrophe. The initial pulse of extinction in the Late *rhenana* conodont zone may have been due to hypothermia, perhaps triggered by carbon dioxide depletion in the atmosphere as predicted by the chemical weathering model. The major sea-level fall that is proposed to have occurred in the succeeding early part of the *linguiformis* conodont zone may be evidence of a short-lived pulse of glaciation in Gondwana, triggered by this global cooling event (but the magnitude of this sea-level fall remains in dispute; see Fact 7). This initial hypothermic pulse may have killed plants on land and the temperature-sensitive reefs in the oceans. The severe pulses of extinction that then occurred in the oceans in the later part of the *linguiformis* conodont zone may have been triggered primarily by hypoxia, as no further losses in land plant diversity appear to be associated with this event, thus providing no evidence for a second, more severe hypothermic killing pulse on land (although the oxygen isotope evidence suggests that a second major drop in global temperature did occur at this time; see Fact 6). It is a curious fact that both the placoderm and acanthodian fishes suffered more severe diversity losses out in the open oceans than in the freshwater rivers and lakes during the Late *linguiformis* conodont zone extinction pulses.[126] This pattern of extinction is also compatible with hypoxia as a primary kill mechanism for the marine fishes, which would not have been able to escape the anoxic waters like the freshwater fishes did by moving onto land during the Late *linguiformis* conodont zone.

Although this last scenario proposes two kill mechanisms rather than one, they both could potentially be produced by the same phenomenon: the chemical weathering of silicate rocks in uplifted mountain chains located in the equatorial zone of the Earth. That is, the proposed carbon dioxide depletion of the atmosphere in the chemical weathering model may produce the global cooling that leads to hypothermia, and the proposed increased influx of nutrients into the shallow seas in the chemical weathering model may produce the eutrophication and depletion of oxygen in marine waters that leads to hypoxia.

Was the end-Frasnian extinction really the result of tectonic chance? A bad period in Earth's history when numerous continental crustal blocks happened to be colliding, terranes accreting in subduction zones at continental margins, and the chance paleogeographic positioning of most of the upthrust mountain chains in the equatorial zone of the planet? Was that really sufficient? Or was there an additional catastrophic component—impacting asteroids scattered across the surface of the Earth or huge plumes of molten rock rising from the mantle to pour vast sheets of lava across the surface of the Earth—that tipped the scales from bad times to lethal times? To return to the courtroom analogy used at the beginning of this section, can it be proved beyond the shadow of a reasonable doubt that any of these scenarios actually was guilty for the loss of so many animal and plant species on Earth that occurred 374.5 million years ago?

The Mystery of the End-Frasnian Loss in Biodiversity

There remains one additional puzzling biological fact about the diversity loss that occurred on Earth at the end of the Frasnian: much of it was due to a decrease in the rate of speciation, in concert with the increase in the rate of extinction (Fact 8). Both of the proposed hypoxic and hypothermic kill mechanisms are just that—kill mechanisms that can explain increased rates of extinction. What, however, can explain the observed suppression in speciation rates that took place in the late Frasnian (table 4.4)?

For some unknown reason, species lineages present in the late Frasnian nearly stopped splitting into new species lineages, the normal process of evolution. Ever since Darwin proposed the theory of natural selection, phylogenetic lineage fragmentation, or speciation, has been seen to be a function of habitat fragmentation. A species lineage may produce several geographic races, each evolving via natural selection to adapt to their local habitat conditions. If these geographic races accumulate sufficient genetic differentiation, then they will no longer be able to interbreed successfully, and they will become new separate species.

Thus, from the expectations of the theory of natural selection, the near cessation of speciation that occurred in the late Frasnian could be a function of the near elimination of habitat fragmentation; that is, habitat homogenization. Reef ecosystems were hit hard at the very beginning of the end-Frasnian extinction: extinction pulse 1 in the Late *rhenana* conodont zone, as discussed at the beginning of the chapter. Modern reefs provide an incredible diversity of habitats and thus are sites of very high species diversity. Elimination of the huge reef ecosystems in the late Frasnian thus eliminated a major potent source area for new species originations.

A volcanic ash bed located just above the Lower Kellwasser Horizon in German strata, which occurred in the earliest part of the Late *rhenana* conodont zone, has been radiometrically dated to be 377.2 ± 1.7 million years old.[127] The age of the Frasnian–Famennian boundary is currently estimated to be 374.5 ± 2.6 million years old.[128] Taking the two age assignments at face value, the collapse of reef ecosystems occurred 2.7 million years before the Frasnian–Famennian boundary. However, the age assignment for the Lower Kellwasser Horizon has a ±1.7 million year uncertainty, and that for the Frasnian–Famennian boundary has a ±2.6 million year uncertainty. Taking those two uncertainties into account, the maximum amount of time that could have elapsed from the collapse of reef ecosystems to the end of the Frasnian is 7.0 million years,[129] and the minimum amount of time is 1.6 million years.[130]

Thus the marine environments of the last 7.0, or 2.7, or 1.6 million years of the Frasnian were characterized by an impoverishment of habitats due to the destruction of the reefs in extinction pulse 1. When the severest pulses of extinction occurred in the *linguiformis* conodont zone in the latest Frasnian, not only did marine species go extinct, but they were not replaced by any new species originations in the homogeneous marine environments of the late Frasnian, and the resulting diversity drop is one of the five most severe losses of biodiversity to have occurred in Earth's history.

Alycia Stigall also links the cessation of vicariant speciation (Fact 8) to the spread of invasive marine species in the habitat-impoverished seas of the late Frasnian. The progressively rising sea levels of the Frasnian breached many marine geographic barriers, allowing the widespread migration of species, previously isolated from one another, in the "Great Devonian Interchange."[131] Stigall argues that the "establishment of broadly adapted, geographically widespread invasive species was the likely trigger for speciation depression. Allopatric speciation by vicariance requires passive isolation of previously adjacent populations. The numerous range expansion events during this interval would have prohibited sustained geographic isolation, thereby cutting off the primary mechanism of vicariant speciation."[132] The effect of modern invasive species in producing biotic homogenization with "a few winners replacing many losers" is also a major focus of the study of human-induced extinctions in our present world.[133]

The problem with these marine scenarios is the fact that the very same phenomenon—a strange near-cessation in new species origination—also occurred in land plants in the terrestrial realm (Fact 8). The collapse of reef ecosystems in the oceans has no effect on habitat diversity on the land. A key to understanding what happened on land may be provided by the previously described work of Scott and Glasspool (see Fact 2 above). To repeat part of one of their observations: "By the Mid-Late Frasnian, *monospecific archaeopterid forests* dominated lowland areas and coastal settings over a *vast geographic range*"[134] (emphases mine). These late Frasnian forests were not like modern forests, which are composed of many species of trees. In contrast, the late Frasnian forests were more like modern monocultural forests that are grown in forestry agriculture, which are composed of a large numbers of a single species of tree. The Smithsonian paleobotanist William DiMichele and his colleague Robert Hook have previously commented on the potential consequences of the evolution of a floral monoculture in the Late Devonian: "Plant communities of the Early and Middle Devonian had relatively high guild depth (the number of different species with similar ecological strategies and resource requirements)."[135] By the Late Devonian, however, the flora had changed:

> During the Late Devonian, habitat or niche partitioning appears to have been well established. Guild depth, however, was low. The ecomorphotype-to-species ratio would have been very close to one. As a consequence, the options for successional recovery were few. Over short time spans (thousands of years), such systems may have been relatively stable with higher levels of species–species connectedness because of greater species resource specificity than in earlier Devonian ecosystems. Local and regional patterns of disturbance and recovery may have been comparable in their degree of regularity to modern systems in similar physical settings. However, over millions of years the low guild depth should have made these communities *very susceptible to dramatic changes* in both dominance–diversity hierarchy and structure.[136] (emphasis mine)

In the Late Devonian flora, almost every floral ecomorphotype was represented by a single species. That is, there was one species of towering tall trees, one species of medium-sized trees, and so on, unlike the modern flora, where a given ecomorphotype may have been independently evolved by numerous different species from different phylogenetic lineages.

The dangers inherent in growing floral monocultures are well known in modern agriculture: a single new disease, or an unexpected change in the climate, can trigger a catastrophic die-off in plants. When extinction struck land plants in the Early *rhenana* conodont zone of the Frasnian, it hit a homogeneous, monocultural flora that was not producing new species. The end result was the same as in the marine realm: not only did land plant species go extinct, but they were not replaced by any new species originations. The post-extinction flora maintained the same low level of diversity that remained unchanged through the time span of the latest Frasnian and the Famennian Gap.

In summary, in addition to the two proposed kill mechanisms of hypothermia and hypoxia that we considered in detail in this chapter, we must consider the effects of the loss of habitat and ecological diversity that occurred in the late Frasnian. The late Frasnian ecosystems of the Earth were on the edge of collapse: animal species and plant species were no longer producing new species replacements, and ecological niches were occupied by single species rather than a diversity of different species. When those ecologically monospecific species began to go extinct, whether by freezing to death or suffocating to death or both, there were no other differently adapted species to replace them and take over the ecological roles or modes of life they had lived. The ecosystems of the Frasnian collapsed, and a "Big Five" diversity loss was the result.

As we shall see in the next chapter, the effects of the end-Frasnian biodiversity crisis are greater than the distinction of having been one of the five most severe losses of biodiversity to have occurred in Earth's history. The course of evolution on land was changed, redirected. The legacy of the end-Frasnian catastrophe extends all the way to today: our modern world would not be the same if that catastrophe had not occurred.

CHAPTER 5

The Second Animal Invasion

The End-Frasnian Bottleneck

An evolutionary or genetic bottleneck is produced when the population sizes of a given species shrink almost to the critical minimum level from which a species cannot recover. Such a critical reduction in species' population sizes can be produced either by the abrupt death of the majority of the individuals in the populations or by the failure of the individuals in one generation of the populations to reproduce enough offspring to ensure the survival of the subsequent generation—or by both processes acting together simultaneously.

One of the immediately observable consequences of a species surviving an evolutionary bottleneck is the sharp reduction of genetic diversity seen in the population of survivors. Numerous examples of this phenomenon are known from nature—the cheetahs in Africa survived a severe evolutionary bottleneck, but the remaining cheetahs have such low genetic diversity that they are still in danger of extinction. They came very close indeed to the minimum viable population limit. European and Asian populations of humans, *Homo sapiens,* also show the characteristic low genetic diversity that results from passing through an evolutionary bottleneck. It is estimated that all European and Asian humans could be the descendants of as few as 160 people, possibly a single clan of hunter-gatherers, who managed to exit Africa by crossing the Bab al-Mandab

(Gate of Grief) of the Red Sea about 50,000 years ago. Previous groups of *Homo sapiens* had attempted to spread out of Africa via present-day Israel about 80,000 years ago, but these migrants did not make it. Apparently they were exterminated by another group of humans then living in southern Europe: the Neanderthal people, *Homo neanderthalensis*.[1]

Famennian tetrapods show a sharp reduction in morphological diversity relative to the Frasnian tetrapods, a low morphological diversity that reflects their underlying low genetic diversity. In 2002, Cambridge University tetrapod specialist Jennifer Clack remarked on the pronounced reduction in morphological variance that is seen in tetrapods that evolved after the Famennian Gap:

> The tetrapodomorphs of the Frasnian, *Obruchevichthys*, *Elginerpeton*, and *Livoniana*, although not known from extensive material, hint at a level of diversity that exceeds that found in tetrapods known from the Famennian. This is shown by the differences among their dentitions, with patterns not seen anywhere else among the lineage. The tetrapods of the Famennian, however, show a conservatism in jaw, palate, skull roof, and girdle structure *that is quite striking*. They are all about the same size and have similar broad, flattened head shapes.[2] (emphasis mine)

Livoniana is a tetrapodomorph fish that is no longer thought to be ancestral to tetrapods (as discussed in chapter 3), but Clack's observations in 2002 regarding *Elginerpeton pancheni* (figs. 3.4 and 3.5) and *Obruchevichthys gracilis* still stand, and to these observations we can now add the additional morphologies seen in the Frasnian tetrapod species *Sinostega pani* and *Metaxygnathus denticulus* (fig. 3.7).

Uppsala University tetrapod specialist Per Ahlberg also notes the striking lack of variation in skull morphology in the new Famennian tetrapod genera, and he takes it as evidence that these genera are conservative survivors of more diverse ancestors:

> The genera display considerable postcranial variation, but their jaw morphology and head shape *are strikingly uniform*; these common features must have characterized the actual stem lineage, and may be

> evidence for a conserved mode of prey capture. . . . This suggests that *Ventastega, Acanthostega,* and *Ichthyostega* are *conservative survivors* from a Lower Famennian (or earlier) radiation.[3] (emphasis mine)

Ventastega (figs. 3.4 and 3.6) is indeed thought to belong to a ghost lineage that survived from the Frasnian, and *Acanthostega* and *Ichthyostega* are new Famennian genera of tetrapods that we will encounter in this chapter. In 2012, Clack further notes that both of the Frasnian species *Sinostega pani* and *Metaxygnathus denticulus* "seem to resemble Famennian forms" and agrees with Ahlberg in that "this could be taken to mean that the specialized Frasnian forms suffered during the Frasnian–Famennian extinction event, to be wiped out *in favor of the more conservative forms*"[4] (emphasis mine).

After the end-Frasnian extinction, only a single species is known from the fossil record in the Famennian Gap: *Jakubsonia livnensis,* which was found in Russia, part of the territorial range of the extinct Frasnian species *Obruchevichthys gracilis* in southeast Laurussia (color plate 8). In fact, of the new tetrapod species seen in the second invasion of land, all appear only in Laurussia and only two species are known outside of the tropical region of Laurussia. The worldwide geographic range seen in Frasnian tetrapod species was gone (color plate 8), and the new tetrapod species appearing in the late Famennian was markedly restricted in their geographic range (color plate 11).

The reductions in morphological variance and geographic range seen in the tetrapod species appearing in the fossil record in and after the Famennian Gap are classic evolutionary bottleneck phenomena, and I refer to the cause of these effects as the "End-Frasnian Bottleneck."[5] The legacy of the End-Frasnian Bottleneck in shaping the subsequent evolutionary history of terrestrial vertebrates, our ancestors, will be explored in more detail in chapter 8.

The Invasion Begins

The first Famennian tetrapod to appear in the fossil record after the end-Frasnian extinction is the Russian species *Jakubsonia livnensis* (color plate 11). The strata containing this species have been dated to occur somewhere in the interval from the Late *triangularis* to Latest *crepida* conodont zones,[6] thus this species appears in the Famennian Gap itself (table 4.2). It is the sole fossil species known from the Famennian Gap. However, we know that two ghost lineages from the Frasnian also survived through the Famennian Gap, as species in these lineages are found later in the Famennian fossil record. Curiously, this phenomenon (one fossil species, two ghost lineages) will be repeated in the Early Carboniferous, following the end-Famennian extinction.

After the Famennian Gap, numerous tetrapod species begin to appear in the fossil record. Some of these new species are found in habitats that were far from marine strata, thus terrestrial spore zonations are used to locate them in time. The terrestrial spore zonation of the Famennian Age is given in table 5.1, from which it can be seen that the Famennian is subdivided into nine spore zones, as opposed to the 21 marine conodont zones we considered in chapter 4 (table 4.2). The correlation of the spore zones with the conodont zones in the Famennian is given in table 5.2; notice the curious "logarithmic" effect in the duration of the terrestrial spore zones with reference to the marine conodont zones. In the earlier Famennian, a single spore zone can span the time of up to five conodont zones, whereas in the latest Famennian a spore zone has about the same time span as a conodont zone. In essence, this logarithmic-like scaling

TABLE 5.1 Terrestrial spore zonation of the Famennian Age for the subtropical region of ancient Laurussia (modern western Europe).

Geologic Age	Spore Zones
FAMENNIAN	**VI** Zone, pars (lowermost *V. verrucosus–R. incohatus* Zone)
	LN Zone (*R. lepidophyta–V. nitidus* Zone)
	LE Zone (*R. lepidophyta–I. explanatus* Zone)
	LL Zone (*R. lepidophyta–K. literatus* Zone)
	VH Zone (*A. verrucosa–V. hystricosus* Zone)
	VCo Zone (*D. versabilis–G. cornuta* Zone)
	GF Zone (*G. gracilis–G. famenensis* Zone)
	DV Zone (*K. dedaleus–D. versabilis* Zone)
	BA Zone, pars (uppermost *R. bricei–C. acanthaceus* Zone)

Source: Spore zonation modified from Streel et al. (1987), Maziane et al. (1999), Streel et al. (2000), Filipiak (2004), Streel (2009), and Blieck et al. (2010).

TABLE 5.2 Correlation of the terrestrial spore zones with the marine conodont zones in the Famennian.

Geologic Age	Spore Zones	Conodont Zones	
FAMENNIAN	**VI** Zone, pars	Late *praesulcata* Zone	
	LN Zone	Late *praesulcata* Zone	
	LE Zone	Middle *praesulcata* Zone	
	LL Zone	Early *praesulcata* Zone	
	LL Zone	Late *expansa* Zone	
	VH Zone	Middle *expansa* Zone	
	VCo Zone	Early *expansa* Zone	
	VCo Zone	Late *postera* Zone	
	GF Zone	Early *postera* Zone	
	GF Zone	Late *trachytera* Zone	
	GF Zone	Early *trachytera* Zone	
	GF Zone	Latest *marginifera* Zone	
	GF Zone	Late *marginifera* Zone	
	DV Zone	Early *marginifera* Zone	Famennian Gap
	DV Zone	Late *rhomboidea* Zone	Famennian Gap
	DV Zone	Early *rhomboidea* Zone	Famennian Gap
	DV Zone	Latest *crepida* Zone	Famennian Gap
	DV Zone	Late *crepida* Zone	Famennian Gap
	DV Zone	Middle *crepida* Zone	Famennian Gap
	DV Zone	Early *crepida* Zone	Famennian Gap
	DV Zone	Late *triangularis* Zone	Famennian Gap
	BA Zone, pars	Middle *triangularis* Zone	Famennian Gap
	BA Zone, pars	Early *triangularis* Zone	Famennian Gap

Source: Zonation correlations modified from Ziegler and Sandberg (1990), Streel et al. (2000), Filipiak (2004), Streel (2009), and Blieck et al. (2010).

Note: Some correlations are approximate; for example, the base of the Famennian **GF** spore zone is known to be diachronous, being equivalent to the Late *marginifera* condont zone in some areas but extending down to the Late *rhomboidea* condont zone in others.

reflects the evolutionary increase in speciation rates of the land plants as they adapted to evermore different habitats, and spread to ever larger geographic regions of the Earth, during the time span of the Famennian.

No less than five new tetrapod species appeared in Greenland after the Famennian Gap: *Ichthyostega stensioei, Ichthyostega watsoni, Ichthyostega eigili, Acanthostega gunnari,* and *Ymeria denticulata* (color plate 11).[7] The oldest strata containing these ichthyostegid, acanthostegid, and ymerid species have been dated in the lower part of the **GF** spore zone and in the interval from the Late *marginifera* to Early *trachytera* conodont zones.[8] Thus these five new species appeared at most two conodont zones after, if not immediately after, the end of the Famennian Gap (table 5.2).

The Frasnian ghost lineage species *Ventastega curonica* appeared in Latvia (color plate 11), also part of the territorial range of the extinct Frasnian species *Obruchevichthys gracilis*, and has been dated in the Early *expansa* conodont zone.[9] This fossil species is thus four conodont zones younger than the five Greenland species.

Two new tetrapod species appeared in Pennsylvania in the United States: *Hynerpeton bassetti* and the Frasnian ghost lineage species *Densignathus rowei* (color plate 11). The strata containing these species have been dated in the **VH** spore zone,[10] or Middle *expansa* conodont zone (table 5.2), hence these species are one conodont zone younger than *Ventastega curonica*.

Last, the species *Tulerpeton curtum* also appeared in Russia (color plate 11), part of the territorial range of the early Famennian species *Jakubsonia livnensis*, but in strata dated in the very latest Famennian, the interval from the Early to Late *praesulcata* conodont zones.[11] This last new species appears in the fossil record sometime between one and three conodont zones after *Hynerpeton bassetti* and *Densignathus rowei* (table 5.2).

Thus, in the Famennian fossil record, eight new genera of tetrapods appear in the following temporal sequence: first *Jakubsonia*, then *Ichthyostega*, *Acanthostega*, and *Ymeria*, next *Ventastega*, then *Hynerpeton* and *Densignathus*, and last *Tulerpeton*. Phylogenetic analyses reveal, however, that they did not evolve in that sequence. *Densignathus* and *Ventastega* (figs. 3.4 and 3.6) are species within old lineages that extend back in time into the Frasnian, lineages that are much more plesiomorphic than the newly evolved Famennian lineages (table 5.3). Within the coexisting Greenland genera, phylogenetic analyses reveal that *Ichthyostega* was more derived than *Ymeria*, and *Ymeria* was more derived than *Acanthostega*.[12] *Hynerpeton* was more derived than *Ichthyostega*,[13] but less so than *Tulerpeton*. *Tulerpeton* was clearly the most derived of the Famennian genera thus far discovered (table 5.3). Last, *Jakubsonia* was very fragmentary and had few diagnostic traits for phylogenetic analysis, although aspects of its shoulder girdle suggest similarity to

TABLE 5.3 The evolution of Famennian tetrapods in the second invasion of land.

Sarcopterygii (lobe-finned fishes + descendants)
– Crossopterygii (coelacanths)
– Dipnoi (lungfishes)
– **Tetrapodomorpha** (tetrapod-like fishes + descendants)
– – basal tetrapodomorphs
– – – *Kenichthys campbelli*
– – Rhizodontia
– – Osteolepidiformes
– – – Osteolepididae
– – – Megalichthyidae
– – – Eotetrapodiformes
– – – – Tristichopteridae
– – – – *Gogonasus andrewsae* †Frasnian
– – – – unnamed clade
– – – – – Elpistostegalia †Frasnian
– – – – – **Tetrapoda** (limbed vertebrates)
– – – – – – Family incertae sedis †Frasnian?
– – – – – – – Unknown early Eifelian tetrapod species in Zachelmie, Poland
– – – – – – Family incertae sedis †Frasnian?
– – – – – – – Unknown early Frasnian tetrapod species in Valentia Island, Ireland
– – – – – – Family Elginerpetontidae †Frasnian
– – – – – – – *Elginerpeton pancheni* †Frasnian
– – – – – – – *Obruchevichthys gracilis* †Frasnian
– – – – – – Family incertae sedis †Frasnian
– – – – – – – *Sinostega pani* †Frasnian
– – – – – – unnamed clade
– – – – – – – Family incertae sedis
– – – – – – – – *Densignathus rowei*
– – – – – – – unnamed clade
– – – – – – – – Family incertae sedis
– – – – – – – – – *Ventastega curonica*
– – – – – – – – unnamed clade †Frasnian
– – – – – – – – – Family incertae sedis †Frasnian
– – – – – – – – – – *Metaxygnathus denticulus* †Frasnian
New Famennian Invader Lineages:
– – – – – – – – Family incertae sedis
– – – – – – – – – *Jakubsonia livnensis*
– – – – – – – – unnamed clade
– – – – – – – – – Family Acanthostegidae
– – – – – – – – – – *Acanthostega gunnari*
– – – – – – – – – unnamed clade
– – – – – – – – – – Family incertae sedis
– – – – – – – – – – – *Ymeria denticulata*

(continued)

TABLE 5.3 (*continued*)

---------- unnamed clade
----------- Family Ichthyostegidae
------------ *Ichthyostega stensioei*
------------ *Ichthyostega watsoni*
------------ *Ichthyostega eigili*
----------- unnamed clade
------------ Family incertae sedis
------------- *Hynerpeton bassetti*
------------ unnamed clade
------------- Family Tulerpetontidae
-------------- *Tulerpeton curtum*

Source: Phylogenetic data modified from Benton (2005), Lecointre and Le Guyader (2006), Ahlberg et al. (2008), and Clack et al. (2012).

Note: Major vertebrate clades of land invaders are marked in bold-faced type.

† extinct taxa. Where a period is given, taxa went extinct in that specific geologic period.

Ventastega.[14] Thus *Jakubsonia* is shown in table 5.3 in a sister Family to that of *Ventastega,* and in an unresolved quadrichotomy with the extinct clade holding *Metaxygnathus denticulus* and the new clade holding the Family Acanthostegidae. In summary, in terms of degree of evolutionary derivation, the eight Famennian tetrapod genera are arranged in the following phylogenetic sequence, from least to most derived: *Densignathus, Ventastega, Jakubsonia, Acanthostega, Ymeria, Ichthyostega, Hynerpeton,* and *Tulerpeton* (table 5.3).

While still preliminary, the phylogenetic analysis of Famennian tetrapod species has some interesting logical consequences. First, from the fossil record alone, it would appear that five new species appeared very soon after the Famennian Gap: *Acanthostega gunnari, Ymeria denticulata, Ichthyostega stensioei, Ichthyostega watsoni,* and *Ichthyostega eigili.* From phylogenetic analyses, that number rises to seven new species, in that the lineages containing the two species *Densignathus rowei* and *Ventastega curonica* must also be included. Species of these two lineages had to be present somewhere on Earth prior to the evolution of the more derived species of *Acanthastega, Ymeria,* and *Ichthyostega.*

Second, the plesiomorphic species *Densignathus rowei* and *Ventastega curonica* (figs. 3.4 and 3.6) would have to have belonged to very old lineages, lineages spanning at the very minimum the time interval from the late Frasnian to the Early *expansa* conodont zone for *Ventastega curonica* and to the Middle *expansa* conodont zone for *Densignathus rowei* (table 5.2). As we saw in chapter 3, Per Ahlberg and colleagues argue "not only that *Ventastega* represents a lineage of Frasnian origin, but that a substantial part of the Devonian tetrapod radiation occurred during the Frasnian."[15] These logical consequences must follow from the phylogenetic analysis. The only other logical possibility is that *Densignathus rowei* and *Ventastega curonica* are the products of reverse evolution: that is, they are the descendants of more derived Famennian tetrapod species that have secondarily, convergently, re-evolved plesiomorphic traits like coronoid fangs and internal gills (traits that we will consider in more detail below). That scenario, while not impossible, is unlikely.[16]

The species *Densignathus rowei* and *Ventastega curonica* represent Frasnian survivor lineages. Given the similarity of *Jakubsonia livnensis* to *Ventastega curonica,* did the lineage of *Jakubsonia livnensis* really evolve during the Famennian Gap, or is this species also a member of a survivor lineage from the late Frasnian? The argument could be made that the true second vertebrate invasion was not mounted by these plesiomorphic species, species from lineages of Frasnian origin. Rather, the second vertebrate invasion was mounted by their more derived descendants; that is, the true new burst of evolution of Famennian tetrapods actually begins with the appearance of *Acanthostega gunnari, Ymeria denticulata, Ichthyostega stensioei, Ichthyostega watsoni,* and *Ichthyostega eigili* shortly after the end of the Famennian Gap.

Something clearly happened in terrestrial environments after the Famennian Gap. Fossil fragments of only one tetrapod species appear in the fossil record of the Famennian Gap. However, perhaps as many as three species existed, species that were survivors of the End-Frasnian Bottleneck, but species that had such small population sizes that they had a very low probability of preservation in the fossil record. Shortly after the end of the Famennian Gap, an explosion of species diversification and an increase in population sizes occurred in Famennian tetrapods: in Greenland, more than 110 skulls of individuals of the species of *Ichthyostega* have been collected thus far from the fossil record![17] That is, even the fragmentary fossil record reveals a post-Famennian Gap population size of more than 110 individuals of *Ichthyostega* species found in strata stretching along a 100-kilometer (62-mile) outcrop belt in Greenland. Such fossils have not been found only in Greenland—a fragment of a lower right jaw of an "*Icthyostega*-like tetrapod"[18] has also been discovered in Belgium, a jaw that has been dated to the lower part of the **GF** spore zone, and thus it is of exactly the same age as the *Ichthyostega* species and *Acanthostega gunnari* in Greenland.[19]

Times had changed. Whatever environmental conditions that had persisted through the time span of the Famennian Gap had abated. A minimum of five newly evolved species of tetrapod had flourished in Greenland, attaining large population sizes and spreading as far as 1,500 kilometers (930 miles) to the east in southeastern Laurussia (present-day Belgium).[20]

The New Vertebrate Invaders

What did these new Famennian land invaders look like? Let us first consider *Jakubsonia livnensis,* even though it may represent a Frasnian survivor lineage rather than a new Famennian invader. The early Famennian tetrapod species *Jakubsonia livnensis* is known only from a few fragments of the skull roof, a fragment of the lower left jaw, two bones from the shoulder girdle (cleithrum and scapulocoracoid), and a femur from a hind limb. The fragments of the skull roof contain the frontal bones and indicate that the animal had a short snout and large eyes. The morphology of the femur indicates that *Jakubsonia livnensis* was an aquatic tetrapod and primarily used its hind limbs for paddling.[21] The whole animal may have been about one meter (3.3 feet) in length.

Numerous fossils of *Acanthostega gunnari* have been discovered in Greenland, including an exceptional find of three individuals that been swept down a river and buried together in point-bar sands in a bend in the river. *Acanthostega gunnari* had a flattish skull with a rounded snout (fig. 3.4), with jaws about 200 millimeters (about eight inches) long. The jaws contained two rows of numerous small, sharp teeth: an outer row of larger teeth along the jaw margin, and an inner row of much smaller teeth located on the palate in the roof of the mouth and on the coronoids in the lower jaw. The palatal teeth include a pair of fangs, but the dentition is more derived in that no coronoid fangs are present whereas they still are in *Ventastega curonica*.

The well-preserved nature of the fossil skull material reveals that *Acanthostega gunnari* still possessed a lateral-line system, a plesiomorphic trait possessed by fish that use the organ system to detect pressure waves in water that might indicate an approaching predator to be avoided, or approaching prey to be seized. In addition, *Acanthostega gunnari* also possessed internal gills, another plesiomorphic, fish-like trait. The delicate gill bars are still preserved in the posterior region of the lower jaw and skull, as is the post-branchial lamina of the shoulder girdle that forms the posterior margin of the chamber holding the internal gills.

Almost the entire skeleton of the animal has been discovered, including the delicate bones of the digits on its feet. To the great surprise of paleontologists, it was discovered that *Acanthostega gunnari* had eight digits, not the expected five, on both the forelimb and hind limb. In the forelimb all of the digits were of different sizes, from small to larger to smaller again around the distal margin of the foot, whereas in the hind limb the leading two digits were small, the following five digits were stouter, and the last digit was tiny. The feet of the animal were wider than they were long. The tail of the animal possessed a fin on both the dorsal and ventral surface, clearly indicating that the animal spent a great deal of time swimming in water, during which it probably used its broad feet as paddles (fig. 5.1). The total animal was about one meter (3.3 feet) in length.[22]

Only two individuals of the species *Ymeria denticulata* have been discovered thus far. One individual is represented by a partially preserved fossil skull that includes both lower jaws, part of the palate, the maxillae and premaxillae, and a mold of parts of the shoulder girdle bones.[23] The other individual is represented by a fossil lower jaw. The jaws of *Ymeria denticulata* were somewhat smaller than those of *Acanthostega gunnari*, being about 120 millimeters (about 5 inches) long, and the whole animal was probably about 600 millimeters (two feet) in length. Because little of the post-cranial skeleton is preserved, we do not know at present how many digits the animals possessed on their feet.

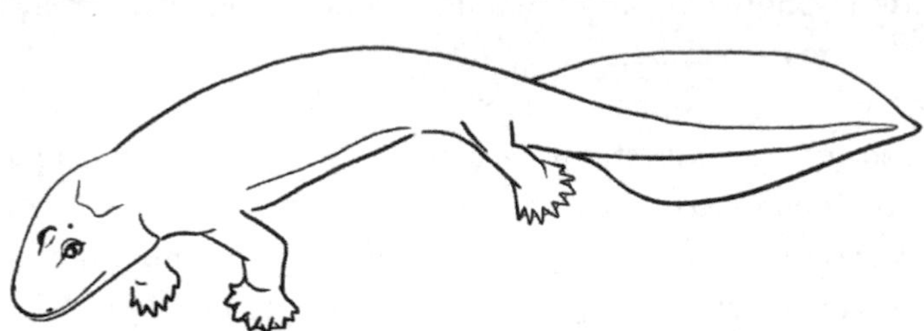

FIGURE 5.1 Reconstruction of the Famennian tetrapod *Acanthostega gunnari* from Greenland. Note the large tail fin, used for swimming, and the broad, paddle-like feet with eight toes. The living animal was about one meter (3.3 feet) in length. *Credit*: Illustration by Kalliopi Monoyios.

Three species of *Ichthyostega* have been discovered in Greenland: *Ichthyostega stensioei, I. watsoni,* and *I. eigili*. The three species did not coexist in the same area in Greenland, which would be ecologically unusual, but rather are separated in time. Stratigraphically, *Ichthyostega stensioei* is the oldest species, *I. watsoni* is younger, and *I. eigili* is the youngest.[24] The quality of preservation of the Greenland *Ichthyostega* fossils is exceptional and, as with *Acanthostega gunnari*, this includes the delicate bones of the digits on the animals' feet. Surprisingly, *Ichthyostega stensioei* possessed more than five digits on its feet—but not eight like the related species *Acanthostega gunnari*. Instead, *Ichthyostega stensioei* had seven digits on its feet, three small digits on the leading edge of the foot, followed by four larger and stouter digits.

The head of the animal was flattish with a blunt, rounded snout (fig. 3.4), and the largest animal found had jaws about 250 millimeters (9.8 inches) long. Similar to *Acanthostega gunnari*, the jaws contained two rows of numerous sharp, conical teeth, but the outer row of around 30 teeth were much larger than those of *Acanthostega gunnari*. Unlike *Acanthostega gunnari*, no palatal fangs are present. *Ichthyostega stensioei* still possessed a lateral-line system, though not as well developed as that seen in *Acanthostega gunnari*, and it also still possessed internal gills.

FIGURE 5.2 Reconstruction of the Famennian tetrapod *Ichthyostega stensioei* from Greenland. Note the smaller tail fin in this species compared to *Acanthostega gunnari* (fig. 5.1), and only seven toes in the paddle-like foot. The living animal was about 1.5 meters (five feet) in length. *Credit*: Illustration by Kalliopi Monoyios.

Curiously, the forelimbs of the animal were longer and more robust than the hind limbs. Jennifer Clack points out that this is also a feature of living seals, modern amphibious mammals, and suggests that *Ichthyostega stensioei* may have locomoted in a fashion similar to a seal: on land the seal primarily uses its large forelimbs to drag its body along, whereas in the water the seal tucks its limbs to its sides and swims with its tail.[25] The tail of *Ichthyostega stensioei* possessed a well-developed fin on the dorsal surface with only a fringe of a fin on the ventral surface (fig. 5.2), unlike the larger, well-developed fin on both the dorsal and ventral surfaces of the tail seen in *Acanthostega gunnari* (fig. 5.1). Clearly *Ichthyostega stensioei* spent more time out of the water than *Acanthostega gunnari*, but its finned tail and paddle-like feet also indicate aquatic adaptations. The whole animal was about 1.5 meters (five feet) in length.

Body fossils of *Hynerpeton bassetti* (color plate 5) include two shoulder girdles and one posterior fragment of the right lower jaw. Importantly, the shoulder girdle of the species lacks a post-branchial lamina, which was a newly derived trait and indicates that *Hynerpeton bassetti* had lost the internal gills that are present in more plesiomorphic tetrapod species. In addition, the size and shape of the shoulder girdle "indicates that the muscles for pectoral support and mobility were more developed in *H. bassetti* than in other known Devonian tetrapods. The morphology of the shoulder girdle of *H. bassetti* also suggests that it was capable of powerful protraction, retraction, and elevation of the forelimb."[26] The jaw fragment is 107 millimeters (4.2 inches) long and does not reveal any information concerning the dentition of the animal. The jaw is smaller and much more gracile than that of the coexisting Frasnian ghost lineage species *Densignathus rowei,* and the animal itself was probably smaller than the estimated one meter length of *Densignathus rowei.*

Last, body fossils of the Russian tetrapod species *Tulerpeton curtum* include the fragments of the skull, right forelimb, left shoulder girdle, fragments of vertebrae and ribs, and the right hind limb. The shoulder girdle of *Tulerpeton curtum* lacks the post-branchial lamina, like *Hynerpeton bassetti,* thus the species also had lost the use of internal gills for breathing.

Of particular significance is the fact that the bones in both of the limb fossils are still articulated and include the delicate digital bones. Surprisingly, *Tulerpeton curtum* did not possess either eight or seven digits on its feet—it possessed six. Thus a clear evolutionary trend in the reduction of the number of digits in the feet of tetrapods appears to have occurred in the Famennian: from eight digits in *Acanthostega gunnari* to seven digits in *Ichthyostega stensioei* to six digits in *Tulerpeton curtum.* The next step in this evolutionary sequence would be the evolution of pentadactyly—the possession of five digits on the hand and foot, which became the standard morphological condition in the evolution of subsequent land vertebrates. Thus far, no pentadactyl tetrapods have been found in Famennian strata, and that last evolutionary step appears to have occurred in the Carboniferous. Assembling the various fragments of the skeleton of *Tulerpeton curtum* reveals that the animal was about 1.5 meters (five feet) in length.[27]

In summary, known Famennian tetrapod species ranged in size from *Ymeria denticulata,* which was about 0.6 meters (two feet) in length, to *Ichthyostega stensioei* and *Tulerpeton curtum,* at about 1.5 meters (five feet) in length. In contrast, known Frasnian tetrapod species ranged in size from *Metaxygnathus denticulatus* (0.5 meters [1.5 feet long]) to

Elginerpeton pancheni and *Obruchevichthys gracilis* (1.5 meters [five feet] long) to the unknown species that made the Eifelian trackways in Poland (2.5 meters [8.2 feet] long). Variation in size in the older Frasnian tetrapods is much greater than that seen in tetrapod faunas after the End-Frasnian Bottleneck. In addition, the Frasnian *Elginerpeton pancheni* and *Obruchevichthys gracilis* had large, narrow, triangular-shaped skulls (fig. 3.4), whereas all of the Famennian tetrapods have small, broad, spade-shaped skulls.[28]

Ecology and Biogeography of the New Vertebrate Invaders

Reduction in morphological variation is not the only effect of the End-Frasnian Bottleneck on tetrapod evolution. A sharp reduction in geographic range is also evident. The Frasnian tetrapod species had a worldwide geographic range, from Laurussia to the isle of North China, far away to the east, and even further away to the eastern margin of the great southern supercontinent Gondwana (present-day Australia; color plate 8). In contrast, all of the known Famennian tetrapod species are confined to Laurussia (color plate 11). Eight of those species are found exclusively in the tropical region of southeastern Laurussia; only *Densignathus rowei* and *Hynerpeton bassetti* (color plate 5) have been found in higher latitude regions of southern Laurussia, at about 25° south of the equator (color plate 11), but *Densignathus rowei* represents a Frasnian survivor lineage rather than a new Famennian invader lineage.

It is interesting to note that the oldest yet discovered Famennian fossil species, *Jakubsonia livnensis*, has been found in near-shore marine strata. Thus the Famennian species *Jakubsonia livnensis* lived in an environment similar to that of the oldest known tetrapod, the species that made the trackways in Poland in the Eifelian Age. In addition, the Frasnian ghost lineage species *Ventastega curonica* (fig. 3.6) still lived in these near-shore marine habitats in the later Famennian. Is this mere coincidence, or is it a consequence of the end-Frasnian Bottleneck? Were the tetrapod survivors of the end-Frasnian extinction the more primitive marginal-marine species and had all of the more advanced freshwater species perished?

The *Jakubsonia livnensis* fossils are found in coarse-grained sandstones, possibly intertidal beach or stream sands, that are overlain and underlain by strata that contain marine brachiopods. Other vertebrate species found with *Jakubsonia livnensis* include bottom-dwelling detritivores, the antiarch placoderms (color plate 9) *Bothriolepis sosnensis, B.* cf. *leptocheira, "B." zadonica, Remigolepis* sp., and the ptyctodontid placoderm *Chelyophorus* sp.; fishes specializing in eating small aquatic arthropods and molluscs, the acanthodian *Devononchus* cf. *laevis*; the smaller predatory lungfishes, *Conchodus* sp., *Holodipterus* sp., and *Dipterus* sp., and the osteolepiform tetrapodomorphs *Megapomus markovksyi* and *Glyptopomus* sp.; and a large predator, the porolepiform crossopterygian *Holoptychius* cf. *nobilissimus*[29] (for the phylogenetic relationships of the various fish groups, see table 4.1). Note that species of the placoderm genus *Remigolepis,* which coexisted with the Australian tetrapod *Metaxygnathus denticulus* in the Frasnian, had arrived in Laurussia in the Famennian.

In contrast to *Jakubsonia livensis* and *Ventastega curonica,* the fossils of the species *Acanthostega gunnari* (fig. 5.1), *Ymeria denticulata, Ichthyostega stensioei* (fig. 5.2), *I. watsoni,* and *I. eigili* are all found in freshwater river deposits, far from the coastline. The region of Greenland where these species are found was once a great river basin fed by streams flowing northward from the towering Caledonide mountain chain to the south.

Acanthostega gunnari probably lived in actively flowing river channels. Several articulated skeletons of this species are found in river point-bar deposits, and the fact that they are still articulated indicates that they were not transported very far. The large ray fin located on the tail of *Acanthostega gunnari* indicates that this species was an active-swimming ambush predator, preying on fish living in the rivers of tropical Laurussia. Fossils of the large predatory freshwater porolepiform crossopterygian fish *Holoptychius* sp. and the lungfishes *Soederberghia groenlandica* and *Oervigia nordica* are found in the same strata as *Acanthostega gunnari.*[30] The detritivorous placoderms (color plate 9) *Bothriolopis nielseni, Phyllolepis neilseni, Remigolepis* cf. *kullingi,* and *R.* cf. *kochi* are also occasionally found,[31] but, curiously, no acanthodian fishes are found.[32]

In contrast, the species of *Ichthyostega* lived on the floodplains adjacent to the rivers, with a mode of life perhaps similar to modern crocodiles. Their tails had much smaller ray fins than *Acanthostega gunnari*, and their lateral-line systems were less well developed, indicating that they did not spend as much time in the water as that coexisting species. The monsoonal rains of the tropical climate that these species lived in produced periodic floods, and most of the *Ichthyostega* fossils are found in these flood deposits. The fossils are mostly disarticulated; thus they were transported a considerable distance from the original habitat of the animals.[33] We now know that an ichthyostegid-like species of tetrapod also lived about 1,500 kilometers (930 miles) away in present-day Belgium,[34] on the north side of the Variscian mountain chain (color plate 11). Presumably these animals migrated around the Caledonide mountains, probably along the coast to the east rather than by climbing over the Caledonides!

Hynerpeton bassetti (color plate 5) lived in the coastal delta plain habitats of southern Laurussia, located 25° south of the equator (color plate 11); thus the new Famennian tetrapods had once again managed to migrate out of the tropics of Laurussia and into higher latitude regions of the Earth. The coastal delta plains inhabited by *Hynerpeton bassetti* were adjacent to the Appalachian Sea to the northwest and were fed by numerous meandering streams flowing down from the young Appalachian mountain chain to the southwest. The Frasnian ghost lineage species *Densignathus rowei* coexisted with *Hynerpeton bassetti* in time, but it probably lived in the rivers rather than on the floodplain. Unlike *Designathus rowei*, *Hynerpeton bassetti* was an exclusive air breather because this species no longer possessed internal gills.

Numerous fish species inhabited the rivers and lakes frequented by *Hynerpeton bassetti*: the bottom-dwelling placoderm detritivores *Groenlandaspis pennsylvanica*, *Turriaspis elektor*, and *Phyllolepis rosimontina* but apparently no species of *Bothriolepis* (color plate 9) or *Remigolepis*; the acanthodian *Gyracanthus* cf. *sherwoodi*, which was an open-water filter feeder; the paleoniscoid actinopterygian *Limnomis delaneyi* and two unidentified species of tetrapodomorph fishes, a rhizodontid and a megalichthyid, which were smaller predators; and the tristichopterid tetrapodomorph fish *Hyneria lindae* (color plate 5), which was a large predator.[35] The usual large predator found in the same habitat as tetrapods, the porolepiform crossopterygian *Holoptychius,* was apparently not present, but the carnivore *Hyneria lindae* was fully three meters (ten feet) long and thus posed a serious danger to the much smaller *Densignathus rowei* and its still smaller cousin *Hynerpeton bassetti* (color plate 5). In addition to preying on smaller fish in the rivers, the exclusively air-breathing *Hynerpeton bassetti* also probably spent a lot of time preying on the land-dwelling arthropods (and thereby remained out of reach of *Hyneria lindae* in the rivers).

Curiously, the most derived of the Famennian tetrapod species, *Tulerpeton curtum,* is found in sediments deposited in estuarine to brackish conditions, evidenced by co-occurring fossils of marine algae and serpulid worms.[36] Paleogeographic reconstructions suggest that *Tulerpeton curtum* lived on and around islands in a shallow sea, and that the nearest large landmass was almost 200 kilometers (124 miles) away.[37] This species had lost the use of internal gills and was thus an exclusive air breather, yet it still seems to have been able to swim great distances given its island habitat far from the mainland.

Fauna coexisting with *Tulerpeton curtum* include the usual association of mostly freshwater fishes: the detritivorous antiarch placoderms *Remigolepis armata* and *Bothriolepis* sp.; small prey–gathering acanthodian species *Devononchus concinnus, D. laevis,* and *Cheiracanthus* sp.; four smaller carnivores, the paleoniscid actinopterygian *Moythomasia* sp., the lungfish *Andreyevichthys epitomus,* the osteolepiform tetrapodomorph *Strunius* sp., and the tristichopterid tetrapodomorph *Eusthenopteron* sp; and the large predatory porolepiform crossopterygian *Holoptychius* cf. *nobilissimus.* However, fragmentary fossils of unidentified sharks also occur, giving evidence of the proximity of the habitat to the sea. Further evidence for marine incursions and brackish conditions in the region can be seen by the presence of the serpulid worm *Serpula vipera,* numerous stromatolites (mound structures formed by marine cyanobacteria), fragmentary fossils of the marine charophyte alga *Quasiumbella* sp., and seven species of ostracodes in overlying strata.[38]

In summary, the late Famennian aquatic ecosystems favored by the tetrapods managed to recover much of the ecological diversity that had been present in Frasnian ecosystems. The bottom-dwelling detritivore niche was filled by the placoderms, and the spectrum of food size–determined niches was filled by a spectrum of small to mid-sized acanthodians, lungfishes, and osteolepiform tetrapodomorphs. The large, top-predator niche was filled by the porolepiform crossopterygians, tristichopterid tetrapodomorphs, and the tetrapods themselves (table 5.3).

However, there was one large hole in the Famennian aquatic ecosystem tapestry: the elpistostegalian tetrapodomorph fishes were gone (table 5.3). Instead, new and larger rhizodont tetrapodomorph species began to appear, large predatory fishes that moved into the ecological niche left vacant by the extinction of the elpistostegalians.

New Arthropod Invaders?

The fossil record is silent with respect to the arthropod invaders. At first glance, it would seem that there were no new arthropod invaders. That is, unlike the tetrapods, the arthropods did not experience an expansion of diversity after the Famennian Gap. Not a single new clade of arthropod is known from the Famennian Age.

Is this real, or is it just an artifact of preservation in the fossil record? Almost all that we know about arthropod species in the Early and Middle Devonian comes from the Rhynie Chert, the Alken-an-der Mosel Shale, the Gilboa Shale, and the South Mountain Shale *Lagerstätten* (chapter 3). No such equivalent *Lagerstätten* have yet been found in the Famennian. Some arthropod fossils are found at the late Famennian Red Hill site in Pennsylvania, along with the tetrapods *Designathus rowei* and *Hynerpeton bassetti* (color plate 5). These are very fragmentary, but the millipede species *Orsadesmus rubecollus* and trigonotarbid arachnid *Gigantocharinus szatmaryi* have been identified.[39] Neither of these arthropod groups are new, however; they are known to have existed on Earth since the Late Silurian (chapter 3). In conclusion, as far as we can tell, the arthropod invaders appeared not to have diversified further after the Famennian Gap.

Prelude to Disaster

Life was good, at least for our ancestors. The tetrapods had survived the diversity constriction of the End-Frasnian Bottleneck—perhaps just barely—and they rebounded in diversity after the Famennian Gap. Population densities bloomed, as witnessed by the hundreds of fossils of individuals of the *Ichthyostega* (fig. 5.2) species found in Greenland. Geographic dispersal began again, as tetrapods ventured out once more from their tropical refuges into the higher-latitude regions of the Earth in which their ancestors had perished. Evolutionary advancement took place in their evolving lineages: their ancestral fish-like dentitions, lateral-line systems, and internal gills were abandoned as they became more adapted to life on dry land.

Famennian tetrapods were finally poised to become lords of the land; they needed only to make a few more improvements in their limbs for efficient walking—but they never made it. The environmental catastrophe that triggered the end-Famennian extinction began instead.

CHAPTER 6

The Second Catastrophe and Retreat

The End-Famennian Catastrophe

The end-Devonian biodiversity crisis struck the Earth.[1] In the waning years of the Famennian, vertebrates and arthropods once again began to die in greater numbers. Population sizes once again dwindled, became smaller and smaller, and vanished. By the end of the Famennian, all of the known Famennian tetrapod species had perished.

On land, the huge forests of the Earth began to die. The trunks of the great lignophyte tree *Archaeopteris hibernica* (color plate 5) still towered 30 meters (100 feet) into the skies—but their leaves were gone, they were dead. One by one, the great trunks fell. Only a few of the *Archaeopteris* trees survived into the Early Carboniferous, only to die soon thereafter as well. The paleobotanists William DiMichele and Robert Hook, whose work we have previously encountered in chapter 4 in the end-Frasnian crisis, write of this second crisis in the forests of the Earth:

> There are major differences between Late Devonian and Early Carboniferous floras. . . . Because both *Archaeopteris* and *Rhacophyton* were extinct by the Early Carboniferous, Tournaisian landscapes were markedly different. Areas that were forest dominated for the previous 15 million years gave way to an entirely new plant-community

> reorganization. *No tree-sized counterparts* of *Archaeopteris* capable of forming dense, shady forests are known prior to the end of the Tournaisian and beginning of the Visean, when the gymnosperm[2] *Pitus* appears. The remaining vegetation, including newly evolved forms growing in floodplains and other lowland environments, appears to have been enriched in species more to the r-selected end of the life-history spectrum; these were largely shrubby pteridosperms,[3] *less than 2 m[eters] in height*."[4] (emphases mine)

To modern ecologists, plants that have an r-selected life-history strategy are commonly referred to as *weeds*: they grow rapidly, produce large numbers of seeds and offspring, and die quickly. This type of a life-history strategy is very useful to modern disaster species, enabling them to rapidly colonize areas of the Earth where the normal vegetation has been destroyed, such as in a forest fire. In the end-Famennian extinction, the Devonian-style forests were destroyed forever.

Out in the oceans, the twilight of the Devonian sounded the death knell for the ancient jawless fishes and the great armored fishes, the Placodermi (color plates 7 and 9). Both of these groups had suffered the loss of half of their phylogenetic lineages in the end-Frasnian extinction, but they had survived. They did not survive the end-Famennian extinction (table 6.1); all of their genetic lineages were terminated—they would evolve no further.

These swimming fishes were not alone; ecologically, almost all of the organisms that were part of the nekton or plankton suffered severe diversity losses in the end-Famennian extinction. An astonishing 26 families (96 percent) of the swimming ammonoid cephalopods perished; only a single family—represented by only two genera—survived into the Early Carboniferous.[5] Their swimming cousins, the nautiloid cephalopods, fared a little better: they lost 58 percent of their families, but that still was one of the worst losses of diversity in their evolutionary history.[6] In the phytoplankton, the tiny floating acritarchs came close to extermination and are unequivocally known from only one site on Earth in the Early Carboniferous.[7] In the zooplankton, the once numerous chitinozoans and tentaculitoids suffered total extinction.[8]

TABLE 6.1 Effect of the end-Famennian extinction on the fishes.

Vertebrata (vertebrate animals)
– Petromyzontiformes (lampreys; living jawless fishes)
– unnamed clade
– – Pteraspidomorphi **†Famennian**
– – unnamed clade
– – – Anaspida **†Famennian**
– – – unnamed clade
– – – – Thelodonti †Frasnian
– – – – unnamed clade
– – – – – unnamed clade **†Famennian**
– – – – – – Osteostraci **†Famennian**
– – – – – – Galeaspida †Frasnian
– – – – – – Pituriaspida †Eifelian
– – – – – Gnathostomata (jawed vertebrates)
– – – – – – Placodermi **†Famennian** (armored fishes)
– – – – – – – Acanthothoraci †Frasnian
– – – – – – – unnamed clade **†Famennian**
– – – – – – – – unnamed clade **†Famennian**
– – – – – – – – – Rhenanida †Frasnian
– – – – – – – – – Antiarchi **†Famennian**
– – – – – – – – unnamed clade **†Famennian**
– – – – – – – – – Arthrodira **†Famennian**
– – – – – – – – – unnamed clade **†Famennian**
– – – – – – – – – – Petalichthyida †Frasnian
– – – – – – – – – – Ptyctodontida **†Famennian**
– – – – – – unnamed clade
– – – – – – – Chondrichthyes (cartilaginous fishes)
– – – – – – – unnamed clade
– – – – – – – – Acanthodii
– – – – – – – – – Climatiiformes **†Famennian**
– – – – – – – – – unnamed clade
– – – – – – – – – – Ischnacanthiformes **†Famennian**
– – – – – – – – – – Acanthodiformes
– – – – – – – – Osteichthyes (bony fishes + descendants)
– – – – – – – – – Actinopterygii (ray-finned fishes)
– – – – – – – – – **Sarcopterygii** (lobe-finned fishes + descendants)

Source: Modified from the phylogenetic classification of Benton (2005) for extinct fishes, and the phylogenetic classification of Lecointre and Le Guyader (2006) for living fishes.

Note: Major vertebrate clades of land invaders are marked in bold-faced type.

† extinct taxa. Where a period is given, taxa went extinct in that specific geologic period. Fish groups that perished in the Famennian are marked **†Fammenian**.

Thus the end-Famennian extinction had a curiously different effect in the oceans of the Earth than did the end-Frasnian. In the end-Frasnian extinction, just about every ecological type of marine life suffered diversity losses. In the end-Famennian extinction, life in the oceanic water column was decimated, but animals that lived on the bottoms of the seas, the benthos, did not suffer such severe diversity losses. For example, the remaining stromatoporoid sponges were exterminated, and the trilobites and rugose corals suffered renewed extinctions,[9] but the brachiopod shellfish, tabulate corals, and bryozoans crossed into the Early Carboniferous relatively unscathed, unlike at the end of the Frasnian.[10] On land, the diversity loss in land-plant macrofossil genera was major but not as severe as in the end-Frasnian: a drop from 33 genera in the late Famennian to 22 in the Tournaisian, a 33 percent reduction in standing diversity as opposed to the 46 percent reduction in standing diversity that occurred in the late Frasnian.[11]

The End-Frasnian Bottleneck in species diversity was a product both of the increase in extinction rates and the precipitous drop in speciation rates that occurred in the late Frasnian, as discussed in chapter 4 (Fact 8). In contrast, speciation rates in the marine realm rebounded in the Famennian, and the loss of biodiversity that occurred in the late Famennian was almost entirely due to an elevation in extinction rates acting alone.[12]

The timing of the end-Famennian extinction is also curiously different from that of the end-Frasnian. In the end-Frasnian extinction, land plants suffered their maximum diversity loss at the beginning of the extinction event (in the Late *rhenana* condont zone), whereas marine organisms suffered their maximum diversity loss near its end (in the *linguiformis* condont zone). That pattern is reversed in the end-Famennian extinction. In the oceans, marine species perished in the latest part of the Middle *praesulcata* condont zone in the late Famennian (see table 6.2 for the zonal timescale; the Tournaisian timescale will be discussed in the next section of the chapter). On land, in the equivalent **LE** spore zone, terrestrial plants were unaffected. The land plants then suffered a major loss of diversity in the following **LN** spore zone, in the last full zone of the Famennian (table 6.2).[13] In the oceans, in the equivalent Late *praesulcata* condont zone, the marine organisms that had survived the earlier pulse of extinction were unaffected.

TABLE 6.2 Zonal timescales for the Famennian, Tournaisian, and Visean.

Geologic Age	Spore Zones	Conodont Zones	
VISEAN		Lower *girtyi–collinsoni* Zone	
		mononodosa Zone	
		Uppermost *bilineatus* Zone	
		Upper *bilineatus* Zone	
		Middle *bilineatus* Zone	
		Lower *bilineatus* Zone	
		Upper *commutata* Zone	
		Middle *commutata* Zone	
		Lower *commutata* Zone	
		texanus-homopunctatus Zone	
TOURNAISIAN		*anchoralis–latus* Zone	Tournaisian Gap
		Upper *typicus* Zone	Tournaisian Gap
		Lower *typicus* Zone	Tournaisian Gap
		isosticha Zone	Tournaisian Gap
		Upper *crenulata* Zone	Tournaisian Gap
		Lower *crenulata* Zone	Tournaisian Gap
		sandbergi Zone	Tournaisian Gap
		Upper *duplicata* Zone	Tournaisian Gap
		Lower *duplicata* Zone	Tournaisian Gap
		sulcata Zone	Tournaisian Gap
FAMENNIAN	**VI** Zone, pars	Late *praesulcata* Zone	
	LN Zone	Late *praesulcata* Zone	
	LE Zone	Middle *praesulcata* Zone	
	LL Zone	Early *praesulcata* Zone	
	LL Zone	Late *expansa* Zone	
	VH Zone	Middle *expansa* Zone	
	VCo Zone	Early *expansa* Zone	
	VCo Zone	Late *postera* Zone	
	GF Zone	Early *postera* Zone	
	GF Zone	Late *trachytera* Zone	
	GF Zone	Early *trachytera* Zone	
	GF Zone	Latest *marginifera* Zone	
	GF Zone	Late *marginifera* Zone	
	DV Zone	Early *marginifera* Zone	Famennian Gap
	DV Zone	Late *rhomboidea* Zone	Famennian Gap
	DV Zone	Early *rhomboidea* Zone	Famennian Gap
	DV Zone	Latest *crepida* Zone	Famennian Gap
	DV Zone	Late *crepida* Zone	Famennian Gap
	DV Zone	Middle *crepida* Zone	Famennian Gap
	DV Zone	Early *crepida* Zone	Famennian Gap
	DV Zone, pars	Late *triangularis* Zone	Famennian Gap
	BA Zone, pars	Middle *triangularis* Zone	Famennian Gap
	BA Zone, pars	Early *triangularis* Zone	Famennian Gap

Source: Tournaisian and Visean conodont zones modified from the zonal timescale of Filipiak (2004). Famennian spore and conodont zonations are from table 5.2.

Note: The "Tournaisian Gap" spans the entire Tournaisian Age of the Early Carboniferous; see text for discussion.

This disjunct between extinction severities in the oceans and on the land is perhaps seen in the biostratigraphic zonal correlations themselves (table 6.2). For many years it was thought that the Devonian–Carboniferous boundary corresponded perfectly to the **LN**-*sulcata* zonal boundary; that is, the base of the *sulcata* conodont zone in the oceans corresponded to the base of the **VI** spore zone on land. Such a correspondence would be expected if some environmental trigger affected life both in the oceans and on the land at the same time. However, it now appears that the base of the *sulcata* zone occurs some 14 centimeters (5.5 inches) higher than the base of the **VI** spore zone in Hangenberg strata in Saurland, Germany.[14] This would mean that the biotic changeover in land plants, from the **LN** to the **VI** zone, occurred slightly earlier in time than the changeover in marine conodonts, from the Late *praesulcata* to the *sulcata* zone (tables 5.1, 5.2, and 6.2). The time disjunct is slight, but it provides evidence for the different biotic responses of the life in the seas and on the land to the end-Famennian extinction.

Although different in these ways from the overall more severe end-Frasnian extinction, the results of the twin pulses of the end-Famennian extinction were profound. The Devonian Period of geologic time came to an end, and the Early Carboniferous was a very different world.

The Tournaisian Gap

When the sun rose in the sky at the dawn of the Tournaisian Age, 359.2 million years ago, the land was once again strangely silent. The great forests were gone. The great armored fishes were gone from the rivers, lakes, and seas. On land, the second valiant attempt of the vertebrates to invade the terrestrial realm had failed—the tetrapods had not survived the end-Famennian extinction (table 6.3).

Obviously, as with the first failed invasion, some few tetrapods did manage to survive somewhere on the Earth, else we would not be here today. It is clear, however, that the tetrapod species that are known to have mounted the second terrestrial invasion of land in the Famennian failed, and they were all driven to extinction. In acknowledgment of the fact that we, the tetrapods, still exist, the series of unnamed clades

TABLE 6.3 Effect of the end-Famennian extinction on the tetrapods and tetrapodomorph fishes.

Sarcopterygii (lobe-finned fishes + descendants)
– Crossopterygii
– – Porolepiformes **†Famennian**
– – unnamed clade
– – – Onychodontida **†Famennian**
– – – Actinistia (living coelacanths)
– Dipnoi (lungfishes)
– **Tetrapodomorpha** (tetrapod-like fishes + descendants)
– – basal tetrapodomorphs
– – – *Kenichthys campbelli*
– – Rhizodontia
– – Osteolepidiformes
– – – Osteolepididae **†Famennian**
– – – Megalichthyidae
– – – Eotetrapodiformes
– – – – Tristichopteridae **†Famennian**
– – – – *Gogonasus andrewsae* †Frasnian
– – – – unnamed clade
– – – – – Elpistostegalia †Frasnian
– – – – – **Tetrapoda** (limbed vertebrates)
– – – – – – Family incertae sedis †Frasnian?
– – – – – – – Unknown early Eifelian tetrapod species in Zachelmie, Poland
– – – – – – Family incertae sedis †Frasnian?
– – – – – – – Unknown early Frasnian tetrapod species in Valentia Island, Ireland
– – – – – – Family Elginerpetontidae †Frasnian
– – – – – – – *Elginerpeton pancheni* †Frasnian
– – – – – – – *Obruchevichthys gracilis* †Frasnian
– – – – – – Family incertae sedis †Frasnian
– – – – – – – *Sinostega pani* †Frasnian
– – – – – – unnamed clade
– – – – – – – Family incertae sedis **†Famennian**
– – – – – – – – *Densignathus rowei* **†Famennian**
– – – – – – – unnamed clade
– – – – – – – – Family incertae sedis **†Famennian**
– – – – – – – – – *Ventastega curonica* **†Famennian**
– – – – – – – – unnamed clade †Frasnian
– – – – – – – – – Family incertae sedis †Frasnian
– – – – – – – – – – *Metaxygnathus denticulus* †Frasnian
New Famennian Invaders:
– – – – – – – – Family incertae sedis **†Famennian**
– – – – – – – – – *Jakubsonia livnensis* **†Famennian**
– – – – – – – – unnamed clade
– – – – – – – – – Family Acanthostegidae **†Famennian**

(continued)

TABLE 6.3 *(continued)*

----------- *Acanthostega gunnari* **†Famennian**
---------- unnamed clade
----------- Family incertae sedis **†Famennian**
------------ *Ymeria denticulata* **†Famennian**
----------- unnamed clade
------------ Family Ichthyostegidae **†Famennian**
------------- *Ichthyostega stensioei* **†Famennian**
------------- *Ichthyostega watsoni* **†Famennian**
------------- *Ichthyostega eigili* **†Famennian**
------------ unnamed clade
------------- Family incertae sedis **†Famennian**
-------------- *Hynerpeton bassetti* **†Famennian**
------------- unnamed clade
-------------- Family Tulerpetontidae **†Famennian**
--------------- *Tulerpeton curtum* **†Famennian**

Source: Phylogenetic data modified from Benton (2005), Lecointre and Le Guyader (2006), Long et al. (2006), Ahlberg et al. (2008), Friedman and Sallan (2012), and Clack et al. (2012).

Note: Major vertebrate clades of land invaders are given in bold.

† extinct taxa. Where a period is given, taxa went extinct in that specific geologic period. Tetrapodomorph groups that perished in the Famennian are marked **†Fammenian**.

produced during Famennian tetrapod evolution are not shown to have gone extinct in table 6.3. If that lineage had indeed terminated, I would not be writing this book.

Only in the Visean Age of the Early Carboniferous do numerous tetrapod species once again appear in the fossil record, and I refer to the near absence of any tetrapod fossils in Tournaisian-age strata at the beginning of the Early Carboniferous as the "Tournaisian Gap" (table 6.2). The Tournaisian Age spanned from 359.2 million years ago to 345.3 million years ago (table 2.2), lasting some 13.9 million years. Thus the Tournaisian Gap is, curiously, almost exactly twice as long as the Famennian Gap. As with the preceding Famennian Gap, in this critical interval of 13.9 million years something went seriously wrong with the terrestrial environments of the Earth, something that only began to abate in the early Visean Age, when tetrapods once again recovered sufficient population sizes to appear in the fossil record.

University of Chicago paleontologist Michael Coates and his colleague, Cambridge paleontologist Jennifer Clack, proposed designating the interval of time spanning the Tournaisian and most of the Visean Ages as "Romer's Gap" in 1995,[15] in honor of Harvard paleontologist Alfred Sherwood Romer (1894–1973), who first commented on the strange absence of tetrapod fossils in the Early Carboniferous. Clack notes: "After the end of the Devonian, the fossil record of tetrapods becomes very sparse . . . only a few fragments of limb bones had been found from rocks representing the succeeding 30 million years. The significance of this gap in the record was noted during the early 1950s by paleontologist A. S. Romer, who pointed out that much of the evolution of terrestriality probably occurred during this period for which there is almost no information" and that this "interval covers the whole of the Tournaisian and most of the Viséan."[16]

However, as will be shown in chapter 7, phylogenetic analyses prove that significant numbers of tetrapod species must have existed in the early Visean; quite a few fossil tetrapod species have now been dated to the middle Visean, and one has been discovered in strata dated to the very end of the Tournaisian itself. In addition, one family of tetrapods (the whatcheeriids) is now known to have achieved a worldwide distribution by the mid-Visean, clearly demonstrating that the severe environmental conditions of the Tournaisian Age had waned, and that tetrapods were once again able to migrate over wide areas of the Earth in the early Visean Age. In contrast, only three basal, plesiomorphic families of tetrapods are known to have existed on the Earth in the Tournaisian. The entire 13.9 million-year span of the Tournaisian apparently produced very little evolutionary innovation in the tetrapod clade.

For these reasons, many researchers refer to the "Tournaisian Gap" (13.9 million years) rather than to Romer's Gap (30 million years). The critical interval of time in tetrapod evolution in the Early Carboniferous is the Tournaisian Age. By the early Visean Age, tetrapods were recovering their population numbers, were producing new evolutionary innovations and clades, and were geographically on the move.

The Tournaisian Gap is also clearly recognizable in the pattern of evolution of land plants. The end-Famennian extinction destroyed the Devonian-style forests, with their towering, spore-reproducing lignophyte trees. In contrast, the tallest land plants outside of the peat swamps during the Tournaisian Gap were only two meters (6.6 feet) high.[17]

In the peat swamps themselves, monospecific stands of the taller lycophyte tree *Lepidodendropsis hirmeri* could be found. The wetland trees of the Tournaisian "*Lepidodendropsis* flora" were around 10 meters (33 feet) tall, with slender, pole-like trunks only about 10 to 20 centimeters (3.9 to 7.8 inches) in diameter. Only in the Visean, following the Tournaisian Gap, did abundant, new, large lignophyte trees re-evolve outside of the swamps, such as the lyginopterid spermatophyte tree *Pitus primaeva*, with heights up to 40 meters (131 feet), and the forests of the Earth began to recover. As with the Famennian Gap, both terrestrial and marine ecosystems show the effects of the Tournaisian Gap.[18]

What Went Wrong?

Once again, we are faced with the very same question that we asked back in chapter 4, but focused on a period some 15 million years later: what are the empirical observations—the facts—about environmental conditions on Earth during the late Famennian and the 13.9 million years of the Tournaisian Gap? Here, at least, we have much more definitive data to answer that question.

FACT 1: *Massive continental glaciers formed on the Earth in the latest Famennian.* Unlike the mysterious global cooling hypothesized for the latest Frasnian and in the Famennian Gap in chapter 4, empirical evidence for global cooling in the latest Famennian is unequivocal: massive deposits of glacial tillite sediments are found in present-day Brazil, Bolivia, Peru, Central African Republic, and Niger, and in the mountains of eastern North America.[19] Thus continental ice sheets covered huge areas of Gondwana (fig. 6.1, color plate 11), and lowland glaciers were even found in temperate regions of Laurussia.

The oldest of the glacial tillites are found in the Amazonas Basin in Brazil and are dated to the **VCo** spore zone, just four conodont zones after the end of the Famennian Gap (table 6.3). The most significant phase of the global cooling occurred in the latest Famennian: glacial tillites dated to the **LE** and **LN** spore zones are found in the Brazilian Solimões and Parnaíba basins, in Peru and Bolivia, and in the Appalachian basin in North America. Thus ice sheets existed in Gondwana

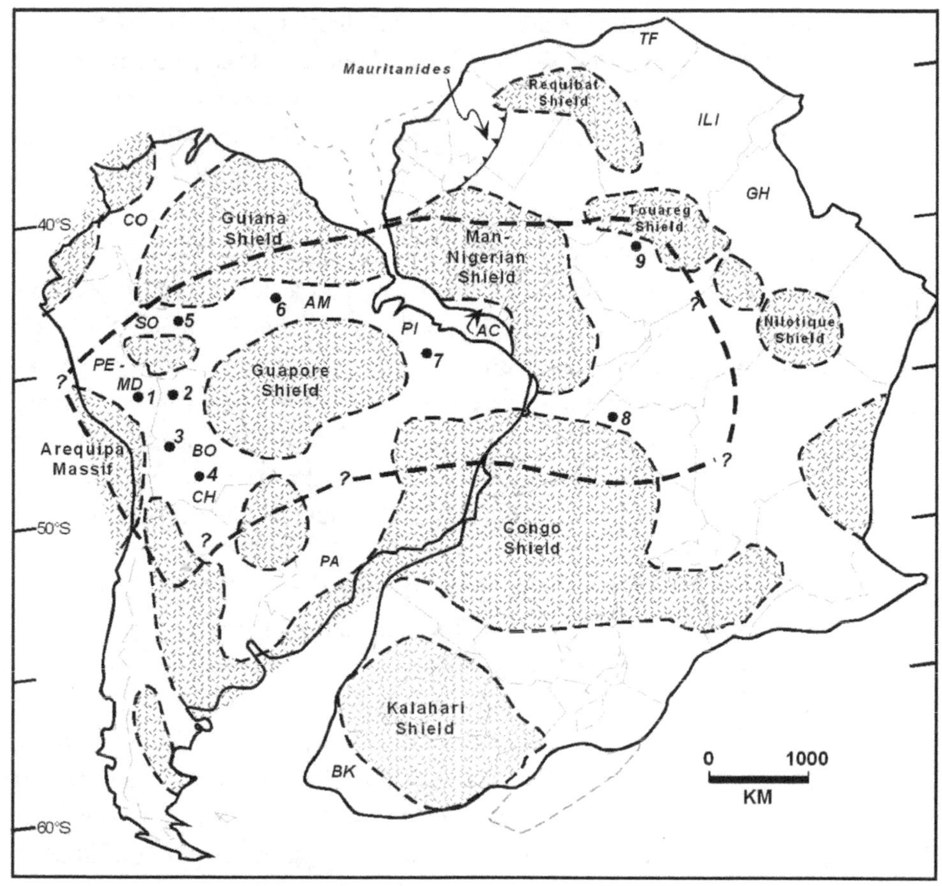

FIGURE 6.1 Paleogeographic extent of the Famennian ice sheet (heavy dashed line) on the continent of Gondwana, which covered parts of present-day South America and Africa. Lighter dashed lines indicate continental shield regions, and numbered points indicate sedimentary basins containing glacial strata. *Credit*: From *Palaeogeography, Palaeoclimatology, Palaeoecology*, Volume 268, by P. E. Isaacson, E. Díaz-Martínez, G. W. Grader, J. Kalvoda, O. Babek, and F. X. Devuyst, "Late Devonian–earliest Mississippian glaciation in Gondwanaland and its biogeographic consequences," pp. 126–142, copyright © 2008 Elsevier. Reprinted with permission.

during the last seven conodont zones in the Famennian (table 6.2), a time span of approximately five million years,[20] during which continental ice sheets eventually covered at least 16 million square kilometers (9.9 million square miles) of land[21] and lowland glaciers existed in Laurussia, within 30° of the equator.[22]

Of particular interest are the glacial tillites found in the **LN** spore zone in the Appalachian basin in North America. These were produced by coalescing alpine glaciers in the ancient Appalachian mountain chain, and it is surprising that they survived later erosion and still exist in the geological record—an unusual "sedimentary *Lagerstätte*" comparable to the fossil *Lagerstätten* that we have considered in previous chapters. Ice-borne dropstone boulders, rafted away from the fronts of these lowland glaciers, have even been found in marine strata to the west in present-day Kentucky.[23] These deposits prove that the Earth was cold enough in the late Famennian to produce massive valley and coalescing lowland glaciers in temperate zones within 30° of the equator. These were not just high-altitude alpine glaciers, such as the Pleistocene remnants that still survive today in equatorial regions above the altitude of 4,000 meters (13,000 feet) in Kenya and Ecuador, but glaciers that extended all the way down to sea level.

FACT 2: *Major drops in global sea level occurred during the glacial periods in the latest Famennian.* There is nothing surprising in this phenomenon—just as in our own Pleistocene glaciations, the buildup of glaciers on the land produced major drops in global sea level. That is, if water leaves the oceans by evaporation and is prevented from returning to the oceans by freezing on land in massive ice sheets, then the amount of water in the oceans decreases and the altitudinal level of the sea surface falls. The estimated size of the Gondwanan glacier that existed in the **LN** spore zone in the late Famennian, some 16 million square kilometers (9.9 million square miles), is predicted to have produced a produced a sea-level drop of at least 60 meters (197 feet).[24]

It can be proved that the actual drop in sea level that occurred in the late Famennian glaciation was much larger than 60 meters. When the sea level falls, the coastlines of the oceans retreat from the land. Land that once was under water is now exposed and, more importantly, the mouths of rivers that used to empty into the ocean at the former coastline are now far from the new coastline and are now at an altitudinal level that is higher than that of the new sea level. As a result, the rivers cut downward in their river beds until they reach the altitudinal level of the new sea level, producing deeply incised valleys. In the recent Pleistocene glaciations, valley incision depths approached 100 meters (328 feet) indicating, of course, that the level of the sea fell by 100 meters during the period of maximum ice buildup on the land.

The late Famennian glaciation occurred over 359 million years ago, but geological evidence of valley incision still exists in isolated regions of the world. Deeply eroded river-channel facies dated to the **LN** spore zone exist for the Berea, Cussewango, and Gay-Fink rivers in the Appalachian basin in North America and for the Seiler river in Germany. The valley incision depths in these ancient river valleys range from 75 to 90 meters (246 to 295 feet), almost as much as that seen in the Pleistocene glaciation.[25] Thus the severity of global cooling in the late Famennian approached that seen in the Pleistocene in terms of the amount of water frozen on the land.

FACT 3: *Rainfall greatly increased in temperate and tropical climates during the glacial periods in the late Famennian and Tournaisian Gap.* Data concerning rainfall amounts in ancient terrestrial environments come from the analysis of fossil soil horizons. The Red Hill site in Pennsylvania has been extensively studied by paleopedologists Gregory Retallack and colleagues[26] chiefly because this site has yielded fossils of two separate species of Famennian tetrapods, *Densignathus rowei* and *Hynerpeton basseti* (color plate 5), and the environment that these two species inhabited is of interest. Analyses of fossil soil horizons at Red Hill reveal a cyclic alternation of four environments: a subhumid warm phase, followed by a semiarid warm phase, followed by a semiarid cool phase, and ending with an arid cool phase. Thus both rainfall and temperature appear to have been cyclically alternating during this interval of time—the **VH** spore zone, some 4.9 million years after the Famennian Gap and 3.5 million years before the end of the Famennian.[27]

Both *Densignathus rowei* and *Hynerpeton basseti* are found only in strata deposited in the subhumid warm phase, the wettest part of the climate cycle, indicating that these animals liked moisture. This is not surprising, as these animals were still aquatic tetrapods and lacked several of the adaptations needed for full terrestrial existence, as discussed in chapter 5. Retallack and colleagues conclude: "If these early tetrapods lived on land at all, it was during wet seasons of feeding and breeding, rather than fatally dry seasons."[28]

Further analyses by Maryland Geological Survey geologist David Brezinski and colleagues of fossil soil horizons distributed throughout the remainder of the Famennian in the central Appalachians of North America indicate that climatic conditions became increasingly humid and wet, evidenced by the marked change in the rock record from arid-deposited, red-colored strata containing numerous thin layers of calcareous fossil soils but few plant remains, to moist-deposited, gray-colored strata with abundant coaly plant fragments and no calcareous paleosols. Calcareous paleosols indicate the existence of evaporative, arid conditions on land and do not form in wet environments. Not only do these stratigraphic changes record a major increase in the abundance of plants in terrestrial ecosystems, but offshore marine strata record an increase in organic carbon content from both marine phytoplankton productivity and from the increased transport of abundant plant material from the land. Land plants, just like the tetrapods, apparently liked moisture.

This trend in increased rainfall and climatic change to wetter conditions in eastern North America began in the **VCo** spore zone with the onset of continental glaciation in Gondwana. Brezinski and colleagues argue that the trend to "humid conditions that began to develop in the late Famennian persisted into the early Visean,"[29] that is, through the Tournaisian Gap. They note that there were drier intervals, just as there were interglacial intervals, in the Tournaisian Gap:

> A relatively short-term climatic excursion to somewhat drier conditions within the Appalachian basin is indicated by the Tournaisian Patton Red Shale. . . . This apparent dry stage lasted up to 100 Ky [thousand years], but overlying coaly strata suggest that the late Tournaisian climate quickly returned to the humid conditions seen at the beginning of the Tournaisian. Subsequently, *early Visean strata of the Appalachian basin suggest there was an abrupt shift to aridity* with the onset of an early Carboniferous warm period. This dry period is indicated by the development of thick calcic soils and evaporites within the Maccrady Formation in Virginia, and calcic-inceptisols within the Mauch Chunk Formation in Pennsylvania and Maryland.[30] (emphasis mine)

Closer to the equator on the Laurussian continent, Visean climates in present-day Scotland, Ireland, England, and northern Canada were also arid on a seasonal basis. This is evidenced by widespread charcoal-bearing strata found in these localities, indicating frequent wildfires and dry conditions. At any single site, wildfires appear to have occurred every three to 35 years, indicating a fire-prone flora. Evidence for seasonality comes from growth rings found in arborescent spermatophytes, indicating an annual rainfall pattern that occurred in alternating wet and dry periods, thus producing alternating growth periods in woody plant tissue. These two pieces of evidence led paleobotanist Howard Falcon-Lang to conclude that Visean climates in tropical eastern Laurussia were savanna-like and monsoonal, with one dry season alternating with one monsoonal rainy season on an annual basis.[31]

FACT 4: *A progressive increase in the amount of terrestrially derived organic carbon in marine black shales occurred throughout the span of the Famennian and into the early Tournaisian Gap.* In the Famennian, the center of the Appalachian basin of North America lay in eastern Ohio and western Pennsylvania, and in this region the organic carbon content of the early Famennian Huron Shale was almost totally derived from marine phytoplankton. However, by the late Famennian Early *expansa* condont zone, the amount of terrestrially derived organic carbon in the Cleveland Shale had increased to almost 60 percent of the total organic carbon.[32]

The Early *expansa* condont zone in the marine realm correlates with the **VCo** spore zone on land (table 6.2), the same interval of time that saw the development of continental ice sheets on Gondwana (Fact 1) and increased rainfall in temperate and tropical regions of the Earth (Fact 3). The greatly increased amount of terrestrially derived organic carbon deposited in marine black shales clearly indicates greatly increased productivity of land plants after the Famennian Gap, which persisted throughout the remainder of the Famennian.

FACT 5: *Oxygen levels in the Earth's atmosphere in the late Famennian and Tournaisian Gap were higher than 13 percent, and may have been higher than the present-day level of 21 perecent.* As we saw in chapter 4, charcoal deposits exist in strata dated to the **VH** spore zone and provide empirical evidence that oxygen levels in the atmosphere of the

Earth had risen to at least 13 percent in the late Famennian. Younger charcoal deposits found around the world prove that wildfires frequently occurred in widely separated regions of the Earth in the time span represented by the last five conodont zones in the Famennian.[33] Thus hard empirical data exist to prove that the atmosphere of the Earth contained 13 percent oxygen, or more, in the last 3.5 million years or so of the Famennian. As also discussed in chapter 4, empirical models suggest that even higher levels of oxygen existed in the atmosphere in this interval of time: 21 percent to 23.5 percent in the Rock-Abundance model and 17 percent to 18 percent in the Geocarbsulf model. Independent data from the reflectance analysis of the charcoals from the **VH** spore zone in the Red Hill tetrapod site in Pennsylvania indicate that the temperature of the wildfires that produced these charcoals was about 575°C (1,067°F), similar to the temperature range found in modern wildfires with our present atmospheric O_2 level of 21 percent.[34] Regardless of the actual figure for atmospheric oxygen levels in the late Famennian, hypoxia is clearly ruled out as a potential kill mechanism for land animals in the end-Famennian extinction.

FACT 6: *The Famennian was punctuated by episodes of extensive and apparently abrupt increases in the geographic extent of black shale depositional regions in the shallow seas of the Earth, the largest of which occurred in the latest Famennian and correlates with marine extinction events. The causes of these episodes of black shale expansion remain unknown.* Even though the atmosphere of the Earth became progressively more oxygenated during the Famennian, episodes of black shale expansion and related marine anoxia continued out in the oceans. Six major episodes are known from the European stratigraphic record: the Nehden, Condroz, Enkeberg, Lower Annulata, Upper Annulata, and Hangenberg Black Shales.[35]

In Europe, the late Famennian extinction horizon for marine organisms correlates with the base of the Hangenberg Black Shale, which is dated to the latest part of the Middle *praesulcata* condont zone in the oceans and to the **LE** spore zone on land (table 6.2). At this time, deposition of normal oxygenated marine limestones abruptly ceased and was replaced by the deposition of black anoxic sediments. Like the twin Kellwasser Horizons in the late Frasnian, a heavy-carbon anomaly also exists at the time horizon of the Hangenberg Black Shale in the late Famennian.[36] Interpretations of the Hangenberg Black Shale with respect to changes in both sea level and global temperatures are disputed, just as we saw for the Kellwasser Horizons in chapter 4. The abrupt switch from oxic to anoxic bottom conditions seen with the onset of Hangenberg Black Shale deposition has been traditionally assumed to be associated with an equally abrupt rise in sea level and replacement of shallow water facies by deep water facies. Alternatively, it has been proposed that sea level and water depth remained the same and that the Hangenberg Black Shale represents a major eutrophication episode in the shallow seas of the Earth. The Hangenberg Black Shale could represent a simultaneous combination of a eutrophication episode and a rise in sea level.

Geochemical analyses of the oxygen isotope ratios in conodont dentary elements indicate that sea-surface temperatures were high in the early part of the Early *praesulcata* Zone, around 27°C to 29°C (80°F to 84°F), but that they dropped by 4°C (by 8.5°F) in the Middle *praesulcata* Zone shortly before the deposition of the Hangenberg Black Shale began.[37] Such a sharp drop in sea-surface temperature could be attributed to the expansion of continental glaciers that are known to have existed on land during the time-equivalent **LE** spore zone (table 6.3), as discussed above. However, expansion of continental glaciers should have resulted in a drop in sea level, not the rise in sea level that the Hangenberg Black Shale is proposed to represent.

Contradicting the geochemical analyses, the pedostratigraphic correlations of Retallack and colleagues indicate that the marine Hangenberg Black Shale occurred at the same time as the terrestrial "deep-calcic paleosol horizon H" in Pennsylvania.[38] All of the deep-calcic paleosol horizons are proposed to represent unusually warm and humid climatic episodes, with increased rainfall and higher atmospheric carbon dioxide levels. Thus a global warming trend and increase in nutrient runoff from land is proposed to have triggered a global eutrophication episode in the shallow seas of the Earth. In addition, a global warming trend would also have triggered an equivalent glacial melting episode, which could have produced the proposed rise in sea level that the Hangenberg Black Shale is thought to represent. Thus, as for the Kellwasser Horizons in the latest Frasnian, the Hangenberg Black Shale in the latest Famennian is predicted to have been deposited during a period of global cooling by one data set, and during a period of global warming by another. Whether sea level rose, fell, or remained the same is also disputed.

FACT 7: *Data concerning the presence and extent of glaciation during the Tournaisian Gap are disputed.* As outlined in Fact 1, the geological data unequivocally demonstrate that massive glacial ice sheets existed on Gondwana during the span of the last seven conodont zones of the Famennian, reaching their maximum areal coverage at the very end of the Famennian in the Late *praesulcata* condont zone. However, once we cross the boundary into the Early Carboniferous, the data become disputed. In a comprehensive 2003 summary of the geological evidence for glaciation, University of Wisconsin geologist John Isbell and colleagues argued that three distinct glacial phases could be recognized in the late Paleozoic: Phase GI, the time span from the Frasnian through the Tournaisian; Phase GII, the time span from the Serpukhovian through the Bashkirian; and Phase GIII, the time span from the late Moscovian to the middle of the Sakmarian in the Permian. Isbell and colleagues argue that glacial strata of Famennian and Tournaisian ages (table 6.4), their proposed glacial phase GI, are present in the northern Gondwana basins in Brazil and Bolivia in South America, and the Tim Mersoï basin in north-central Africa.[39] Later glacial phases were to spread across Gondwana from the west (present-day South America) to the east (present-day Australia).

The fact that extensive ice sheets existed in the Famennian is not disputed (see Fact 1). However, University of Pará geologist Mário Caputo and colleagues argue that glacial strata only from the middle Touraisian can be reliably dated, and that the early Tournaisian and late Tournaisian may have been cold but free of ice (table 6.4). They note that, during glacial periods with lowered sea levels, older strata are exposed and thus eroded and redeposited in glacial strata. Thus these younger glacial strata frequently contain older spores that have been reworked from older rocks. Careful examination of the preservation condition of the spores can reveal whether they are primary, thus reveal the age of the rock that contains them, or if they are secondary, i.e., have been reworked and redeposited, and thus give a false older date to the strata that contain them. Caputo and colleagues thus argue that only glacial

TABLE 6.4 Conflicting data concerning the extent of glaciation in the Early Carboniferous. See text for discussion.

Geologic Age	Conodont Zones	Early Carboniferous Glacial Data (1)	(2)	(3)	(4)
VISEAN	Lower *girtyi–collinsoni* Zone				
	mononodosa Zone				
	Uppermost *bilineatus* Zone				
	Upper *bilineatus* Zone			ICE	ICE
	Middle *bilineatus* Zone		ICE	ICE	ICE
	Lower *bilineatus* Zone		ICE	ICE	ICE
	Upper *commutata* Zone		ICE	ICE	ICE
	Middle *commutata* Zone			ICE	
	Lower *commutata* Zone			ICE	
	texanus–homopunctatus Zone			ICE	ICE
TOURNAISIAN	*anchoralis–latus* Zone	ICE		ICE	ICE
	Upper *typicus* Zone	ICE		ICE	ICE
	Lower *typicus* Zone	ICE		ICE	ICE
	isosticha Zone	ICE	ICE	ICE	ICE
	Upper *crenulata* Zone	ICE	ICE	ICE	ICE
	Lower *crenulata* Zone	ICE	ICE	ICE	ICE
	sandbergi Zone	ICE		ICE	ICE
	Upper *duplicata* Zone	ICE			
	Lower *duplicata* Zone	ICE			
	sulcata Zone	ICE			ICE
FAMENNIAN (pars)	Late *praesulcata* Zone	ICE	ICE	ICE	ICE
	Middle *praesulcata* Zone	ICE	ICE	ICE	ICE
	Early *praesulcata* Zone	ICE	ICE	ICE	ICE
	Late *expansa* Zone	ICE	ICE	ICE	ICE
	Middle *expansa* Zone	ICE	ICE	ICE	ICE
	Early *expansa* Zone	ICE	ICE	ICE	ICE
	Late *postera* Zone	ICE	ICE	ICE	ICE
	Early *postera* Zone				
	Late *trachytera* Zone				

Source: Data source indicated by numbers: (1) Isbell et al. (2003), (2) Caputo et al. (2008), (3) Mii et al. (1999), (4) Frank et al. (2008).

strata from the time interval of the Lower *crenulata* to the *isosticha* condont zones can be reliably dated in the middle Tournaisian.[40] Possible corroboration of this assessment is the documented presence of a large-magnitude heavy-carbon anomaly in strata dating from the

Upper *crenulata* and *isosticha* condont zones in both North America and Europe in equatorial Laurussia. University of Iowa geochemist Matthew Saltzman and his colleagues argue that this heavy-carbon anomaly, "one of the largest known Phanerozoic $\delta^{13}C$ events," corresponds to an episode of high rates of carbon-12 burial and atmospheric carbon dioxide downdraw, and that the resulting reverse-greenhouse cooling effect "is consistent with a late Kinderhookian (Tournaisian) glacial episode."[41] Also in North America, a major fall in sea level occurred in the middle Tournaisian, leading to erosional valley incision as deep as 60 meters (almost 200 feet) in some areas.[42] Thus the geological and geochemical evidence all point to the formation of extensive ice sheets in Gondwana in the middle Tournaisian and a consequent drop of at least 60 meters in sea level.

In contrast to Isbell and colleagues, who considered the Visean to have been ice-free, Caputo and colleagues report the existence of glacial strata in Brazil and Argentina that have been dated to the time interval from the Upper *commutata* to the Middle *bilineatus* condont zones in the middle Visean (table 6.4).[43] This age assignment has been corroborated by radiometric dating of volcanic strata that occur with the glacial strata in Argentina. Uranium-lead dating gives an age of 335.99 ± 0.06 million years for these strata, a date that falls in the middle Visean.[44]

As discussed in chapter 4, oxygen isotope ratios preserved in organic skeletal material are often used to reconstruct ancient temperatures (see Fact 6 in chapter 4). As also discussed in chapter 4, oxygen isotope ratios obtained from Devonian brachiopod seashells consistently yield temperature predictions that are clearly too hot, but that temperature predictions yielded by Carboniferous brachiopod seashell analyses are considered to be accurate! Having given this caveat, two brachiopod-based oxygen isotope analyses have proposed slightly different temperature predictions for the Tournaisian–Visean interval in the Early Carboniferous. Texas A&M University geochemist Horng-sheng Mii and colleagues argue that oxygen isotope ratios in early Tournaisian brachiopod fossils predict warmer conditions and that the Earth was ice-free at this time, but that the Earth cooled and glaciers returned in the middle Tournaisian and that the glaciers persisted into the late Visean (table 6.4).[45] The data that Mii and colleagues used in their 1999 study were obtained solely from North American fossils. In contrast, a larger data base, containing brachiopod isotopic analyses from North America, Europe, Iran, and China, was used in 2008 by University of Nebraska geologist Tracy Frank and colleagues to predict cold temperatures and glaciation in the earliest Tournaisian, a subsequent warming episode and ice-free conditions until the middle Tournaisian, global cooling and glaciation from the middle Tournaisian through the earliest Visean, another warming episode and ice-free conditions in the early Visean, and a return to glacial conditions in the middle Visean (table 6.4).[46]

In summary, hard empirical data from dated glacial strata in Brazil and Argentina prove that Gondwana was glaciated in the middle Tournaisian and the middle Visean. Oxygen isotope data suggest that the extent of the glaciation was more widespread in time. The analyses of Frank and colleagues expand range of time for the two glacial pulses of the middle Tournaisian and middle Visean, whereas the analyses of Mii and colleagues would indicate a single glacial interval that spanned the entire middle Tournaisian to late Visean interval of time.

The Kill Mechanism

Hypothermia was clearly the kill mechanism for many of the species populations that perished in the end-Famennian extinctions (table 6.5). The glacial evidence is unequivocal (Facts 1 and 2)—the Earth froze, just as it did in the Pleistocene, and the biological results were the same: extinction both on the land and in the sea, both in animal and in plant species. The charcoal evidence is also unequivocal (Fact 5)—the atmosphere of the Earth was well oxygenated, so hypoxia is clearly ruled out as a potential kill mechanism for land animals in the end-Famennian extinction.

Thus, unlike the mysterious end-Frasnian extinctions, we have a definitive answer to the question of the kill mechanism for the end-Famennian extinction: beyond any shadow of doubt, it was hypothermia, yes? Unfortunately, for those who prefer single-cause mechanisms, that is not the case. Hypoxia cannot be ruled out as a contributing factor to the end-Famennian extinction in the oceans. Extinction in the marine

TABLE 6.5 Summary of the kill-mechanism models that have been proposed for the end-Famennian extinctions.

I. Hypothermic Kill Mechanism
  A. Glaciation model
  B. Biological weathering model
  C. Equatorial convergence model
II. Hypoxic Kill Mechanism
  A. Marine eutrophication model
III. Combination Kill Mechanism
  A. Hypoxia followed by hypothermia model

Note: The proposed combination kill mechanism model is the reverse of the model proposed for the end-Frasnian extinctions (table 4.4).

realm clearly began before the extinction began on the land, as outlined at the beginning of the chapter. Marine animals began to go extinct in the latest Middle *praesulcata* condont zone, whereas maximum glacial expansion and sea-level fall did not occur until the following Late *praesulcata* condont zone. Land plants began to go extinct only in the **LN** spore zone, the terrestrial time-equivalent of the marine Late *preasulcata* condont zone (table 6.2), during the maximum glacial expansion.

Of course, it could be argued that the marine species were more sensitive to temperature changes than land plants and that they began to go extinct in the earlier phases of the glaciation, whereas the land plants succumbed only when land temperatures reached their most frigid levels and the ice sheets covered the most extensive area of the Earth's land masses. Yet the observed correlation of the onset of the extinction of marine species with the onset of the Hangenberg black-shale expansion episode cannot be dismissed as irrelevant. Marine waters were depleted of oxygen in wide geographic expanses of the Earth's shallow seas in the latest Middle *preaesulcata* condont zone, and marine species began to go extinct at the same time horizon. Correlation does not prove causality, but it is suggestive that many marine species populations succumbed to hypoxia at this time horizon. In other words, these species may have suffocated to death before they got the chance to freeze to death.

It is interesting to note that the oldest glacial tillites in the Famennian are dated to the **VCo** spore zone (Fact 1 above), and the oldest charcoal deposits in the Famennian are dated to the following **VH** spore zone (Fact 5). Oxygen concentrations in the atmosphere of the Earth had risen above 13 percent by the **VH** spore zone, and carbon dioxide concentrations had decreased. Was the initiation of glaciation in the preceding **VCo** spore zone in the southern continent of Gondwana triggered by the onset of this long-term reciprocal trend of oxygen increase and carbon dioxide decrease in the atmosphere during the late Famennian? That is, with the reduction of the concentration of the greenhouse gas carbon dioxide in the atmosphere in the late Famennian, did the atmosphere of the Earth became more prone to heat loss and thus trigger the glaciation?

A much stronger case may be made for global cooling triggered by biological weathering due to the proliferation of land plants, the model of Algeo and colleagues[47] discussed in chapter 4, for the end-Famennian extinction than the end-Frasnian one. It is a fact that seed plants evolved in the **VCo** spore zone of the Famennian (chapter 2) and, being free from needing water in reproduction, these plants could colonize the dry highlands and mountains that were out of reach for non-spermatophyte plants. Thus huge areas of exposed silicate rock were now within the reach of the seed plants for potential biological weathering. The massive increase of land plant–derived carbon that was transported out into the shallow seas (Fact 4) and the increase in the oxygen content of the atmosphere (Fact 5) are empirical evidence of vast geographic coverage of land by plants in the late Famennian. The production of all that organic carbon, and the resultant release of all that oxygen into the atmosphere, could only have been produced by one phenomenon: the removal of carbon dioxide from the atmosphere of the Earth in the process of terrestrial photosynthesis.[48] Added to that phenomenon would be the additional carbon dioxide depletion effect of the biological weathering of the silicate highlands and mountains in the process of carbonate formation. Was global cooling in the late Frasnian primarily produced by chemical weathering, a scenario discussed in chapter 4, and global cooling in the late Famennian primarily produced by biological weathering? Regardless of the particular type of weathering most prominent on land in the late Famennian, an increase in both nutrient and organic carbon (Fact 4) influx into the shallow seas of the Earth is evidenced in the eutrophication event that produced the Hangenberg Black Shale. Hypoxia in the shallow seas is the end result in both weathering models.

In summary, a scenario with two kill mechanisms cannot be ruled out for the end-Famennian extinction. Enhanced biological weathering and ongoing chemical weathering of silicate rocks on land may have triggered widespread eutrophication and resultant hypoxia in the shallow seas of the Earth, triggering the marine extinction pulse that occurred in the Middle *praesulcata* condont zone. Continued carbon dioxide depletion in the atmosphere produced by this weathering effect, combined with the carbon dioxide depletion effect of enhanced photosynthetic productivity by the geographic spread of seed plants into the highlands and mountains of the Earth, may then have triggered the reverse-greenhouse cooling that led to the major continental glaciers that formed in the Late *praesulcata* condont zone, and thus the hypothermia that killed the land plants and animals.

The end-Frasnian extinction remains mysterious—the "two kill mechanisms" scenario for that catastrophe discussed in chapter 4 (table 4.4) is the reverse of that discussed here for the end-Famennian (table 6.5). That is, in the end-Frasnian extinction hypothermia is proposed to have killed land plants and animals first, in the Late *rhenana* condont zone, and hypoxia is proposed to have killed marine animals second, in the *linguiformis* condont zone. The production of major marine eutrophication and hypoxia in the *linguiformis* condont zone, after the sharp reduction in terrestrial plant diversity that occurred in the preceding Late *rhenana* condont zone, therefore could not have been due to the effects of biological weathering by plants and had to have been produced by chemical weathering alone. Is the absence of significant biological weathering in the late Frasnian, and the onset of significant biological weathering by seed plants in the late Famennian, the cause of this peculiar reversal in the timing of hypoxia and hypothermia in the late Famennian versus the late Frasnian? To return to the courtroom analogy that I introduced in chapter 4, can it be proved beyond the shadow of a reasonable doubt that any of these scenarios actually was guilty for the loss of so many animal and plant species on Earth that occurred in the twin Devonian extinctions?

In the late Famennian, one thing has been proved beyond the shadow of any reasonable doubt: massive continental ice sheets formed on the Earth. The Earth froze, just like it did in the Pleistocene. Therefore it might be of interest to compare what we know about what happened during the 50-million-year Cenozoic global cooling trend, which is about 300,000,000 years closer to us in time than the ancient Late Devonian, to what little we definitively know about what happened in the Devonian.

A Comparison of the Late Devonian and Cenozoic Glaciations

In this section, we will examine the possibility that both the end-Frasnian and end-Famennian extinctions were triggered by the onset and intensification of the Late Paleozoic Ice Age. It is well established that the Late Paleozoic Ice Age spanned the interval of time from the late Famennian Age in the Late Devonian through the Capitanian Age in the Middle Permian.[49] This period of global cooling and glaciation lasted for some 67 million years and was the longest-lived icehouse interval in the entire Phanerozoic.

It is also well established that several major biodiversity crises in the Phanerozoic were directly associated with the Late Paleozoic Ice Age: crises in the Famennian at the onset of the glaciation, in the Serpukhovian when the glaciation became much more severe, and in the Capitanian when the glaciation ended during the volcanics of the Emeishan large igneous province in China.[50] If these biodiversity crises are ranked in order of their ecological impact, they all cluster together as the seventh, sixth, and fifth most ecologically severe of the eleven major biodiversity crises that occurred in the Phanerozoic since the beginning of the Ordovician (table 6.6). The ecological crisis in the Frasnian is ranked fourth, just ahead of the cluster of biodiversity crises that are known to have been associated with the Late Paleozoic Ice Age (table 6.6). Was the Frasnian crisis an isolated, unrelated event, or was it also related to this great glacial period in Earth history? Was the Famennian crisis, now thought to have been triggered by the onset of the Late Paleozoic Ice Age, not "ecologically severe enough" to have heralded the onset of such a massive change in the Earth's climate? Could the end-Frasnian fossil record instead have captured the initial ecological shock experienced by

TABLE 6.6 Ecological-severity ranking of the eleven largest Phanerozoic biodiversity crises since the beginning of the Ordovician.

1. Changhsingian (end-Permian)
2. Maastrichtian (end-Cretaceous)
3. Rhaetian (end-Triassic)
4. **Frasnian** (Late Devonian)
5. **Capitanian**
6. **Serpukhovian**
7. **Famennian**, Hirnantian (end-Ordovician)
8. Givetian
9. Eifelian, Ludfordian

Source: From McGhee et al. (2013).

The traditional "Big Five" biodiversity crises (Raup and Sepkoski [1982]) are listed in parentheses. The four biodiversity crises that are thought to be related to the Late Paleozoic Ice Age are marked in bold-face type; see text for discussion.

the hothouse fauna confronted with the onset of an icehouse stage in the Earth's climate, and the Famennian fossil record instead document the lesser ecological effect of the intensification of glaciation in the progressively cooling global climate?

What are the facts? A summary of some key geologic and biotic events that took place in the Late Devonian and Early Carboniferous are listed in table 6.7. The first pulse of extinction in the late Frasnian occurred 377 million years ago, as did the heavy-carbon anomaly that is associated with the Lower Kellwasser horizon. Three additional pulses of extinction then took place in the latest Frasnian, as did a second heavy-carbon anomaly that is associated with the Upper Kellwasser horizon. One final pulse of extinction took place in the earliest Famennian and marks the beginning of the Famennian Gap, which lasted seven million years.

About four million years after the end of the Famennian Gap, the first known ice sheets formed on the southern supercontinent of Gondwana (table 6.7). The glaciers reached their maximum expansion, at least 16 million square kilometers (9.9 million square miles) in area, within the last two million years of the Famennian.[51] The first pulse of extinction in the late Famennian occurred 361 million years ago, as did the heavy-carbon anomaly that is associated with the Hangenberg Black Shale horizon. The second pulse of extinction took place in the latest Famennian, coinciding with the maximum expansion of the ice sheets on land,

TABLE 6.7 Late Devonian and Early Carboniferous geologic and biotic events.

Geologic Time:		Geologic and Biotic Events:	
Visean	343		
(pars)	344		
	345		
Tournaisian	346 Cold	Tournaisian Gap	
	347 Cold	Tournaisian Gap	
	348 Cold	Tournaisian Gap	
	349 Cold	Tournaisian Gap	
	350 ICE	Tournaisian Gap	
	351 ICE	Tournaisian Gap	
	352 ICE	Tournaisian Gap	
	353 ICE	Tournaisian Gap	
	354 Cold	Tournaisian Gap	
	355 Cold	Tournaisian Gap	
	356 Cold	Tournaisian Gap	
	357 Cold	Tournaisian Gap	
	358 Cold	Tournaisian Gap	
	359 Cold	Tournaisian Gap	
Famennian	360 ICE		← land extinction; pulse 2
	361 ICE	← heavy-carbon anomaly	← marine extinction; pulse 1
	362 ICE		
	363 ICE		
	364 ICE		
	365		
	366		
	367		
	368		
	369	Famennian Gap	
	370	Famennian Gap	
	371	Famennian Gap	
	372	Famennian Gap	
	373	Famennian Gap	
	374	Famennian Gap	
	375	Famennian Gap	← marine extinction; pulse 5
Frasnian	376	← heavy-carbon anomaly	← marine extinction; pulses 2, 3, 4
(pars)	377	← heavy-carbon anomaly	← marine + land extinction; pulse 1
	378		

and marks the beginning of the Tournaisian Gap, which lasted for 13.9 million years. A second pulse of glaciation is known to have taken place in the Tournaisian Gap, in the Lower *crenulata* to *isosticha* conodont zonal interval, and lasted about four million years.[52]

For comparison, a summary of some key geologic and biotic events that took place in the Cenozoic are listed in table 6.8. The development of glaciations in the Cenozoic took place in three steps in geologic time: the Early Oligocene, Middle Miocene, and late Pliocene. The Early Oligocene Oi-1 glacial pulse took place 34 million years ago, lasted for 400,000 years, and was associated with a heavy-carbon anomaly.[53] This cold pulse initiated the formation of ice sheets in eastern Antarctica, glaciers that had about half the ice mass of glaciers existing at the present day. These ice sheets persisted for about eight million years before retreating during a warming trend that began in the Late Oligocene and extended into the Middle Miocene (table 6.8). The onset of the Early Oligocene glaciations coincided with two separate pulses of extinction that occurred in the oceans about 33 million years ago and a third pulse of extinction that occurred on land about 32 million years ago.[54]

A second glacial pulse, Mi-1, took place in the earliest Miocene, lasted for 200,000 years, and was also associated with a heavy-carbon anomaly.[55] However, it only initiated a series of brief and smaller glaciations in Antarctica, unlike the Oi-1 glacial pulse. The Earth began to cool rapidly in the Middle Miocene, about 14 million years ago, and a pulse of extinction occurred in both the marine and terrestrial realms at this same time.[56] Glaciological modeling suggests that the mean summer temperature in Antarctica dropped by at least 8°C (by 17°F) in the 220,000-year interval of time from 14.07 million to 13.85 million years ago, bringing a full polar climate to the region.[57] Massive ice sheets formed in eastern Antarctica 12 million years ago and persist to the present day (table 6.8). The last step in the Cenozoic glaciation took in the late Pliocene, about three million years ago, with the formation of ice sheets in the Northern Hemisphere of the Earth. The onset of glaciation in the Northern Hemisphere coincided with extinction pulses first in the marine realm[58] and then later in the terrestrial realm (table 6.8).

A comparison of the temporal scaling of the geologic and biotic events that took place in the Cenozoic, vis à vis those that took place in the Late Devonian and Early Carboniferous, is given in table 6.9. If the first

TABLE 6.8 Cenozoic geologic and biotic events.

Geologic Time:	Ice Sheets: *SH*: *NH*:	Geologic and Biotic Events:
Holocene	00 ICE ICE	
Pleistocene	01 ICE ICE	← land extinction
	02 ICE ICE	← marine extinction
Pliocene	03 ICE ICE	← marine extinction
	04 ICE	
	05 ICE	
Late Miocene	06 ICE	
	07 ICE	
	08 ICE	
	09 ICE	
	10 ICE	
	11 ICE	
	12 ICE	
Middle Miocene	13 Cold	
	14 Cold	← marine + land extinction
	15	
	16	
Early Miocene	17	
	18	
	19	
	20	
	21	
	22	
	23	← Mi-1 Glacial Pulse + heavy-carbon anomaly
Late Oligocene	24	
	25	
	26	
	27 ICE	
	28 ICE	
Early Oligocene	29 ICE	
	30 ICE	
	31 ICE	
	32 ICE	← land extinction; pulse 3
	33 ICE	← marine extinction; pulses 1, 2
	34 ICE	← Oi-1 Glacial Pulse + heavy-carbon anomaly
Late Eocene	35	
	36	

Abbreviations: *SH* = Southern Hemisphere; *NH* = Northern Hemisphere.

TABLE 6.9 A temporal comparison of Cenozoic and Devono-Carboniferous geologic and biotic events (formats modified from tables 6.6 and 6.7).

CENOZOIC		DEVONO-CARBONIFEROUS	
Geologic Time:	Geologic and Biotic Events:	Geologic Time:	Geologic and Biotic Events:
HOL	00 ICE ICE	VIS	344
PLE	01 ICE ICE ← land †	(pars)	345
	02 ICE ICE ← marine †	TOU	346 Cold (Tour. Gap)
PLI	03 ICE ICE ← marine †		347 Cold (Tour. Gap)
	04 ICE		348 Cold (Tour. Gap)
	05 ICE		349 Cold (Tour. Gap)
Late	06 ICE		350 ICE (Tour. Gap)
MIO	07 ICE		351 ICE (Tour. Gap)
	08 ICE		352 ICE (Tour. Gap)
	09 ICE		353 ICE (Tour. Gap)
	10 ICE		354 Cold (Tour. Gap)
	11 ICE		355 Cold (Tour. Gap)
	12 ICE		356 Cold (Tour. Gap)
Mid	13 Cold		357 Cold (Tour. Gap)
MIO	14 Cold ← marine + land†		358 Cold (Tour. Gap)
	15		359 Cold (Tour. Gap)
	16	FAM	360 ICE ← land†
Early	17		361 ICE ← marine†
MIO	18		362 ICE
	19		363 ICE
	20		364 ICE
	21		365
	22		366
	23		367
Late	24		368
OLI	25		369 *Cold?* (Fam. Gap)
	26		370 *Cold?* (Fam. Gap)
	27 ICE		371 *Cold?* (Fam. Gap)
	28 ICE		372 *Cold?* (Fam. Gap)
Early	29 ICE		373 *Cold?* (Fam. Gap)
OLI	30 ICE		374 *Cold?* (Fam. Gap)
	31 ICE		375 *ICE?* ← marine † (Fam. Gap)
	32 ICE ← land †	FRA	376 *ICE?* ← marine †
	33 ICE ← marine †	(pars)	377 *ICE?* ← marine + land †
	34 ICE		378
Late	35		379
EOC	36		380
(pars)	37		381

Note: Geologic events that remain to be proved are given in italics with question marks.

† extinction events.

extinction pulses that took place in the Early Oligocene are aligned with those that took place in the late Frasnian, it is interesting to note that the second extinction pulses that occurred in the Middle Miocene and the late Famennian almost align as well. That is, the Middle Miocene extinction occurred 19 million years after the onset of the Early Oligocene extinction, and the late Famennian extinction occurred 16 million years after the onset of the late Frasnian extinction. If we assume that the Earth cooled at a similar rate in the Cenozoic and the Late Devonian, then the temporal spacing of these extinctions may not be accidental—they may reflect the timing of the step-wise glaciation of the planet.

Step 1 of the Cenozoic glaciation occurred in the Early Oligocene. It is characterized by an initial glacial pulse and heavy-carbon anomaly, the formation of ice sheets that persisted for eight million years, and extinctions in both marine and terrestrial realms (table 6.8). The last two million years of the late Frasnian were also characterized by heavy-carbon anomalies and extinctions in both marine and terrestrial realms, followed by the seven-million-year long Famennian Gap (table 6.7). Was the nine-million-year span of the late Frasnian and Famennian Gap the first step in the Late Devonian glaciation, equivalent to the eight-million-year span of the ice sheets in the Oligocene during the first step of the Cenozoic glaciation? If so, one would predict the formation of ice sheets in Gondwana at least in the last three million years of the late Frasnian and earliest Famennian, if not for the entire nine-million-year span of the late Frasnian and the Famennian Gap (table 6.9).

Step 2 of the Cenozoic glaciation occurred in the Middle Miocene. Its onset was characterized by rapid cooling and extinctions in both the marine and terrestrial realms, as well as the renewed formation of ice sheets in Antarctica (table 6.8). Ice sheets are known to have formed in Gondwana in the late Famennian, and extinctions occurred in both the marine and terrestrial realms (table 6.7). Ice sheets are also known to have been present for at least four million years in the middle of the Tournaisian Gap, if not for the entire 13.9 million years of the Tournaisian Gap (table 6.9). Is the 13.9-million-year Tournaisian Gap the equivalent of the 12-million-year span of the ice sheets that formed in the second step of the Cenozoic glaciation?

In the Cenozoic, the extinction that occurred at the onset of the glaciation was more severe than the second extinction that occurred within the glaciation; that is, the Late Oligocene extinction was more severe than the Middle Miocene extinction. The pattern is the same in the Late Devonian: the late Frasnian extinction was more severe than the late Famennian extinction. Thus was the late Frasnian extinction triggered by the onset of the Late Devonian glaciations, and did the late Famennian extinction occur within the glacial period (table 6.9)?

The proposal that the patterns of extinction seen in the Late Devonian and Cenozoic (table 6.9) both reflect the timing of the step-wise glaciation of the Earth during these two separate periods in geologic time runs into serious difficulty when step 3 of the Cenozoic glaciation is considered. The late Pliocene formation of glaciers in the Northern Hemisphere of the Earth (table 6.8) clearly had no equivalent in the early Visean Age (table 6.9). The glaciers that did form in the middle Visean (table 6.4) are not of that magnitude, and they are still in the Southern Hemisphere of the Earth rather than the Northern. Rather, it is not until the Serpukhovian Age, some 17 million years after the end of the Tournaisian Gap, that major ice sheets formed in the Serpukhovian-Bashkirian glaciation that straddled the Early-Late Carboniferous boundary.[59] In addition, it is only in the Serpukhovian that the next major extinction event takes place in geologic time.[60]

Another serious difference between the Late Devonian and Cenozoic glaciations is the magnitude of the geologic and biotic events that occurred in those time intervals—both were much larger in the Late Devonian than the Cenozoic! For example, the magnitude of the heavy-carbon anomalies that occurred in the late Frasnian was a factor of 3.75 times larger than those that occurred in the Early Oligocene.[61] The magnitude of the generic diversity loss in marine organisms that occurred in the late Frasnian was a factor of 3.5 times larger than the diversity loss that occurred in the oceans in the Early Oligocene: a loss of 35 percent versus a loss of 10 percent.[62] The magnitude of marine generic diversity that was lost in the late Famennian was a factor of 4.8 times larger than the diversity loss that occurred in the Middle Miocene: a 29 percent loss versus a 6 percent loss.[63] Last, the magnitude of the heavy-carbon anomaly was only a factor of 1.5 larger in the Famennian than in the Miocene, but it was still larger.[64]

Why were the heavy-carbon anomalies so much larger in the Late Devonian than in the Cenozoic? Why was so much more carbon-12 buried in black shale and carbonate deposits during the Late Devonian? Was this difference simply due to a higher carbon dioxide content in the atmosphere of the Earth in the Late Devonian than in the Cenozoic; that is, was there simply much more carbon available to be fixed in organics and carbonates in the Late Devonian? The Geocarbsulf empirical model of the Earth's atmosphere that we previously considered (see Fact 2 in chapter 4) predicts the carbon dioxide content of the Frasnian atmosphere to have been between 0.12 percent to 0.15 percent, and that that level fell to around 0.06 percent to 0.09 percent by the end of the Famennian.[65] These Late Devonian levels are much higher than the 0.03 percent carbon dioxide level that was present in the Earth's atmosphere before the Industrial Revolution.[66]

However, the Geocarbsulf model also predicts an atmospheric carbon dioxide level of around 0.06 percent to 0.09 percent at the beginning of the Oligocene, 34 million years ago, similar to the levels predicted for the atmosphere at the end of the Famennian.[67] Other models predict even higher levels of carbon dioxide in the atmosphere in the Early Oligocene: 0.10 percent to 0.15 percent, and those levels sharply fell to 0.03 percent by the end of the Oligocene.[68] These latter Early Oligocene carbon dioxide figures are more similar to the Frasnian levels predicted by the Geocarbsulf model. Thus, in summary, the carbon dioxide content of the Earth's atmosphere in the Late Devonian may not have been greatly different from that of the Cenozoic. Thus the mystery of the large-magnitude heavy-carbon anomalies in the Late Devonian remains a mystery.

Why were the extinctions so much more severe in the Late Devonian than in the Cenozoic if both were triggered by the same phenomenon, the step-wise glaciation of the Earth? University of Hawaii paleontologist Steve Stanley has proposed an answer to that question: the biota of the Late Devonian world was more prone to extinction than the biota of the Cenozoic, a phenomenon revealed in the classic analyses of University of Chicago paleontologists Dave Raup and Jack Sepkoski that prove that the mean extinction rate of marine animals has declined through time; that is, Cambrian marine species had a much higher rate of extinction than Cenozoic marine species.[69] This phenomenon is not unexpected; the theory of natural selection would predict the evolution of increasing extinction resistance with time. In the Late Devonian, the biota was dominated by ancient Paleozoic species with higher extinction rates than modern species. For example, Stanley has shown that in the end-Frasnian extinction the older species lineages suffered extinction at a 20 percent higher rate than the more recently evolved species lineages and, as the majority of the Late Devonian species belonged to these older lineages, the total extinction rate was predictably high.[70] In contrast, Cenozoic marine species are much more resistant to extinction, hence the total extinction rates seen in the Early Oligocene and Middle Miocene are predictably lower than those seen in the Late Devonian.

A final similarity between the Late Devonian and Cenozoic intervals of geologic time is the attribution of the decline in carbon dioxide content in the atmosphere, and the subsequent cooling and eventual glaciation of the planet, to the chemical weathering of silicate rocks exposed by tectonic uplifting of mountain chains, particularly mountain chains located in the tropics.[71] Previous attribution of the Cenozoic glaciations to changing oceanic flow patterns following the opening of the southern ocean gateways and the thermal isolation of the Antarctic continent, or to variations in the amount of heat the Earth receives from the sun during the several Melankovitch cycles of variation in the Earth's orbit, are now argued to be secondary factors superimposed on the primary cause of cooling: carbon dioxide downdraw from the atmosphere. In the Cenozoic, that downdraw is proposed to have been triggered by the weathering of the silicate rocks of the Deccan Traps when the Indian subcontinent was located in the tropics and the subsequent uplift of the Himalayan mountains and Tibetian Plateau when India collided with Asia (the equatorial convergence model of weathering, as discussed in the Kill Mechanism section of chapter 4). In the Devonian, that downdraw is proposed to have been triggered by the weathering of mountain chains uplifted by the collision of the continents of Laurussia with Gondwana, Kazakhstan with Laurussia, Siberia with Kazakhstan, and the fact that the majority of the subsequent uplifted mountain chains were located in the tropics (Fact 1 in chapter 4).

In conclusion, although many similarities exist between the Late Devonian and Cenozoic step-wise glaciations and extinction pulses, atmospheric carbon dioxide depletions, and the uplifting of huge mountain chains, it still has not been geologically proved that ice sheets formed on the Earth in the late Frasnian (hence the question marks in table 6.9). The end-Frasnian extinction remains a mystery.

CHAPTER 7

Victory at Last

The End-Famennian Bottleneck

Just as in the End-Frasnian Bottleneck,[1] the same classic evolutionary bottleneck phenomena of reduction in morphological variance and geographic range are seen in the tetrapod species appearing in the fossil record in and after the Tournaisian Gap, and I refer to this effect as the "End-Famennian Bottleneck." McGill University paleontologist and tetrapod specialist Robert Carroll comments on this reversal of the fortunes of the numerous Famennian tetrapods and the peculiar nature of the amphibians that are found in the Early Carboniferous:

> Fossils from the very end of the Devonian appear on the verge of a new way of life, lacking only a few changes in the limb structure necessary for them to become the lords of the land. However, the Lower Carboniferous beds, where you would expect to find the first fully terrestrial vertebrates, provide frustratingly little evidence of the next step in vertebrate evolution, *but a variety of peculiar amphibians that have left no descendants* and tantalizing fragments of others that may point the way to advanced amphibians.[2] (emphasis mine)

Cambridge University tetrapod specialist Jennifer Clack similarly notes that:

> However, it seems clear that immediately after the Devonian, there was indeed a period of low diversity representing a recovery period from an extinction event. . . . There seems to be little obvious connection between the tetrapods of the Late Devonian and those of the Early Carboniferous. . . . It may nonetheless be that *most of the Devonian forms became extinct at the end of the Devonian.*[3] (emphasis mine)

In addition, the University of Chicago vertebrate paleontologists Lauren Cole Sallan and Michael Coates explicitly refer to the effects of the end-Famennian extinction (which they refer to as the Hangenberg extinction) as an evolutionary bottleneck:

> Major vertebrate clades suffered acute and systematic effects centered on the Hangenberg extinction involving long-term losses of over 50 percent of diversity and the restructuring of vertebrate ecosystems worldwide. Marine and nonmarine faunas were equally affected, precluding the existence of environmental refugia. . . . The Hangenberg event represents a previously unrecognized *bottleneck in the evolutionary history of vertebrates* as a whole and a historical contingency that shaped the roots of modern biodiversity.[4] (emphasis mine)

The fact that a major extinction event in the history of life on Earth occurred at the end of the Famennian was pointed out over 40 years ago by American Museum of Natural History paleontologist Norman Newell in his classic paper "Revolutions in the History of Life,"[5] thus the event is far from being previously unrecognized. The point that Sallan and Coates argue is, however, that the severity of that extinction for the vertebrates in particular has been previously unappreciated.

Curiously, just as in the aftermath of the end-Frasnian extinction, after the end-Famennian extinction only a single tetrapod species is known from the Tournaisian Gap: *Pederpes finneyae,* which is found in Scotland, once again in southeast Laurussia.[6] Of the 23 new species of tetrapods that appear in the Visean fossil record after the Tournaisian Gap, 18 species (78 percent) are found in the restricted geographic region of southeast Laurussia.[7]

TABLE 7.1 Terrestrial spore zonation timescale for the subtropical region of ancient Laurussia (modern western Europe) in the Tournaisian and Visean.

Geologic Age	Spore Zones
VISEAN	**NC** Zone, pars (lower to middle *B. nitidus–R. carnosus* Zone)
	VF Zone (*T. vetustus–R. fracta* Zone)
	ME Zone (*M. margodentata–R. ergonulii* Zone)
	DP Zone (*T. distinctus–M. parthenopia* Zone)
	TC Zone (*P. tessellatus–S. campyloptera* Zone)
	TS Zone (*K.triradiatus–K. stephanephorus* Zone)
	Pu Zone (*Lycospora pusilla* Zone)
TOURNAISIAN	**CM** Zone (*S. claviger–A. macra* Zone)
	PC Zone (*S. pretiosus–R. clavata* Zone)
	BP Zone (*S. balteatus–R. polyptycha* Zone)
	HD Zone (*K. hibernicus–U. distinctus* Zone)
	VI Zone, pars (lower to upper *V. verrucosus–R. incohatus* Zone)

Source: Spore zonations modified from the 2004 zonal timescale of Filipiak.

The Final Assault

The timing of the third, and final, vertebrate invasion of land is measured by reference to the terrestrial spore zonation of the Early Carboniferous (table 7.1). The correlation of the spore zonations with the marine conodont zonations and the Tournaisian Gap that we considered in chapter 6 (table 6.2) are given in table 7.2. The Tournaisian Age lasted 13.9 million years (table 2.2) and is divided into five spore zones, thus a rough estimate of the duration of a spore zone in the Tournaisian is about 2.8 million years. The Visean Age lasted 17 million years and is divided into seven spore zones, thus a rough estimate of the duration of a spore zone in the Visean is about 2.4 million years.

The first tetrapod to appear in the Carboniferous is the Scottish species *Pederpes finneyae* (table 7.3), which will be discussed in detail in the next section of the chapter. The strata containing the fossils of *Pederpes finneyae* are dated to the **CM** spore zone,[8] the very last zone in the Tournaisian (table 7.1) and the very end of the Tournaisian Gap (table 7.2). It is the first known representative to appear of the Family Whatcheeriidae;[9] more members of the family are soon to appear in the following Visean Age of the Early Carboniferous.

TABLE 7.2 Correlation of the terrestrial spore zones with the marine conodont zones in the Tournaisian and Visean.

Geologic Age	Spore Zones	Conodont Zones	
VISEAN	**NC** Zone, pars	Lower *girtyi–collinsoni* Zone	
	NC Zone, pars	*mononodosa* Zone	
	NC Zone, pars	Uppermost *bilineatus* Zone	
	VF Zone	Upper *bilineatus* Zone	
	ME Zone	Middle *bilineatus* Zone	
	DP Zone	Lower *bilineatus* Zone	
	TC Zone	Upper *commutata* Zone	
	TS Zone	Middle *commutata* Zone	
	Pu Zone	Lower *commutata* Zone	
	Pu Zone	*texanus–homopunctatus* Zone	
TOURNAISIAN	**CM** Zone	*anchoralis–latus* Zone	Tournaisian Gap
	CM Zone	Upper *typicus* Zone	Tournaisian Gap
	CM Zone	Lower *typicus* Zone	Tournaisian Gap
	PC Zone	*isosticha* Zone	Tournaisian Gap
	PC Zone	Upper *crenulata* Zone	Tournaisian Gap
	BP Zone	Lower *crenulata* Zone	Tournaisian Gap
	HD Zone	*sandbergi* Zone	Tournaisian Gap
	VI Zone, pars	Upper *duplicata* Zone	Tournaisian Gap
	VI Zone, pars	Lower *duplicata* Zone	Tournaisian Gap
	VI Zone, pars	*sulcata* Zone	Tournaisian Gap

Source: Spore and conodont zonations modified from the 2004 zonal timescale of Filipiak.

Phylogenetic analyses indicate that the whatcheeriids are not the most plesiomorphic of the Carboniferous tetrapods; that is, that they are more derived than the families of the Colosteidae and the Crassigyrinidae (table 7.3). Thus at least two other tetrapod species populations existed somewhere on Earth during the Tournaisian Gap, representing the ghost lineages of the Colosteidae and the Crassigyrinidae. The identity of these species remains unknown. Thus the Tournaisian Gap is curiously similar to the Famennian Gap (chapter 4) in that only one fossil species is known from the Gap itself, yet we know that two ghost lineages spanned the Gap.

The third invasion began in earnest in the Visean Age, following the Tournaisian Gap. The first of these Visean species appears in strata dated to the **TS** spore zone in the early to middle Visean (table 7.1), and no less than 23 new tetrapod species had appeared by the **VF** spore zone

TABLE 7.3 Phylogenetic classification of the Tournaisian tetrapod invaders.

Sarcopterygii (lobe-finned fishes + descendants)
– Crossopterygii (coelacanths)
– Dipnoi (lungfishes)
– **Tetrapodomorpha** (tetrapod-like fishes + descendants)
– – basal tetrapodomorphs
– – – *Kenichthys campbelli*
– – Rhizodontia
– – Osteolepidiformes
– – – Osteolepididae †Famennian
– – – Megalichthyidae
– – – Eotetrapodiformes
– – – – Tristichopteridae †Famennian
– – – – *Gogonasus andrewsae* †Frasnian
– – – – unnamed clade
– – – – – Elpistostegalia †Frasnian
– – – – – **Tetrapoda** (limbed vertebrates)
– – – – – – Family incertae sedis †Frasnian?
– – – – – – – Unknown early Eifelian tetrapod species in Zachelmie, Poland
– – – – – – Family incertae sedis †Frasnian?
– – – – – – – Unknown early Frasnian tetrapod species in Valentia Island, Ireland
– – – – – – Family Elginerpetontidae †Frasnian
– – – – – – – *Elginerpeton pancheni* †Frasnian
– – – – – – – *Obruchevichthys gracilis* †Frasnian
– – – – – – Family incertae sedis †Frasnian
– – – – – – – *Sinostega pani* †Frasnian
– – – – – – unnamed clade
– – – – – – – Family incertae sedis †Famennian
– – – – – – – – *Densignathus rowei* †Famennian
– – – – – – – unnamed clade
– – – – – – – – Family incertae sedis †Famennian
– – – – – – – – – *Ventastega curonica* †Famennian
– – – – – – – – unnamed clade †Frasnian
– – – – – – – – – Family incertae sedis †Frasnian
– – – – – – – – – – *Metaxygnathus denticulus* †Frasnian
– – – – – – – – Family incertae sedis †Famennian
– – – – – – – – – *Jakubsonia livnensis* †Famennian
– – – – – – – – unnamed clade
– – – – – – – – – Family Acanthostegidae †Famennian
– – – – – – – – – – *Acanthostega gunnari* †Famennian
– – – – – – – – – unnamed clade
– – – – – – – – – – Family incertae sedis †Famennian
– – – – – – – – – – – *Ymeria denticulata* †Famennian
– – – – – – – – – – unnamed clade
– – – – – – – – – – – Family Ichthyostegidae †Famennian

(continued)

TABLE 7.3 *(continued)*

– – – – – – – – – – – – – *Ichthyostega stensioei* †Famennian
– – – – – – – – – – – – – *Ichthyostega watsoni* †Famennian
– – – – – – – – – – – – – *Ichthyostega eigili* †Famennian
– – – – – – – – – – – – unnamed clade
– – – – – – – – – – – – – Family incertae sedis †Famennian
– – – – – – – – – – – – – – *Hynerpeton bassetti* †Famennian
– – – – – – – – – – – – – unnamed clade
– – – – – – – – – – – – – – Family Tulerpetontidae †Famennian
– – – – – – – – – – – – – – – *Tulerpeton curtum* †Famennian
New Tournaisian Invader Lineages:
– – – – – – – – – – – – – – unnamed clade
– – – – – – – – – – – – – – – [Family Colosteidae]
– – – – – – – – – – – – – – – unnamed clade
– – – – – – – – – – – – – – – – [Family Crassigyrinidae]
– – – – – – – – – – – – – – – – unnamed clade
– – – – – – – – – – – – – – – – – Family Whatcheeriidae
– – – – – – – – – – – – – – – – – – *Pederpes finneyae*

Source: Phylogenetic data modified from Benton (2005), Lecointre and Le Guyader (2006), Long et al. (2006), Ahlberg et al. (2008), and Clack et al. (2012).

Note: "Ghost" tetrapod lineages are enclosed in brackets (e.g., [Family Colosteidae]); see text for discussion. Major vertebrate clades of land invaders are marked in bold-faced type.

† extinct taxa. Where a period is given, taxa went extinct in that specific geologic period.

near the end of the Visean. This span of time in the Visean represents five spore zones (table 7.1), or approximately 12 million years.

The families of the Colosteidae and the Crassigyrinidae, ghost lineages in the Tournaisian Gap, now appear in the fossil record with the species *Greererpeton burkemorani* from West Virginia in North America and *Crassigyrinus scoticus* in Scotland, respectively (table 7.4). The family of the whatcheeriids, represented by the sole species *Pederpes finneyae* in the Tournaisian Gap (table 7.3), blossomed in diversity, with four new species appearing in the Visean: *Whatcheeria deltae* and *Sigournea multidentata* from Iowa in North America, *Ossinodus pueri* from Australia, and *Occidens portlocki* from Ireland[10] (table 7.4). The Irish species *Occidens portlocki* lived close to the Scottish species *Pederpes finneyae*, but the appearance of whatcheeriids far away to the west in central Laurussia and to the east in the very eastern margin of Gondwana demonstrates that this family of tetrapods had achieved a worldwide distribution. In particular, strata containing the Australian

TABLE 7.4 Phylogenetic classification of the Visean tetrapod invaders.

Tetrapoda (limbed vertebrates)
– Frasnian Families
– – Famennian Families
– – – Family Colosteidae
– – – – *Greererpeton burkemorani*
– – – unnamed clade
– – – – Family Crassigyrinidae
– – – – – *Crassigyrinus scoticus*
– – – – unnamed clade
– – – – – Family Whatcheeriidae
– – – – – – *Whatcheeria deltae*
– – – – – – *Sigournea multidentata*
– – – – – – *Ossinodus pueri*
– – – – – – *Occidens portlocki*
– – – – – unnamed clade
– – – – – – Family Baphetidae
– – – – – – – *Eucritta melanolimnetes*
– – – – – – – *Loxomma allmanni*
– – – – – – unnamed clade (sister group to the baphetids)
– – – – – – – Batrachomorpha (amphibians)
– – – – – – – – basal batrachomorphs ("temnospondyls")
– – – – – – – – – *Balanerpeton woodi*
– – – – – – – unnamed clade (sister group to the batrachomorphs)
– – – – – – – – Lepospondyli
– – – – – – – – – basal lepospondyls ("microsaurs")
– – – – – – – – – – *Acherontiscus caledoniae*
– – – – – – – – – – *Adelogyrinus simorhynchus*
– – – – – – – – – – *Dolichopareias disjectus*
– – – – – – – – – – *Palaeomolgophis scoticus*
– – – – – – – – – Holospondyli
– – – – – – – – – – [Nectridea]
– – – – – – – – – – Aistopoda
– – – – – – – – – – – *Lethiscus stocki*
– – – – – – – – – – – *Ophiderpeton kirtonense*
– – – – – – – – **Reptiliomorpha**
– – – – – – – – – basal reptiliomorphs ("anthracosaurs")
– – – – – – – – – – *Silvanerpeton miripedes*
– – – – – – – – – – *Eldeceeon rolfei*
– – – – – – – – – – *Eoherpeton watsoni*
– – – – – – – – – – *Antlerpeton clarkii*
– – – – – – – – – – *Doragnathus woodi*
– – – – – – – – – – *Pholidogaster pisciformis*
– – – – – – – – – – *Westlothiana lizziae*
– – – – – – – – – Batrachosauria

(*continued*)

TABLE 7.4 (*continued*)

- - - - - - - - - - - [Seymouriamorpha]
- - - - - - - - - - - unnamed clade
- - - - - - - - - - - - [Diadectomorpha]
- - - - - - - - - - - - **Amniota** (conquerors of the land!)
- - - - - - - - - - - - - basal amniotes?
- - - - - - - - - - - - - - *Casineria kiddi*

Source: Phylogenetic and stratigraphic data modified from Benton (2005), Lecointre and Le Guyader (2006), Sallan and Coates (2010), and Clack (2012).

Note: "Ghost" tetrapod lineages are enclosed in brackets (e.g., [Nectridia]). Older paraphyletic tetrapod group names are given in quotation marks within the parentheses. Major vertebrate clades of land invaders marked in bold-faced type. See text for discussion.

species *Ossinodus pueri* have been dated to the **TS** spore zone in the early Visean[11] (table 7.1). Clearly, the severe environmental conditions of the Tournaisian Gap had waned, and tetrapods were on the move in the early Visean Age.

The three basal families of Tournaisian Gap tetrapods were followed by an explosion of new tetrapod groups in the mid-to late Visean. In this short period of time, the divergent tetrapod lineages that produced the modern-day amphibians and the ancestors of the modern-day amniotes—if not the amniotes themselves—both appeared on Earth. The Tournaisian colosteids, crassigyrinids, and whatcheeriids are followed by two species of baphetids that appeared in the Visean (table 7.4): *Eucritta melanolimnetes* from Scotland and *Loxomma allmanni* from England. These four families are now extinct, but they gave rise to the batrachomorphs, whose ancestors thrive today.

Modern amphibians belong to the clade of the Batrachomorpha (table 7.4), descendants of the basal batrachomorphs that are referred to as "temnospondyls" in the older literature (they are a paraphyletic grade rather than a clade). In the Visean, however, the batrachomorphs also gave rise to the clade of the lepospondyls, now extinct, and the clade of the reptiliomorphs, our ancestors. The basal batrachomorphs are represented in the Visean by a single species, *Balanerpeton woodi* from Scotland. We know there had to be many other basal species present at this time as well, because the more derived batrachomorph clades of the lepospondyls and reptiliomorphs have numerous Visean species present in the fossil record. No less than four species of basal lepospondyls or "microsaurs" (non-holospondyl lepospondyls, a paraphyletic older taxon) are present in the Visean, as well as two species of highly derived aistopods. In addition, one ghost lineage, the nectrideans, had to exist in the Visean as well because they are not as derived as the aistopods, which existed in the Visean (table 7.4). Some of these lepospondyl species are very peculiar—some have even lost their key tetrapod trait, namely, they are legless! The morphologies of these animals will be considered in more detail in the next section of the chapter.

The other major batrachomorph clade to appear in the Visean are the reptiliomorphs, ancestors of modern-day amniotes: reptiles, birds, and mammals. The amniotes evolved the final suite of adaptations needed to finally become the lords of the land, the conquerors of the terrestrial realm of the Earth. No less than seven basal reptiliomorph species, or "anthracosaurs" (non-batrachosaurian reptiliomorphs, a paraphyletic older taxon) appeared in the Visean (table 7.4).

Finally, there is one enigmatic Visean species that may in fact be the very first amniote, *Casineria kiddi* from Scotland (table 7.4). The body of this little animal was only 80 millimeters (three inches) long from shoulder to hip, and the fossil skeleton is missing its head, but it clearly was a highly derived reptiliomorph.[12] How highly derived remains controversial, as will be discussed in the next section of the chapter when we consider its anatomy in more detail. I have provisionally included it in table 7.4 as a possible basal amniote, possibly the first of the animals that were the true conquerors of the land.

All of the divergent species and clades listed in table 7.4 appeared in the fossil record on Earth within approximately 12 million years, the span of time from the **TS** through the **VF** spore zones in the Visean (table 7.1). It could be argued that many of the divergent speciation events that produced so many separate clades in the mid-to late Visean must have occurred in species existing in the early Visean. Then where are the early Visean fossils of these species? One possibility is that these probable early Visean species had population sizes that were still as yet too small to have had a very high probability of preservation. The severe environmental conditions of the Tournaisian Gap had abated, but it took time for tetrapods to recover sufficient numbers to appear in the fossil record.

A second possibility is that the probable early Visean tetrapod species do not appear in the fossil record because favorable preservational conditions were not present. If fact, six (26 percent) of the 23 Visean species listed in table 7.4 come from one locality: the East Kirkton Limestone *Lagerstätte* in Scotland.[13] This late Visean *Lagerstätte,* like the Rhynie Chert *Lagerstätte* in the Early Devonian Pragian, is another hot springs deposit, and it preserved organisms that lived in a series of habitats around thermal springs and volcanic fumaroles that existed in eastern Scotland. Thus, just as for the Rhynie Chert, we are fortunate that the silica deposits and volcanic ashes of the East Kirkton strata give us a glimpse into the diversity of tetrapod species present in this unique window of time, preserved by the unusual environmental conditions present in the *Lagerstätte.*

The strata of the East Kirkton *Lagerstätte* are dated to the **VF** spore zone, near the very end of the Visean[14] (table 7.1). Thus it could be alternatively argued that all of Visean tetrapod evolution indeed took place in the mid-to late Visean, and that probably the only early Visean species present on Earth belonged to the basal Tournaisian families of the colosteids, crassigyrinids, and whatcheeriids. That is, there are no early Visean fossils of tetrapods more derived than these three basal families because no more derived species of tetrapods had yet evolved.

We know that this argument can be disproved because fossils of the highly derived Scottish species *Casineria kiddi* (table 7.4) are not from the late Visean East Kirkton *Lagerstätte.* Instead, they are found in strata from the Cheese Bay in Scotland, strata that have been dated to the late **TS** spore zone[15] or to the mid-Visean (table 7.1). This species is the most derived of all of the Visean tetrapods, thus proving that all of the divergent evolutionary events that produced the clades shown in table 7.4 must have occurred *by the mid-Visean.* Thus it is clear that numerous *early Visean* tetrapods must have existed, but they have not been discovered (as yet) in the fossil record.

The fossil record in last age of the Early Carboniferous, the Serpukhovian, merely evidences a continued diversification within the major groups of tetrapods that had appeared in the Visean (table 7.4); no new clades of tetrapods appeared. Instead, three new species of baphetids appeared (*Spathicephalus pereger, Spathicephalus mirus,* and *Megalocephalus pachycephalus*),[16] two new species of basal lepospondyls (*Utaherpeton franklini* and *Adelospondylus watsoni*),[17] and five new species of basal reptiliomorphs (*Proterogyrinus scheelie, Proterogyrinus pancheni, Pholiderpeton(?) bretonense, Caerorhachis bairdi,* and *Papposaurus traquairi*).[18]

Last, the arthropod invaders once again appear in the fossil record in the Early Carboniferous. Similar to the single fossil species of tetrapod that appeared in the Tournaisian Gap, a single new fossil species of the trigonotarbid *Pocononia whitei* appeared at the very end of the Tournaisian Gap. Arthropod specialists William Shear and Paul Selden note the presence of the Tournaisian Gap in arthropod evolution as well as in vertebrate evolution:

> A 23-million-year gap separates the modest Devonian terrestrial [arthropod] faunas from the extensive and complex ones of the Upper Carboniferous. At this time, there are no known Tournaisian records for terrestrial [arthropods]; but a single trigonotarbid specimen (*Pocononia whitei*) has been collected from the Pocono Mountain Formation of Virginia, probably located near the Tournaisian/Visean boundary. A significant Visean site for terrestrial animals, including arachnids, myriapods, and a rich fauna of tetrapods is found at East Kirkton, West Lothian, Scotland.[19]

The Visean East Kirkton *Lagerstätte* does indeed preserve some fascinating arthropods, including the gigantic scorpion *Pulmonoscorpius kirktonensis,* which was 700 millimeters (2.3 feet) long! Some of the millipede fossil fragments show the characteristic pores that accompany the poison glands found in modern living millipedes, thus these Early Carboniferous species had evolved this chemical defense against predators.[20] However, the arthropod species that do appear in the fossil record in the Early Carboniferous Visean are representatives of the clades established back in the Devonian (table 3.5) and do not evidence the evolution of any major new clades. The same is not true for the very last Age of the Early Carboniferous, the Serpukhovian, as will be considered when we examine the evolution of the flying insects.

The Strange Early Carboniferous Vertebrates

Some of the new vertebrate invaders that appear in the fossil record in the Early Carboniferous were very peculiar animals indeed. Rather than exhibiting the expected new adaptations for terrestrial existence, some were secondarily aquatic forms that had clearly abandoned the land entirely. Others did not possess the defining trait of the tetrapods themselves—they had secondarily lost all four of their limbs!

Phylogenetic analyses indicate that three clades of tetrapods existed on Earth during the dawn of the Carboniferous (table 7.3), but fossils of only one of these have been discovered thus far. *Pederpes finneyae*, the only fossil tetrapod species yet known from the Tournaisian Gap, was about one meter (3.3 feet) in total length, similar in size to the extinct Famennian species. It is very similar to four other species that appeared in the Visean, *Whatcheeria deltae, Sigournea multidentata, Ossinodus pueri*, and *Occidens portlocki*, thus they all are included in the family Whatcheeriidae (tables 7.3 and 7.4). These animals had skulls around 200 to 250 millimeters (eight to ten inches) in length and, curiously, possessed some plesiomorphic traits that had been lost by the more advanced Famennian species: it had teeth on its palate and on the coronoid in the lower jaw, as well as traces of a lateral line system. The animals also had an ilium in the hind limb more similar to that seen in *Acanthostega gunnari* than in the more derived Famennian tetrapods. The shape of the skull itself is more derived than the Famennian tetrapods, however, being narrow and tall rather than broad and flat, and their eyes faced sideways rather than upward (fig. 7.1). The hind feet of the animals were also more derived: they had only five toes, the now-standard pentadactylus condition for tetrapods. The forefeet, however, remained polydactylus and possessed a small sixth toe. Clearly, however, the whatcheeriids were more adapted to walking on land than swimming in water.[21]

The colosteids evolved before the whatcheeriids (tables 7.3 and 7.4) but remain as yet a ghost lineage in the Tournaisian Gap, thus we do not know what the first species of this family looked like. The colosteid species *Greererpeton burkemorani* is found in the Visean fossil record and is surprising in that the animal was clearly fully aquatic, with an

FIGURE 7.1 Reconstruction of the whatcheeriid tetrapod *Pederpes finneyae* from Scotland. Note the presence of six toes in the forefoot and the laterally placed eyes in the head. The living animal was about one meter (3.3 feet) long. *Credit*: Illustration by Kalliopi Monoyios.

ecology perhaps similar to the modern giant salamander *Andrias davidianus,* a 1.5-meter-long (five-foot-long) river-dwelling ambush predator. *Greererpeton burkemorani* had a skull about 150 to 180 millimeters (six to seven inches) in length, had a well developed lateral line system, was broad and flattened, and had eyes that were situated dorsally (fig. 7.2).[22] In short, they were much more plesiomorphic and similar to Famennian tetrapods than the whatcheeriids. However, their forelimbs were different in that they were very small, and their forefeet possessed only four toes. Given their smaller skulls, the early colosteids were probably smaller than the whatcheeriids, thus less than a meter (3.3 feet) in length.

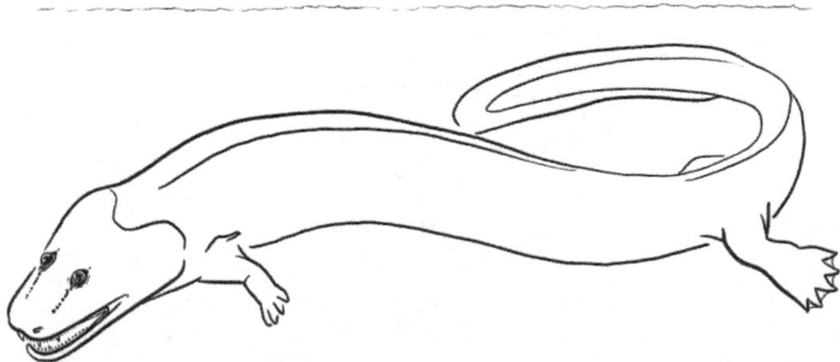

FIGURE 7.2 Reconstruction of the colosteid tetrapod *Greererpeton burkemorani* from West Virginia, North America. Note the presence of only four toes in the forefoot, the reduced size of the limbs relative to the body size, and the tail fin for swimming. The living animal was a little less than one meter (3.3 feet) long. *Credit*: Illustration by Kalliopi Monoyios.

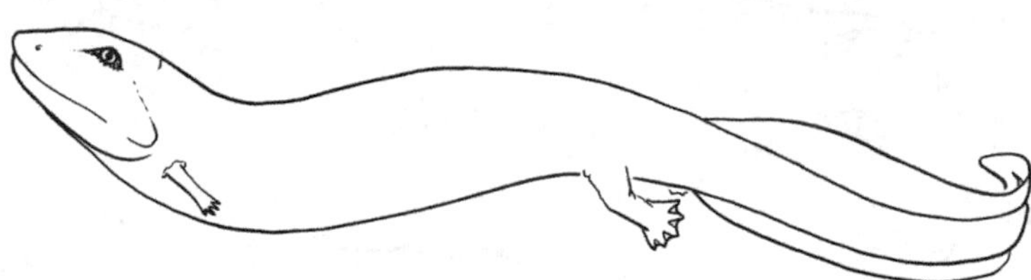

FIGURE 7.3 Reconstruction of the crassigyrinid tetrapod *Crassigyrinus scoticus* from Scotland. Note the almost vestigial forelimbs, the small hind limbs relative to the body size, and the tail fin for swimming. The living animal was large, about two meters (6.6 feet) long. *Credit*: Illustration by Kalliopi Monoyios.

The crassigyrinids occupy a phylogenetic position intermediate between the colosteids and the whatcheeriids (tables 7.3 and 7.4); but again because they are a ghost lineage in the Tournaisian Gap, we do not know what the first species of this family looked like. The crassigyrinid species *Crassigyrinus scoticus* is present in the Visean fossil record, and it is bizarre. The animal was big, around two meters (6.6 feet) in length, and possessed a skull that was 350 millimeters (14 inches) long. It had powerful jaws filled with sharp teeth and multiple fangs, thus the animal is thought to have been a large predator. However, the animal's limbs (fig. 7.3)—the tetrapods—were almost nonexistent! These animals, like

FIGURE 7.4 Reconstruction of the baphetid tetrapod *Eucritta melanolimnetes* from Scotland. Note the short, compact body and large limbs and feet. The living animal was small, about 250 millimeters (ten inches) long. *Credit*: Illustration by Kalliopi Monoyios.

the colosteids, were fully aquatic. The crassigyrinids, however, were secondarily convergent on fish—an evolutionary reversal—rather than evolving in the direction of greater terrestriality. Many aspects of their skulls and dentition are reacquired fish traits, and they were clearly devolving their appendages and returning to limbless, fish-like, swimming morphologies.[23]

In summary, of the three families of tetrapods that existed on Earth during the Tournaisian Gap, only the whatcheeriids represent an evolutionary advancement in the invasion of land. The colosteids were more similar to the aquatic tetrapods of the Famennian, and the crassigyrinids were actually an evolutionary reversion to fish forms, a step backwards away from the land. The whatcheeriids, in contrast, were land walkers; they were no longer primarily aquatic tetrapods. They were also highly successful on land, as witnessed by their worldwide distribution in the early Visean, as discussed in the previous section of the chapter.

The Visean baphetid species *Eucritta melanolimnetes* and *Loxomma allmanni* are phylogenetically close to the clade of the Batrachomorpha and the Reptiliomorpha (table 7.4), a clade that not only produced the clade of the still living amphibians but also that of the amniote conquerors of land as well. *Eucritta melanolimnetes* is the most basal baphetid known at present and, although fossils of five individuals of this species have been found, none is complete. The animal had a short, compact body, with strong hind limbs and large feet (fig. 7.4). Its hind feet, like those of the whatcheeriids, had only five toes. Unfortunately, the bones of the forefeet are not preserved and thus it is not known with certainty if the animals had five fingers (as shown in fig. 7.4) or if they were still polydactylus. Jennifer Clack, who first described the species, noted that the animal had a peculiar "*mélange* of crown-group characters,"[24] and gave it the name *Eucritta melanolimnetes,* which translates as "the true creature from the black lagoon" and which further demonstrates that Clack has a marvelous sense of humor.[25]

The species *Loxomma allmanni* possessed a more derived baphetid trait: a peculiar, forward-directed extension of the socket in the skull that held the eye. Thus, rather than the expected round hole for the eye socket, the eye socket of *Loxomma allmanni* looked more like the keyhole one sees in an antique door. Unfortunately, only the skull of the species is known. Its skull was 250 millimeters (ten inches) long, and its eyes were placed high on the skull such that the animal could see even if the skull was submerged in water, somewhat similar to the condition seen in modern crocodiles and alligators.[26] Its mode of life may have been similar as well, as its numerous large recurved teeth indicate that it was a predator.

The most ancient species of the batrachomorph clade known at present is *Balanerpeton woodi* from the East Kirkton *Lagerstätte.* Individuals of this species are generally small, with skulls 25 millimeters to 48 millimeters (one to two inches) in length, and with total body lengths around 175 millimeters to 440 millimeters (seven inches to a foot and a half). Their limbs were quite sturdy, with the hind limbs larger than the forelimbs. Their hind feet possessed five toes (fig. 7.5), but their forefeet possessed only four! Thus, rather than the expected evolution of the pentadactylus condition in the forefoot as well as the hind foot, *Balanerpeton woodi* had gone one step further and lost two toes on the forefoot, not just one. More importantly, the species no longer possessed any trace of a lateral-line system: it was a fully terrestrial animal. Further evidence of its terrestriality is seen in the fact that *Balanerpeton woodi* had two otic notches in its skull, located behind each cheek region; two slender, rod-like stapedial bones are found in these regions as well. Jennifer Clack thus argues that *Balanerpeton woodi,* like modern-day amphibians, had eardrums in these otic notch regions of the skull, and that the animal had adapted to hear sound in dry air.

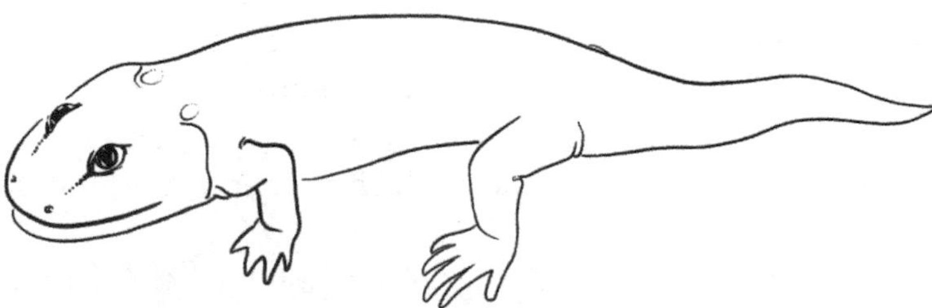

FIGURE 7.5 Reconstruction of the batrachomorph tetrapod *Balanerpeton woodi* from Scotland. Note the presence of only four toes in the forelimb and the somewhat slenderer body than that of *Eucritta melanolimnites* (fig 7.4). The living animal was about 400 millimeters (a foot and a half) long. *Credit:* Illustration by Kalliopi Monoyios.

Further, Jennifer Clack's phylogenetic analyses have revealed that the tympanal ear system evolved convergently in at least five different tetrapod lineages: once in the ancient batrachomorphs and still present in their amphibian descendants (frogs, salamanders, and kin), three additional separate times in the sauropsid amniotes (ancestors of living turtles, lizards, and archosaurs), and independently in the synapsid amniotes (ancestors of mammals). As summarized by Clack: "The evolution of hearing in air was a slow and complicated process; far from being achieved and perfected in the earliest tetrapods, it was separately invented many times by different groups, and rather a long time after tetrapods first gained the land." A natural-selection impetus for the evolution of tympanal ears may have been the evolution of noisy, buzzing insects in the Late Paleozoic: "The timing of the evolution of the characteristics of an ear capable of receiving airborne sound well certainly corresponds to the huge radiation of insects that can be seen in the fossil record."[27] Tetrapods that could hear insects, as well as see them, would certainly be more successful in hunting insects for food than a tetrapod that could not hear. Indeed, the teeth of *Balanerpeton woodi* were numerous but small and undifferentiated; given the small size of the animal itself, this has been taken to indicate that the animal was an insectivore and not a predator on other vertebrates. Clack notes that the final proof that *Balanerpeton woodi* was an insectivore is the fact that some fossil specimens of the animals "have the remains of myriapods preserved in the position where the stomach would have been."[28]

At this point, I would mention that an interesting debate still continues today about the origin of the modern living amphibians: the frogs, salamanders, and caecilians. These are usually considered to constitute a monophyletic clade, the Lissamphibia, although some would dispute even that! The problem is that the lissamphibians share many traits both with the basal batrachomorphs (the paraphyletic "temnospondyls" of older literature) and with the more derived batrachomorph clade of the lepospondyls (table 7.4). One argument maintains that the supposed synapomorphies uniting the lissamphibians with the basal batrachomorphs are in fact basal symplesiomorphies and thus do not reveal any new information about the derivation of the lissamphibians because they are simply ancient traits inherited unchanged from the earliest batrachomorphs. The counter argument maintains that the supposed synapomorphies uniting the lissamphibians with the lepospondyls are convergent traits, independently acquired in these two separate clades as a common functional result of evolutionary miniaturization in both lineages.[29] In table 7.4, I have taken the view that the origin of the lissamphibians lies with the basal batrachomorphs.

The as yet unnamed sister clade to the batrachomorphs contains two separate clades: the extinct lepospondyls[30] and the living clade of the reptiliomorphs (table 7.4). Curiously, the four known Visean species of basal lepospondyls (the paraphyletic "microsaurs" of older literature) are all found in Scotland, but not from the same locality. *Acherontiscus caledoniae* is from Gilmerton, *Adelogyrinus simorhynchus* is found in the Dunnet Shale, *Dolichopareias disjectus* comes from the Burdiehouse Limestone, and *Palaeomolgophis scoticus* is from the Curley Shale.[31] In contrast to the batrachomorph *Balanerpeton woodi* that we just considered, these lepospondyls are decidedly peculiar. *Balanerpeton woodi* was a fully terrestrial hunter of insects, with robust limbs for walking on land and tympanal ears for tracking the sound of insect prey. In contrast, *Adelogyrinus simorhynchus* had no hind limbs at all and its forelimbs were tiny and vestigial. It was clearly an aquatic, eel-like animal that had abandoned the land and returned to the water. The single known fossil of *Acherontiscus caledoniae* is fragmentary but is similar in form to *Adelogyrinus simorhynchus.* These aquatic lepospondyls probably gave rise to the more derived holospondylian clade of the Nectridea, which were also aquatic but are a ghost lineage in the Visean (table 7.4).

The other derived holospondylian clade of the Visean lepospondyls are the aistopods (table 7.4). The two known Visean species of the aistopod clade are also from Scotland: *Ophiderpeton kirtonense* from the East Kirkton *Lagerstätte* and *Lethiscus stocki* from the Wardie Shales.[32] These animals are even more peculiar than the basal lepospondyls—they had no limbs at all, and had very elongated, eel-liked bodies, but they were terrestrial animals rather than aquatic. Some of the aistopods had over 200 vertebrae and were almost a meter in length. It is believed that they filled the ecological niche of modern-day snakes, a group of tetrapods that have also convergently lost their appendages.[33] *Lethiscus stocki* is particularly interesting from an evolutionary point of view because the strata in which it is found, the Wardie Shales, are dated to the **TS** spore zone (tables 7.1 and 7.2) and are thus mid-Visean in age.[34] Thus, by the mid-Visean, a group of presumably land-dwelling tetrapods had already devolved their appendages to fill the burrowing and predatory niche of the modern snake. Once more it is clear that numerous early Visean tetrapods must have existed but have not been discovered (as yet) in the fossil record.

The other major clade in the sister group to the batrachomorphs were the reptiliomorphs (table 7.4). The reptiliomorphs eventually produced the amniotes, the vertebrate conquerors of land and our ancestors. Seven species of basal reptiliomorphs (the paraphyletic "anthracosaurs" of older literature) are known in the Visean (table 7.4).[35] Like the basal batrachomorphs, the basal reptiliomorphs were surprisingly small: their skulls were only 40 millimeters to 50 millimeters (1.6 inches to 2 inches) long, and their body length, from snout to pelvis, was only about 200 millimeters (eight inches). The length of their tails is unknown, but the total length of these early reptiliomorphs was probably only around 300 millimeters (about a foot). *Silvanerpeton miripedes, Eldeceeon rolfei,* and *Westlothiana lizziae* are found in the East Kirkton *Lagerstätte,* thus the fossils of their skeletons are fairly complete, except for their tails. *Antlerpeton clarkii* is known only from fragments of its ribs and limb bones; its primary importance is in demonstrating that the early reptiliomorphs were not confined to Scotland in southeastern Laurussia but had spread west to Nevada in North America.

Unlike the basal batrachomorph *Balanerpeton woodi,* these reptiliomorphs possessed five digits on all of their feet, both in the forelimb and hind limb. *Westlothiana lizziae* appears to be the most derived of the known basal reptiliomorph species; indeed, at one time it was thought to be the earliest known amniote (fig. 7.6). Subsequent examination of its fossil skeleton shows that, although it does possess some amniote traits, such as the fusion of the centra of its vertebrae with their neural arches, it is missing other amniote traits in its skull and ankles. Most interestingly, the animal has no otic notches in the skull, and this is true of other basal reptiliomorphs where the skull is preserved well enough to show this feature if it were present, such as in *Silvanerpeton miripedes.* These animals, although terrestrial (*Eldeceeon rolfei* in particular,

FIGURE 7.6 Reconstruction of the reptiliomorph tetrapod *Westlothiana lizziae* from Scotland. Note the tall, narrow head with laterally placed eyes and the elongated body. The living animal was small, about 200 millimeters (eight inches) long. *Credit*: Illustration by Kalliopi Monoyios.

with its robust limbs; fig. 7.7), apparently did not possess tympanic ear systems like the basal batrachomorph *Balanerpeton woodi*. Instead, tympanic ear systems evolved independently and convergently in the more derived sauropsid and synapsid amniote lineages.[36]

Last, we come to the most derived of the Visean tetrapods, the enigmatic species *Casineria kiddi* (table 7.4). The sedimentary nodule holding the fossil contains most of the body skeleton, but the head and tail are missing. The other Early Carboniferous reptiliomorphs are small, but *Casineria kiddi* can be described as tiny. Its back, from the base of its neck to its pelvic girdle, was only 80 millimeters (3 inches) long, making it the smallest known Early Carboniferous tetrapod.[37]

The vertebrae of *Casineria kiddi* are solidly ossified, and the animal had long, slender curved ribs. The well-preserved forelimb and forefoot of the animal held the greatest surprise (fig. 7.8), as described by Jennifer Clack: "The humerus is much more slender than that of any other Early Carboniferous tetrapod, with an obvious shaft and with

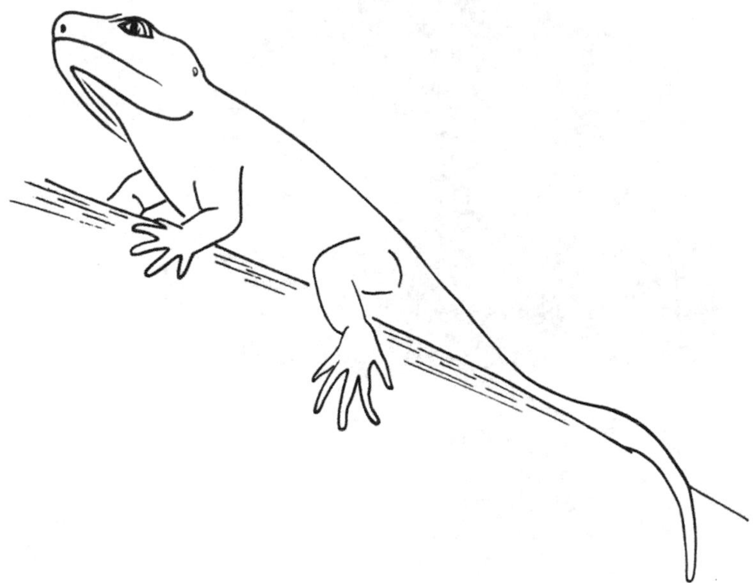

FIGURE 7.7 Reconstruction of the reptiliomorph tetrapod *Eldeceeon rolfei* from Scotland. Note the short, compact body and large limbs. The living animal was about 350 millimeters (14 inches) long. *Credit*: Illustration by Kalliopi Monoyios.

the two ends set at different angles to each other (known as torsion). The radius and ulna are also slender, with an olecranon process on the ulna. . . . These features alone strongly suggest a fully terrestrial animal," and in the forefoot: "the manus has five slender digits. . . . The last phalanx on each digit (the ungual) is noticeably curved, and the whole arrangement suggests a hand capable of grasping. No other Early Carboniferous tetrapod shows digits like this, but they are similar to those found in Late Carboniferous early amniotes."[38] Robert Carroll also remarks that "uniquely, *Casineria* had curved terminal phalanges (unguals) forming claws, as in many early amniotes."[39] In summary, Clack notes that "it has been suggested that the origin of amniotes is connected with an evolutionary step involving small size . . . and here is a specimen that accords well with that theory."[40]

Unfortunately, the head of the animal is missing, thus it is impossible to prove that *Casineria kiddi* was indeed the first known amniote.

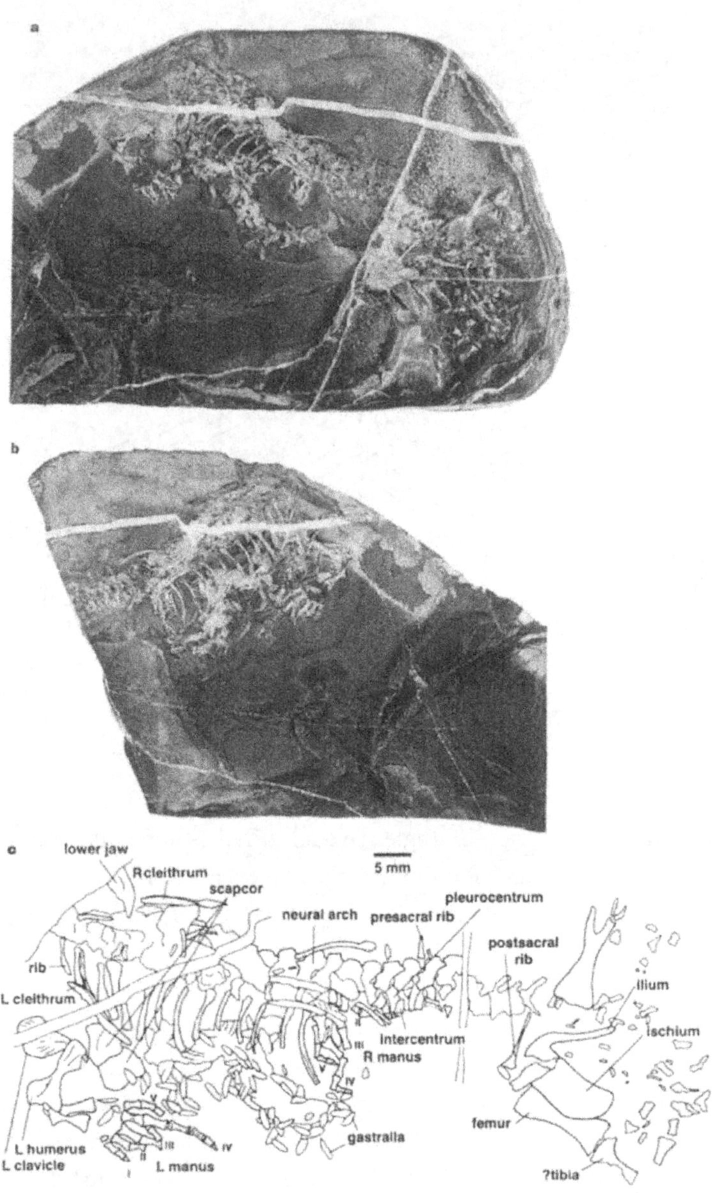

FIGURE 7.8 (a) Part and (b) counterpart of the rock slab holding the fossil skeleton of *Casineria kiddi.* (c) A drawing of the skeleton, with selected bones labeled. Note that the left hand ("L manus") in the lower left of the sketch has five fingers (labeled with Roman numerals). The living animal was a little over 80 millimeters (three inches) long. *Credit*: Macmillan Publishers Ltd: *Nature* (Paton et al., 1999), copyright © 1999. Reprinted with permission.

For this reason I have listed it with a question mark in table 7.4 as a possible basal amniote. Clearly, however, it is the most derived species of tetrapod yet known from the Visean.

In summary, the Early Carboniferous vertebrates were very strange animals. As we saw in chapter 5, Famennian tetrapods had made considerable evolutionary advancement in adapting to life on dry land: they lost their ancestral fish-like dentitions, lateral-line systems, and internal gills, and they progressively reduced the number of digits present in their feet in the move from paddling in water to walking on land. In contrast, after passing through the End-Famennian Bottleneck, we have two basal families of Early Carboniferous tetrapods that had abandoned land entirely and retreated back into the water, the colosteids and the crassigyrinids (table 7.4), and a more derived family, the baphetids (table 7.4), that also seems to have become more aquatic than terrestrial, judging from the morphology of the species *Loxomma allmanni*. We have a more derived clade of batrachomorph tetrapods, the aistopods (table 7.4), that were no longer tetrapodal at all—they had lost their hard-earned four limbs entirely and were snake-like.

Only the early, more plesiomorphic family of the whatcheeriids and the later, more derived clades of the batrachomorphs and reptiliomorphs were land walkers (excluding the aistopod batrachomorphs, which were land "crawlers"). The Tournaisian whatcheeriid tetrapods were about one meter (3.3 feet) in length, thus similar in size to their Famennian ancestors. In contrast, the basal members of the more derived batrachomoph and reptiliomorph clades in the Visean were very peculiar in that they were all very small—the largest batrachomorph was about 450 millimeters (a foot and a half) long, and the largest reptiliomorph was about 300 millimeters (about a foot) long. The most derived of the reptiliomorphs and the possible first amniote tetrapod, *Casineria kiddi*, was the smallest of them all. With a body trunk length of 80 millimeters, the entire animal was probably less than 200 millimeters (eight inches) long. This peculiar trend in the evolution of highly derived traits in animals with small body sizes will continue in the Late Carboniferous, where fossils of undisputed amniote species are found—and they are quite small.

Why did the derived Early Carboniferous vertebrate land dwellers, and the early amniotes in the Late Carboniferous, evolve such small body sizes? Smithsonian Institution arthropod specialist Conrad Labandeira summarizes the traditional view that this evolutionary trend was driven by specialized prey choice in these early predators:

> The timing of the initial tetrapod radiation on land was primarily during the Early Carboniferous. Stem-group tetrapods such as fin-limbed *Eusthenopteron* [fig. 3.1] and ichthyostegalian tetrapods occurred during the last stage of the Devonian, but both were more amphibious than truly land based, based on the body and trace-fossil evidence. Although these taxa were quite large, their dentition suggests that land arthropods and small aquatic vertebrates were probable food sources. By the end of the Early Carboniferous, amniotes had originated, giving rise to sauropsid and synapsid clades, many of which were considerably smaller than their Late Devonian and earlier Early Carboniferous forbearers, and probably occupied the feeding niche of micropredators.[41]

Specifically, these micropredators were seen as specializing in feeding on small arthropod species, particularly the numerous new insect species that evolved during this period of time—that is, they became insectivores.

In contrast, University of California paleontologist Richard Cowen proposes an alternative and novel idea to explain the small body sizes seen in these early terrestrial vertebrates: they were arboreal, that is, they lived in the trees. Here he outlines his view of the evolution of a tree canopy ecosystem during this period of time, where he uses the older paraphyletic taxon Reptilia to refer to the amniote clade:

> The currently favored scenario of reptile [= amniote] evolution, mostly due to Robert Carroll, is that reptiles [amniotes] evolved on a forest floor covered with rotting material, leaf litter, fallen branches, and tree stumps, ideal places for prey to hide and reptiles [amniotes] to search. . . . I suggest instead that the first reptiles [amniotes] evolved either on the river banks or in the canopy ecosystem of the Early Carboniferous, not on the forest floor. Whichever suggestion one prefers, I would argue that reptiles [amniotes] were living in the canopy forest in the

> Late Carboniferous. . . . Vertical climbing is easy with a small body size, so small Carboniferous animals could have been tree dwellers, as many salamanders are today. Trees offer damp places in which to lay eggs, and rich insect life high in the canopy forest provides abundant food. Even today, salamanders (and spiders) are the top carnivores in parts of the Central and South American canopy forest. The rich fossil record of Late Carboniferous insects, scorpions, spiders, and reptiles [amniotes] may portray the ecosystem of the canopy rather than the forest floor.[42]

Support for this alternative scenario comes from the morphological observation that the early amniotes and the enigmatic *Casineria kiddi* (fig. 7.8), which may itself have been the earliest amniote, had evolved claws on their digits. Why did these animals evolve claws? Claws can be used by predators to more efficiently seize prey, but they also can be used to climb trees very effectively, as can be seen in modern clawed predators like the cats.

Whether on the forest floor or in the canopy above, the victorious arthropod and vertebrate invaders of land did indeed evolve in the Late Carboniferous, as we shall see in the next two sections of the chapter.

The Victorious Arthropod Invaders

Although the millipedes, centipedes, and arachnids all evolved clades of new invaders in the Late Carboniferous, it is the clade of the insects that produced the final conquerors in the victorious arthropod invasion of land. That victory was achieved by the evolution of a key new innovation by the insects: the evolution of flight. With the evolution of wings, the insects could take to the skies and disperse over huge areas of the land in a very short time. Although the vertebrates preceded the arthropods in the final invasion of land in the Early Carboniferous Visean, the arthropods quickly surpassed them with the evolution of flight in the Late Carboniferous Bashkirian. Vertebrates remained behind, constrained to life on the ground, their dispersal rates limited by their slow walking speeds. A few reptiles evolved the capability to glide from tree to tree in the Permian, such as the diapsid weigeltisaur *Coelurosauravus jaekeli*,[43] but it is not until the Late Triassic that the first major clade of winged vertebrates, the Pterosauria, evolved.

Much of what we know about the new arthropod invaders comes from fossils found in the Mazon Creek Ironstone *Lagerstätte* in Illinois, North America. It is of Late Carboniferous Moscovian Age, and it is famous for its exceptional preservation of fragile arthropod tissues. For example, the ghost lineage of the delicate dipluran hexapods, which we know had to have evolved back in the Early Devonian before the springtails and the first insects (table 3.4), finally appears in this *Lagerstätte* with the species *Testajapyx thomasi*.[44] No less than eight (57 percent) of the 14 new Late Carboniferous species of millipedes, centipedes, and arachnids known to us are found in the Mazon Creek Ironstone *Lagerstätte*.[45]

The arthropleurids comprise a major new clade of millipedes that appeared in the Moscovian (table 7.5). These millipedes were the most gigantic terrestrial arthropods ever to evolve on Earth; for example, some individuals of the species *Arthropleura armata* and *A. mammata* were 2.5 meters (8 feet) long, 150 millimeters (6 inches) wide, and weighed up to 10 kilograms (22 pounds)![46] Although they were the largest of the Late Carboniferous arthropods, we shall see that they were not alone in their gigantism.

The clade of the pentazonians (table 7.5), a ghost lineage in the Devonian (table 3.4), now appeared in the Late Carboniferous Gzhelian fossil record with the amynilyspedid species *Amynilyspes fatimae* from France.[47] The clade of the euphoberiid millipedes appeared with the species *Euphoberia armigera*, another gigantic arthropod with individuals that attained one meter in length. The pleurojulids, another ghost clade in the Devonian (table 3.4), appeared in the Mazon Creek *Lagerstätte* with the two species *Pleurojulus biornatus* and *P. levis*.[48] Last, the millipede clade of the nematophorans are present also in the Mazon Creek *Lagerstätte*, represented by the species *Hexecontasoma carinatum*.

The centipedes evolved a new group of predators in the Late Carboniferous, the epimorphans, which consist of two clades (table 7.5): the ghost geophilomorph lineage, and the more derived scolopendromophs, represented by the species *Mazoscolopendra richardsoni*[49] in the Mazon Creek *Lagerstätte*.[50]

TABLE 7.5 Phylogenetic classification of the Late Carboniferous arthropod invaders.

Panarthropoda
– Onychophora (velvet worms)
– [Tardigrada]
– Arthropoda (true arthropods)
– – Mandibulata
– – – Myriapoda
– – – – Progoneata
– – – – – [Symphyla]
– – – – – Dignatha
– – – – – – [Pauropoda]
– – – – – – Diplopoda (millipedes)
– – – – – – – Pencillata
– – – – – – – – Arthropleuridea†
– – – – – – – – – Eoarthropleurida†
– – – – – – – – – **Arthropleurida**†
– – – – – – – unnamed clade
– – – – – – – – Microdecemplicida†
– – – – – – – – Chilognatha
– – – – – – – – – Zosterogrammida†
– – – – – – – – – Pentazonia
– – – – – – – – – – **Amynilyspedida**†
– – – – – – – – – unnamed clade
– – – – – – – – – – Archipolypoda†
– – – – – – – – – – – Cowiedesmida†
– – – – – – – – – – – Archidesmida†
– – – – – – – – – – – **Euphoberiida**†
– – – – – – – – – – Helminthomorpha
– – – – – – – – – – – Pleurojulida†
– – – – – – – – – – – – **Pleurojulidae**
– – – – – – – – – – – [Colobognatha]
– – – – – – – – – – – Eugnatha
– – – – – – – – – – – – Juliformia
– – – – – – – – – – – – **Nematophora**
– – – – Chilopoda (centipedes)
– – – – – Scutigeromorpha
– – – – – Pleurostigmorpha
– – – – – – [Lithobiomorpha]
– – – – – – Phylactometria
– – – – – – – [Craterostigmomorpha]
– – – – – – – Devonobiomorpha†
– – – – – – – **Epimorpha**
– – – – – – – – **[Geophilomorpha]**

(continued)

TABLE 7.5 (*continued*)

– – – – – – – – **Scolopendromorpha**
– – – Pancrustacea
– – – – Hexapoda
– – – – – Entognatha
– – – – – – Diplura
– – – – – – Elliplura
– – – – – – – [Protura]
– – – – – – – Collembola (springtails)
– – – – – Insecta (insects)
– – – – – – Rhyniognatha†
– – – – – – Archaeognatha (bristletails)
– – – – – – **Dicondylia**
– – – – – – – **[Zygentoma (silverfish)]**
– – – – – – – **Pterygota (winged insects)**
– – – – – – – – **Ephemeroptera (mayflies)**
– – – – – – – – **Metapterygota**
– – – – – – – – – **Odonatoptera (dragonflies, damselflies)**
– – – – – – – – – – **Geroptera†**
– – – – – – – – – – **Holodonata†**
– – – – – – – – – – – **Meganisoptera†**
– – – – – – – – – **Palaeodictyopterida†**
– – – – – – – – – – **Palaeodictyoptera†**
– – – – – – – – – – **unnamed clade†**
– – – – – – – – – – – **Diaphanopterodea†**
– – – – – – – – – – – **Megasecoptera†**
– – – – – – – – – **Neoptera (folding-wing insects)**
– – – – – – – – – – **Protoptera†**
– – – – – – – – – – – **Paoliidae†**
– – – – – – – – – – **Polyneoptera**
– – – – – – – – – – – **Anartioptera**
– – – – – – – – – – – – **[Plecopterida (stoneflies)]**
– – – – – – – – – – – – **[Dermaptera (earwigs)]**
– – – – – – – – – – – – **Orthopterida**
– – – – – – – – – – – – – **Caloneurodea†**
– – – – – – – – – – – – – **Orthoptera (crickets, grasshoppers)**
– – – – – – – – – – – – – – **Archaeorthoptera†**
– – – – – – – – – – – **Dictyoptera (roaches, temites, mantids)**
– – – – – – – – – – – – **basal dictyopterans ("roachoids")**
– – – – – – – – – – – – – **Archimylacrididae†**
– – – – – – – – – – – – – **Subioblattidae†**
– – – – – – – – – – – – – **Necymylacrididae†**
– – – – – – – – – – – – – **Poroblattinidae†**
– – – – – – – – – – – – – **Spiloblattinidae†**
– – – – – – – – – – – – – **Mylacrididae†**

TABLE 7.5 (*continued*)

– – – – – – – – – – – – **unnamed clade**

– – – – – – – – – – – – – **Phylloblattidae†**

– – Cheliceriformes

– – – Chelicerata

– – – – Arachnida

– – – – – Stomothecata

– – – – – – Scorpiones (scorpions)

– – – – – – Opiliones (harvestmen)

– – – – – Phalangiotarbida†

– – – – – Haplocnemata

– – – – – – Pseudoscorpiones (false scorpions)

– – – – – – **Solifugae (camel spiders)**

– – – – – [Palpigradi]

– – – – – Pantetrapulmonata

– – – – – – Trigonotarbida†

– – – – – – Tetrapulmonata

– – – – – – – Uraraneida†

– – – – – – – unnamed clade

– – – – – – – – **Araneae (spiders)**

– – – – – – – – **Schizotarsata**

– – – – – – – – – **Haptopoda†**

– – – – – – – – – **Pedipalpi**

– – – – – – – – – – **Amblypygi (whip spiders)**

– – – – – – – – – – **Uropygi**

– – – – – – – – – – – **Thelyphonida (whip scorpions)**

– – – – – Acaromorpha

– – – – – – Actinotrichida (mites)

– – – – – – unnamed clade

– – – – – – – **Ricinulei (hooded tick spiders)**

Source: Phylogenetic data modified from Grimaldi and Engel (2005), Lecointre and Le Guyader (2006), Sierwald and Bond (2007), Shear and Edgecombe (2010), and Dunlop (2010).

Note: New arthropod lineages are marked in bold-face type. "Ghost" arthropod lineages in are enclosed in brackets (e.g., [Tardigrada]). See text for discussion.

† extinct taxa.

The arachnid predators were much more prolific than the centipedes during the Late Carboniferous. The first fossil members of the clades of the camel spiders, true spiders, whip spiders, whip scorpions, and hooded tick spiders all appeared in the Late Carboniferous (table 7.5). The solifuge species *Protosolpuga carbonaria* and the thelyphonid species *Geralinura carbonaria* appeared in the familiar Mazon Creek

Lagerstätte. However, the oldest known true spider, the species *Eocteniza silvicola,* the haptopod species *Plesiosiro madeleyi,* the whip spider species *Graeophonus anglicus,* and the hooded tick spider species *Curculioides adompha* all appeared in the Moscovian Coal Measures of Europe rather than in North America.[51]

We now have arrived at the true arthropod conquerors of the terrestrial realm in the Late Carboniferous: the flying insects (table 7.5). The oldest known near-basal member of the clade of the winged insects, the Pterygota, is the geropteran species *Eugeropteron lunatum* (fig. 7.9) from the Late Carboniferous Bashkirian Age in Argentina.[52] However, an even more basal member of the pterygotan clade is the ephemeropteran species *Lithoneura lameerei,* found in the Moscovian Age Mazon Creek *Lagerstätte.* Although *Lithoneura lameerei* is thus younger than *Eugeropteron lunatum,* phylogenetic analyses (table 7.5) prove that the basal clade of the mayflies must have evolved in the Bashkirian as well, since the more derived clade of the geropterans is already present there. The existence of these two species, one basal and the other only slightly more derived, gives definitive proof that the winged insects evolved on Earth at the dawn of the Late Carboniferous, some 318 million years ago.

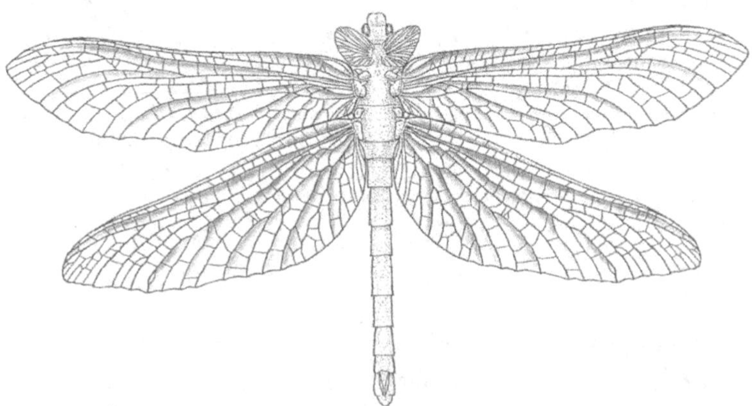

FIGURE 7.9 Reconstruction of the geropteran insect *Eugeropteron lunatum* from the Late Carboniferous of Argentina. The living animal had a wingspan of about 100 millimeters (four inches). *Credit:* From Grimaldi and Engel (2005); copyright © 2005 David Grimaldi and Michael S. Engel. Reprinted with the permission of Cambridge University Press.

Or does it? University of Exeter biologist Robin Wootton and his colleagues note that *Eugeropteron lunatum* was surprisingly advanced in its wing morphology, in that it possessed "features analogous to 'smart' mechanisms in modern dragonflies that are associated with the agile, versatile flight necessary to catch prey in flight. These mechanisms act automatically in flight to depress the trailing edge and to facilitate wing twisting, in response to aerodynamic loading. The presence of similar features suggests that the earliest known odonatoids were already becoming adapted for high-performance flight."[53] Thus these Bashkirian wing types may have had earlier, less high-performance, precursors in the Early Carboniferous. Further, Czech Charles University zoologist Jakub Prokop and colleagues[54] argue that four fossils of even more highly derived pterygotan species have been discovered in strata dated to the last Age of the Early Carboniferous, the Serpukhovian. These four are the palaeodictyopteran species *Delitzschala bitterfeldensis* and the protopteran paoliid species *Kemperala hagensis* from Germany, the archaeorthopteran species *Ampeliptera limburgica* from the Netherlands, and fossil fragments of the wing of another archaeorthopteran species found in the Czech Republic.[55]

The protopteran and archaeorthopteran species in particular are highly derived neopterans (table 7.5), members of the diverse clade of advanced insects that had evolved the capability of folding their wings, unlike the more plesiomorphic dragonflies, which cannot. The delicate wings of a dragonfly always project out at right angles to the length of the dragonfly's body, even when the dragonfly is not using them in flight. Wing damage can be a lethal injury to the dragonfly, thus it clearly would be advantageous to the insect to be able to protect its wings when they are not actually being used to fly. The neopterans solved this problem by evolving the capability to fold their wings against their bodies when they are not in use, and the evolution of wing folding is a defining innovative synapomorphy of this clade of more derived pterygotans.

Thus the presence of highly derived neopteran species in Serpukhovian strata at the end of the Early Carboniferous argues for the evolution of flying insects at some point in time even earlier in the Early Carboniferous. Yet even if the winged insects evolved in the Visean of the Early Carboniferous, they did not become numerous enough to appear in numbers in the fossil record until the Bashkirian Age at the dawn of the Late Carboniferous. In addition to the geropteran species *Eugeropteron lunatum* that we considered previously, the more derived meganisopteran species *Erasipteroides valentini* and *Namurotypus sippeli* also appeared in the Bashkirian. The meganisopterans were gigantic griffenflies, ancient relatives of the living dragonflies. The Late Carboniferous Gzhelian species *Meganura monyi* and the Early Permian species *Meganeuropsis permiana* had wingspans over 700 millimeters (2.3 feet) wide. These gigantic flying predators ruled the skies of the Late Carboniferous and Permian world, but they did not survive the end-Permian mass extinction.[56]

The winged insects also had produced a major new ecological innovation by the dawn of the Late Carboniferous: the evolution of full-scale herbivory, the ability to efficiently use living plants as a food source. This innovation was a key element in the construction of the terrestrial ecosystem familiar to us today, with its energy flow from living plants to herbivores to carnivores, and unlike the ancient arthropod food chains in the Silurian and Devonian, where the energy flow was from dead plants to detritivores to carnivores (as discussed in chapter 3). These early insect herbivores were the palaeodictyopteridans (table 7.5), a peculiar group of beaked sap-suckers. In addition to the long beak on the head of the insect, which it used to pierce plant stems and to suck out sap, the animals had two smaller, wing-like projections located behind their heads but in front of their large wings, giving the animal a six-winged look (fig. 7.10).

The palaeodictyopteridan herbivores were the dominant group of insects in the Late Carboniferous and Permian, constituting over 50 percent of all known insect species, yet they perished in the end-Permian mass extinction and are the only major clade of insects that did not survive to the modern day. Like their griffenfly cousins, many of the palaeodictyopteridan species were gigantic, having wingspans of over 500 millimeters (a foot and a half) wide and beaks over 30 millimeters (1.2 inches) in length.[57]

Last, the folding-wing insects also diversified and flourished in the Late Carboniferous, producing the clades that include modern-day stoneflies, earwigs, crickets, grasshoppers, roaches, termites, and

FIGURE 7.10 Reconstruction of the palaeodictyopteridan insect *Stenodictya lobata* from the Late Carboniferous of France. The living animal had a wingspan of about 130 millimeters (five inches). *Credit*: From Grimaldi and Engel (2005); copyright © 2005 David Grimaldi and Michael S. Engel. Reprinted with the permission of Cambridge University Press.

mantids (table 7.5). The basal neopteran clade of the protopterans is present in the Serpukhovian fossil record, represented by the species *Kemperala hagensis* from Germany. As previously discussed, however, the highly derived polyneopteran clade of the archaeorthopterans was also present in the Serpukhovian, thus all of the winged insect clades listed before the archaeorthopteran clade in table 7.5 may have in fact evolved earlier in the Early Carboniferous Visean, although they as yet have not been found in any Visean *Lagerstätte*. The polyneopteran

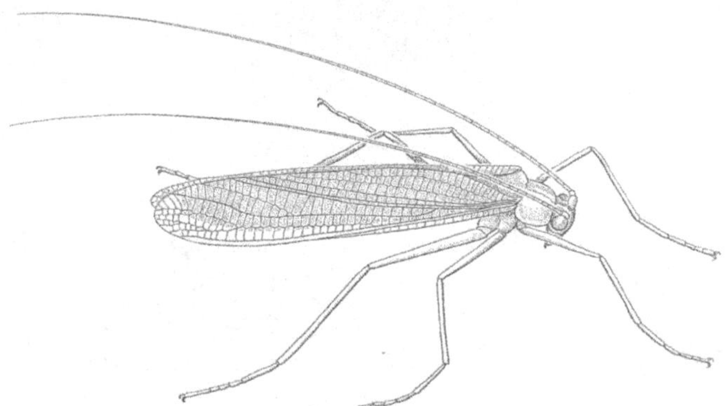

FIGURE 7.11 Reconstruction of the Permian caloneurodean insect *Paleuthygramma tenuicornis* from the Permian of Russia, showing its folded wings. The living animal was about 25 millimeters (one inch) long. *Credit*: From Grimaldi and Engel (2005); copyright © 2005 David Grimaldi and Michael S. Engel. Reprinted with the permission of Cambridge University Press.

clades of the stoneflies, plecopteridans, and the earwigs (dermapterans) are ghost lineages in the Late Carboniferous, but they do appear in the fossil record in the Permian and Triassic, respectively. The basal orthopterid clade of the caloneurodeans (fig. 7.11), represented by the species *Caloneura dawsoni* in the Late Carboniferous Gzhelian of France, is another clade of ancient insects that did not survive the end-Permian mass extinction.[58] Their relatives, the orthopterans, have numerous descendants in the living crickets and grasshoppers. Like these modern kin, the ancient orthopterans were primarily herbivores and scavengers.

The last major clade of folding-wing insects that were present in the Late Carboniferous were the dictyopterans (table 7.5). The basal dictyopterans were roach-like animals that were the most abundant arthropods, in terms of population numbers, in Carboniferous coal swamp ecosystems. No less than six families of basal dictyopteran "roachoids" (fig. 7.12) are known from the Late Carboniferous (table 7.5). The dictyopterans did produce one more derived family at the end of the Late Carboniferous, represented by the Gzhelian species *Phylloblatta gallica* from France. Like other Late Carboniferous insects, some of the roachoids were large,

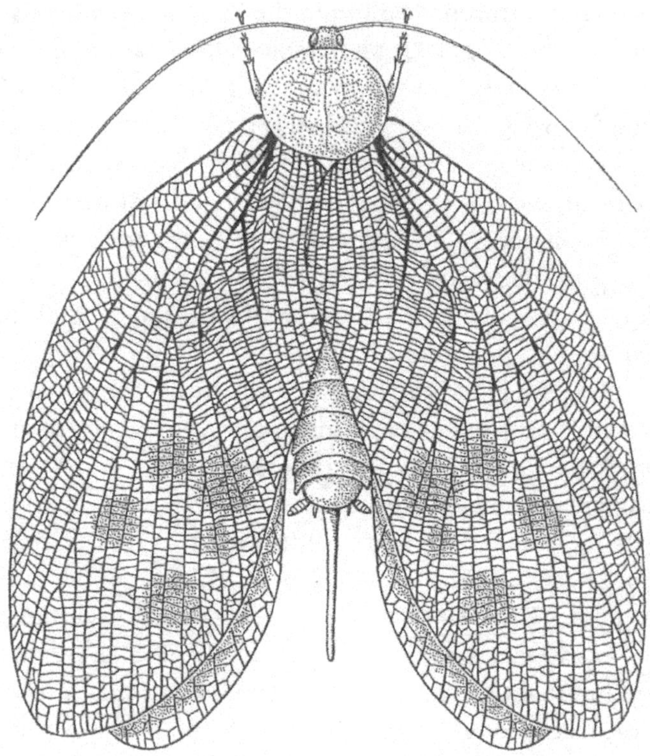

FIGURE 7.12 Reconstruction of the "roachoid" insect *Manoblatta bertrandi* from the Late Carboniferous of France. The living animal was about five centimeters (two inches) long. *Credit*: From Grimaldi and Engel (2005); copyright © 2005 David Grimaldi and Michael S. Engel. Reprinted with the permission of Cambridge University Press.

being over 60 millimeters (2.4 inches) in length.[59] These roachoids, like their modern-day roach kin, were polyphagous, feeding on just about anything edible, both plant and animal, living or dead.

What was the trigger for the evolution of the winged insects, the victorious arthropod conquerors of the land? After two periods of environmental disaster, producing the end-Frasnian and end-Famennian extinctions, nature now may have given the invaders a major environmental advantage in the Visean and Serpukhovian: increased concentrations of oxygen in the atmosphere.[60] Evidence for this environmental trigger may be seen in the gigantic size attained by many of the Carboniferous arthropods. Even in the Early Carboniferous Visean, gigantic ground-dwelling arthropods appeared, such as the scorpion *Pulmonoscorpius kirktonensis*, which was about 700 millimeters (2.3 feet) long. In the following Serpukhovian Age, a few fossils of flying insects are found, such as the palaeodictyopteran *Delitzschala bitterfeldensis*.[61] Flying is a highly energetic activity; it requires a lot of oxygen. Then, in the Late Carboniferous, flying griffenfly predators and flying palaeodictyopteran sap-suckers, with wingspans of 700 millimeters (2.3 feet) and 500 millimeters (a foot and a half), respectively, appeared in the skies. What is wrong with this picture?

Larger, heavier organisms require even more energy to fly than small ones. This problem is made much more serious by the fact that insects breathe with tracheae, a series of small tubes that extend from pore-like openings in their exoskeletons down into their body tissues. The insects depend entirely upon diffusion of oxygen in from these tubules (and reverse diffusion of carbon dioxide out of the tubules) to keep their tissues alive. The size of an insect is strongly limited by this system of breathing. Larger and larger insects have larger and larger volumes of internal tissue, tissues that depend upon the surface areas of the tiny tracheae for gas exchange. With increasing size, the volume of an object increases as a cubic function of linear dimension, whereas surface area increases only as a square function. This constraint—the area—volume effect—means that insects are constrained to sizes small enough to have the proper ratio of internal tissue volume to the surface area of tracheal tubes.

So how could a dragonfly-like insect with a wingspan that is almost three-quarters of a meter in length exist? It should have suffocated under its own mass, its tracheae unable to aerate its large volume of internal tissues. But what if the atmosphere at that time *had more oxygen* than it presently does? An atmosphere with a higher partial pressure of oxygen—more oxygen molecules concentrated in a given volume of air—could potentially lift the constraint of the area–volume effect to higher ratios of tissue volume to tracheal surface area and could make the existence of much larger insects possible.[62] We have previously seen in chapter 4 that two different empirical models predict high oxygen levels in the Earth's atmosphere in the Late Carboniferous (fig. 4.2).

In addition to the appearance of gigantic flying insects in the Late Carboniferous, ground dwelling arthropods became even larger than the 700-millimeter length (2.3-feet) of the Visean scorpion *Pulmonoscorpius kirktonensis*. These were the *Euphoberia armigera* and the *Arthropleura armata,* Moscovian Age millipedes with body lengths of 1,000 millimeters (3.3 feet) and 2,500 millimeters (8 feet), respectively, that we considered at the beginning of this section of the chapter. It is difficult to see how such large arthropods could have existed without having an atmosphere that contained much more oxygen than the present-day atmosphere of the Earth. In addition, increased levels of oxygen in the atmosphere in the early Cenozoic have been linked to the evolution of very large mammals on land during this same interval of time.[63]

The paleoentomologists David Grimaldi and Michael Engel caution against concluding that insect gigantism is indisputable proof of hyperoxic conditions on Earth in the Carboniferous. They point out that there are other evolutionary mechanisms that produce large animals, such as natural selection for defense against predators. Also, our present-day atmosphere with about 21 percent oxygen is not hyperoxic, yet still some very large insects exist on Earth today: the 300-millimeter (one-foot) walking-stick insects and the 150-millimeter (six-inch) beetles, the white-witch moth *Thysania agripinna* with a 250-millimeter (ten-inch) wingspan, and katydids with 200-millimeter (eight-inch) wingspans.[64] Still, however, none of these modern insects approach the size of crawling millipedes that were longer than a bicycle and wider than a truck tire, and dragonfly-like predators with wingspans longer than the length of your arm.

The Victorious Vertebrate Invaders

Just as the winged insects were the final conquerors of the land in the arthropod clade, the amniotes were the victorious conquerors of land in the clade of the vertebrate animals. That victory was achieved by the evolution of a key new innovation by the vertebrates: the evolution of the amniote egg. Prior to this innovation, tetrapods still laid their eggs in water, much like fish.

The amniote egg was an innovative, two-layer adaptation to protect the developing embryo from dehydration in the harsh, dry-air environments of the terrestrial realm. First, the embryo is enclosed in a water-filled region contained by the surrounding amniotic membrane. In essence, rather than floating in an actual pond of water as amphibian embryos do, the amniote embryo floats within its own private amniotic pond. Second, the outside layer of the egg is a tough shell rather than the soft, gelatinous outer covering of the amphibian egg. Both of these layers help protect the embryo from serious water loss, and death by dehydration. Thus the amniotes can lay their eggs on dry land, and the eggs can survive. This is not true of most amphibians. Their eggs will rapidly dehydrate if exposed to dry air—they will shrivel up and shrink as moisture is lost to the atmosphere, and the embryo will die.

Last, a large container of food is enclosed within the amniote egg for the embryo, the yolk sac. Thus the embryo can develop to a considerable degree within the egg itself, feeding on the nutrients from the yolk sac rather than hatching out of the egg at an early growth stage and foraging for food in a free-swimming larval stage, as with many amphibians. In consequence, the amniotes have direct development—there is no larval stage. What emerges from an amniote egg looks like a small, scaled-down version of the adult animal. In contrast, what emerges from a frog egg, a tadpole, looks like a fish, not like a frog. And fish, of course, need water to swim in. A hatchling tetrapod amniote does not.

Similar to the arthropods, the vertebrates experienced an explosion of diversity in the Late Carboniferous. Five new species of basal reptiliomorphs appeared in the Bashkirian Age, the dawn of the Late Carboniferous, seven new species in the following Moscovian, two more in the Kasimovian, and one in the Gzhelian at the close of the Late Carboniferous (table 7.6). Also similar to the arthropods, the first fully herbivorous vertebrates appeared in the fossil record in the Late Carboniferous: the diadectomorphs (table 7.6). In the Batrachosauria, the clade of the seymouriamorphs remained a ghost lineage (they eventually appear in the Permian), but the clade of the diadectomorphs now appeared, represented by the Moscovian species *Limnostygis relictus* and two additional species in the Gzhelian.[65] It is an interesting example of ecological convergence that herbivory evolved in two widely divergent phylogenetic

TABLE 7.6 Phylogenetic classification of the Late Carboniferous reptiliomorph tetrapod invaders, arranged by the geologic age (underlined) in which they are first found.

Reptiliomorpha

– basal reptiliomorphs ("anthracosaurs")

– – Bashkirian: *Anthracosaurus russelli, Eobaphetes kansensis, Pholiderpeton*

– – – *scutigerum, Palaeoherpeton decorum, Calligenethlon watsoni.*

– – Moscovian: *Anthracosaurus lancifer, Pteroplax cornutus, Eogyrinus attheyi,*

– – – *Diplovertebron punctatum, Nummulosaurus kolbii, Spondylerpeton spinatum,*

– – – *Carbonoherpeton carrolli.*

– – Kasimovian: *Neopteroplax conemaughensis, Cricotus heteroclitus.*

– – Gzhelian: *Neopteroplax? relictus.*

– Batrachosauria

– – [Seymouriamorpha]

– – unnamed clade

– – – Diadectomorpha

– – – – Moscovian: *Limnostygis relictus.*

– – – – Gzhelian: *Desmatodon hesperis, D. hollandi.*

– – – **Amniota** (conquerors of the land!)

– – – – basal amniotes?

– – – – – *Casineria kiddi*

– – – – **Sauropsida** (ancestors of reptiles)

– – – – – basal sauropsids

– – – – – – Bashkirian: *Hylonomus lyelli.*

– – – – – – Moscovian: *Cephalerpeton ventriarmatum, Anthracodromeus longipes,*

– – – – – – – *Brouffia orientalis, Coelostegus prothales.*

– – – – – [Anapsida]

– – – – – unnamed clade

– – – – – – Captorhinidae

– – – – – – – Gzhelian: *Concordia cunninghami.*

– – – – – – unnamed clade

– – – – – – – – Moscovian: *Paleothyris acadiana.*

– – – – – – – **Diapsida** (ancestors of lepidosaurs, archosaurs, dinosaurs, birds)

– – – – – – – – Kasimovian: *Petrolacosaurus kansensis.*

– – – – – – – – Gzhelian: *Spinoaequalis schultzei.*

– – – – **Synapsida** (ancestors of mammals)

– – – – – basal synapsids

– – – – – – Bashkirian: *Protoclepsydrops haplous.*

– – – – – unnamed clade ("pelycosaurs")

– – – – – – [Eothyrididae]

– – – – – – [Caseidae]

(continued)

TABLE 7.6 *(continued)*

- - - - - unnamed clade ("pelycosaurs")
- - - - - - Varanopseidae
- - - - - - - Kasimovian: *Milosaurus mccordi.*
- - - - - - - Gzhelian: *Archaeovenator hamiltonensis.*
- - - - - - unnamed clade ("pelycosaurs")
- - - - - - - Ophiacodontidae
- - - - - - - - Moscovian: *Archaeothyris florensis, Clepsydrops collettii, Echinerpeton*
- - - - - - - - - *intermedium.*
- - - - - - - - Gzhelian: *Stereorhachis dominans.*
- - - - - - - unnamed clade ("pelycosaurs")
- - - - - - - - Edaphosauridae
- - - - - - - - - Kasimovian: *Ianthasaurus hardestii, Xyrospondylus ecordi.*
- - - - - - - - - Gzhelian: *Edaphosaurus colohistion.*
- - - - - - - - unnamed clade ("pelycosaurs")
- - - - - - - - - Sphenacodontidae
- - - - - - - - - - Kasimovian: *Haptodus garnettensis.*
- - - - - - - - - **Therapsida** (ancestors of monotremes, marsupials, placentals)
- - - - - - - - - - Biarmosuchia (basal therapsids)
- - - - - - - - - - - Early Permian: *Tetraceratops insignis.*

Source: Phylogenetic and stratigraphic data modified from Benton (2005), Lecointre and Le Guyader (2006), Ward et al. (2006), and Sallan and Coates (2010).

Note: "Ghost" tetrapod lineages are enclosed in brackets (e.g., [Seymouriamorpha]). Older paraphyletic tetrapod group names are given in quotation marks within the parentheses. Major vertebrate clades of land invaders marked in bold-faced type. See text for discussion.

lineages, the arthropods and the vertebrates, at about the same time in the evolution of the terrestrial ecosystem.[66]

The Late Carboniferous diadectomorphs (fig. 7.13) were large animals, ranging from two meters to three meters (six and a half feet to ten feet) in length. They had blunt, peg-like teeth in the front of their mouths for cropping vegetation, and rows of blocky, molar-like cheek teeth for grinding plant material in their mouths.[67] Curiously, these first vertebrate herbivores in the terrestrial ecosystem were closely related to the ultimate conquerors of the land, the amniotes (table 7.6).

Both of the two clades of the amniote vertebrates, the sauropsids and the synapsids, appeared in the Bashkirian Age, the dawn of the Late Carboniferous. The very first amniote species thus must have evolved before the late Bashkirian because the split between the divergent sauropsid and synapsid amniote clades had occurred by then,

FIGURE 7.13 Reconstruction of the diadectomorph advanced reptiliomorph *Diadectes maximus* of the Late Carboniferous and Permian of North America and Europe. Note the short, barrel-chested body and blunt, peg-like teeth in the mouth. The living animal was very large, about three meters (ten feet) long. *Credit*: Illustration by Kalliopi Monoyios.

some 314 million years ago, and these two divergent lineages pursued independent evolutionary pathways forward from this point in time (table 7.6). One of these lineages, the synapsids, produced the lineage that resulted in the human species.

When did the first amniotes evolve? As discussed previously, the very earliest amniotes may have evolved in the mid-Visean late **TS** spore zone (table 7.1) with the appearance of the species *Casineria kiddi* (table 7.4), but this cannot be proved as the fossil skeleton of *Casineria kiddi* is missing its head (fig. 7.8). As will be discussed in detail in a moment, basal sauropsid and synapsid fossil amniote species are known from the late Bashkirian. Trace fossil evidence suggests, however, than more derived species of synapsid amniotes were also present on Earth in the Bashkirian, not just the basal one known from body fossils. This evidence comes from the trackway ichnospecies *Dimetropus*, found in Bashkirian-aged strata in Germany.[68] This trackway was made by a large animal: its hind feet were around 140 millimeters (5.5 inches) long, and its forefeet were around 70 millimeters (2.6 inches) long. The trackway is most similar to those produced by more derived ophiacodontid, edaphosaurid, or sphenacodontid synapsids (table 7.6), all of which were large animals. In contrast to the large *Dimetropus* footprints, the known basal synapsid and sauropsid amniotes were quite small—the entire animal was only 200 millimeters (eight inches) or so in length. If the ichnospecies *Dimetropus* was produced by a highly derived synapsid amniote in the Bashkirian, then the split between the synapsid and saurposid amniotes had to been even older, perhaps occurring in the last Age of the Early Carboniferous, the Serpukhovian. The very first amniote may have evolved even earlier, in the Visean, where *Casineria kiddi* is found.

The earliest known body fossil of a sauropsid amniote species is *Hylonomus lyelli* from Nova Scotia, North America. It was found preserved in the famous Joggins tree-stump *Lagerstätte*: hollow, cylindrical, pitfall traps produced by lepidodendralean scale tree stumps that had rotten away. The stumps of modern lignophyte trees do not produce such hollows because the center of the tree consists of strong heartwood. However, in ancient lepidodendralean scale trees the center of the tree was actually softer than the outside layers of cortex. When the tree died, the trunks of the tree sometimes remained standing, particularly if the base of the trunk had been surrounded by enclosing mud deposited during seasonal flooding of the forest. The inner part of the tree rotted away, leaving a cylindrical hole into which unwitting animals could fall: that is, the rotted out, hollow tree trunk formed a pitfall trap. If the animals were small enough, they could not climb out and escape. This famous *Lagerstätte* was discovered in 1852, and over 30 fossil tree stumps have since been excavated, producing hundreds of skeletons of small tetrapods.[69] The Joggins tree-stump *Lagerstätte* has been dated to the late Bashkirian, about 314 million years ago.[70]

In a previous section of the chapter, we considered the novel proposal by Richard Cowen that the small, early amniotes were tree dwellers. Cowen also proposed an alternative scenario for the Joggins tree-stump *Lagerstätte*. Rather than pitfall traps, hollow tree trunks were the preferred homes of the species found preserved in them:

> The lycopod trees [color plate 5] were fragile and shallow-rooted, and they may have been hollow in life, as many tropical trees are today. After storms or old age felled them, their hollow stumps sometimes remained standing. The fossil forests of Early Pennsylvanian

> (Late Carboniferous) rocks of Nova Scotia are famous because they include many tree stumps and tree trunks fossilized upright in life position. Some early reptiles [amniotes] have been found preserved inside some of the hollow stumps. This may not be a freak of preservation. I think that they *lived* inside the hollow trunks of living trees, as little insectivorous mammals do today in tropical rain forests."[71] (emphasis Cowen's)

Indeed, one of the distinctive morphological characteristics of the amniotes was that they had claws on their digits, a trait that is also present in the digits of the enigmatic Visean species *Casineria kiddi* (fig. 7.8), which may have been the earliest amniote. Would these amniotes really have been unable to climb out of the hollow tree trunks? The inner surface of the tree trunk must have contained irregularities that could have been used as toeholds by an animal with claws, allowing it to climb out.

The sauropsid *Hylonomus lyelli* was a small, slender animal that was only 200 millimeters (eight inches) long from the snout of the head to the tip of the tail. Such an animal today would be considered a medium-sized lizard. The early sauropsids had mouths filled with small, sharp teeth, and they are believed to have been primarily predators of insects and millipedes.[72] Thus the conqueror arthropods became the food source of the conqueror vertebrates.

Four more species of basal sauropsid amniotes are found in the fossil record in the Moscovian, which indicates that the sauropsids were diversifying and solidifying their insectivorous role in the terrestrial ecosystem (table 7.6). Phylogenetic analyses reveal that this diversification was more rapid than might be thought if only the number of species seen in the fossil record is considered. The sauropsid species *Paleothyris acadiana* (fig. 7.14) was also found in the Morien Group tree-stump *Lagerstätte* in Nova Scotia, which is about 10 million years younger than the Joggins *Lagerstätte*.[73] This species is highly derived, very close to the clade of the advanced diapsid sauropsids (table 7.6), yet it is found in strata of the Moscovian Age near the very beginning of the Late Carboniferous. More importantly, this species is more derived than a major group of reptiles still found in the modern world: the anapsids.

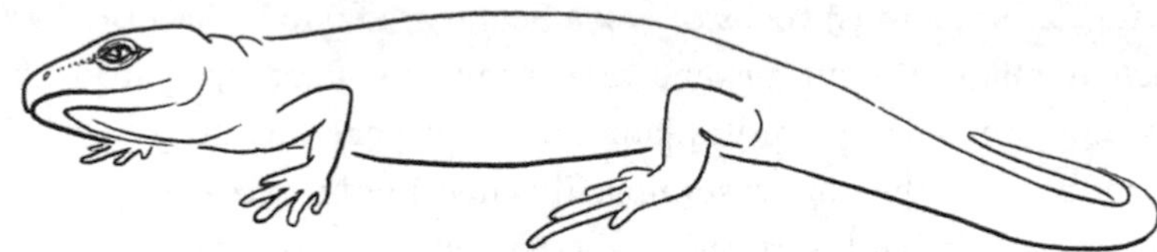

FIGURE 7.14 Reconstruction of the sauropsid amniote *Paleothyris acadiana*, preserved in the Joggins tree stump *Lagerstätte*. Note the gracile body and limbs. The living animal was 400 millimeters (1.3 feet) long. *Credit*: Illustration by Kalliopi Monoyios.

Thus the anapsids, ancestors of modern turtles and tortoises, must have evolved on Earth before the evolution of *Paleothyris acadiana* (table 7.6). The anapsid clade remains a ghost lineage until the Early Permian, when anapsid species finally appear in the fossil record.[74] Likewise, the derived clade of the captorhinid sauropsids also must have evolved before *Paleothyris acadiana*, even though the first known fossil species of the captorhinid clade, the Gzhelian species *Concordia cunninghami*, is not found until the end of the Late Carboniferous (table 7.6).[75] Like the diadectomorphs, the captorhinids filled the herbivore niche in the Late Carboniferous terrestrial ecosystem and are another example of ecological convergence.

Although the Moscovian species *Paleothyris acadiana* is very close to the advanced diapsid clade, the first known diapsid found in the fossil record is the Kasimovian species *Petrolacosaurus kansensis* (fig. 7.15) from Kansas in North America. *Petrolacosaurus kansensis* was a bit over 400 millimeters (1.3 feet) in length from its nose to the tip of its long tail, and it stood about 80 millimeters (three inches) tall on long, slender legs. Its skull had two temporal openings behind the eye, a characteristic diapsid synapomorphy. It had large eyes, a mouth full of small sharp teeth, and was apparently still an insectivore. The only other Carboniferous diapsid known is the Gzhelian species *Spinoaequalis schultzei*, also from Kansas. Curiously, this fossil exhibits morphological changes that indicate it was possibly an aquatic animal rather than fully terrestrial.[76]

The diapsids were to become the most successful clade of the sauropsids, giving rise to all the species of modern lepidosaurian lizards and snakes, archosaurian crocodiles and alligators, and the birds,

FIGURE 7.15 Reconstruction of the diapsid advanced sauropsid *Petrolacosaurus kansensis* from Kansas, North America. Note the even more gracile body and limbs compared to *Paleothyris acadiana* (fig. 7.14). The living animal was 400 millimeters (1.3 feet) long. *Credit*: Illustration by Kalliopi Monoyios.

living members of the great clade of the dinosaurs, most of whom perished in the end-Cretaceous mass extinction. In the Late Carboniferous and Early Permian, however, diapsids were rare, and they only diversified in the Late Permian. Even in the Late Permian, however, they were overshadowed by the other clade of the amniotes: the synapsids, our ancestors.

The earliest known body fossil of a synapsid amniote is the Bashkirian species *Protoclepsydrops haplous*, also from the famous Joggins tree-stump *Lagerstätte* in Nova Scotia.[77] Like the sauropsid *Hylonomus lyelli*, with which it coexisted and probably competed, it was a small, slender insectivore. About 10 million years younger, another synapsid species also found in Nova Scotia in the Morien Group tree-stump *Lagerstätte* is the Moscovian species *Archaeothyris florensis*. In contrast to the basal synapsid *Protoclepsydrops haplous, Archaeothyris florensis* is a highly derived ophiacontid (table 7.6). The existence of *Archaeothyris florensis* in the Moscovian means that three clades of more basal "pelycosaurs" (non-therapsid synapsids, a paraphyletic clade) must have evolved either in the Moscovian or the earlier Bashkirian Age, before the evolution of the clade of the ophiacodontids (table 7.6). These "pelycosaurs" include two ghost families, the Eothyrididae and the Caseidae, that do not appear in the fossil record until the Permian. Another "pelycosaur" family, the Varanopseidae, does appear later in the Late Carboniferous, represented by the Kasimovian species *Milosaurus mccordi* and the Gzhelian species *Archaeovenator hamiltonensis*.[78] The eothyridids and varanopsids were small but probably very aggressive predators; the eothyridids in particular had prominent fang-teeth. In contrast, the caseidids were herbivores with grinding teeth.[79] Thus both lineages of the amniotes, the sauropsids and the synapsids, independently evolved clades to fill the herbivore niche in Late Carboniferous terrestrial ecosystems. Including the diadectomorphs, ecological convergence in the evolution of herbivory occurred in no less than three separate clades of tetrapods during this critical period of terrestrial ecosystem evolution.

The Late Carboniferous ophiacodontid clade (fig. 7.16), represented by the Moscovian species *Archaeothyris florensis, Clepsydrops collettii,* and *Echinerpeton intermedium*, and the Gzhelian species *Stereorhachis dominans*,[80] were larger predators, ranging from 1.5 meters to 3 meters (five feet to ten feet) in length. These were clearly no longer insectivores, but rather carnivores that preyed on other vertebrates. The ophiacodontids were followed by three species of edaphosaurids and one species of sphenacodontids in the Late Carboniferous fossil record (table 7.6).[81] Both of these clades were sail-backed "pelycosaurs"; they possessed large, flat, vertical structures on their backs that looked like the sail on a single-sail sailboat (fig. 7.17). These morphological sails were created by vastly elongating the neural arches of vertebrae located along the animals' backbone, and the vertical spines of the sail were covered by skin tissue in life. It is believed that the surface area of these sails was a

FIGURE 7.16 Reconstruction of the ophiacodontid synapsid *Ophiacodon uniformis* from the Late Carboniferous and Permian of North America. Note the large size of the head relative to that of the body. The living animal was large, about 2.5 meters (eight feet) long. *Credit*: Illustration by Kalliopi Monoyios.

thermoregulatory structure, used by the animals to absorb the heat of the sun when they were cold and to shed heat to the surrounding air (in the shade, away from the sun) when they were too hot.[82] Curiously, although these sail-back synapsids were similar in superficial appearance, their ecologies were very different: the edaphosaurids were herbivores, and the sphenacodontids were carnivores.

FIGURE 7.17 Reconstruction of the sphenacodontid synapsid *Dimetrodon grandis* from the Early Permian of North America and Europe. Note the presence of the large sail-fin on the back. The living animal was very large, about three meters (ten feet) long. *Credit*: Illustration by Kalliopi Monoyios.

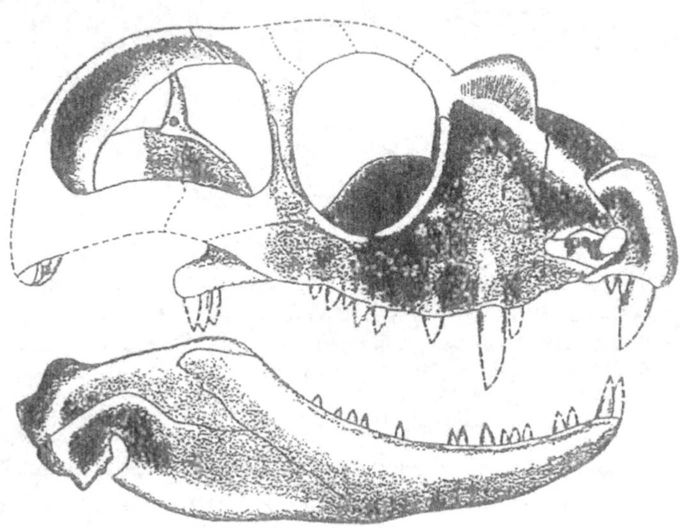

FIGURE 7.18 Skull of the Early Permian therapsid *Tetraceratops insignis* from Texas, North America. The skull is 90 millimeters (3.5 inches) long. *Credit*: Modified from M. J. Benton, *Vertebrate Palaeontology, Third Edition*, © 2005 Blackwell Science Ltd. Used with permission.

The therapsids were to become the most successful clade of the synapsids, giving rise to all of the monotreme, marsupial, and placental mammal species that live today (table 7.6). The most successful clade of the sauropsids, the diapsids, evolved in the Late Carboniferous. As far as we yet know, however, the synapsids did not produce the therapsid clade until the Early Permian: the basal-therapsid, biarmosuchian species *Tetraceratops insignis* (fig. 7.18) from Texas in North America. Advanced therapsids soon followed in the Middle Permian, and true mammals in the Late Triassic (fig. 7.19).

Some of the Late Carboniferous tetrapods were very large animals, like the diadectomorphs, ophiacodontids, edaphosaurids, and sphenacodontids, many of which reached three meters in length. This is in contrast to the basal batrachomorphs and reptiliomorphs of the Visean, most of which were only 300 millimeters (about one foot) in total length. It is possible that the large size attained by some tetrapods in the Late Carboniferous was due to the same environmental trigger that

FIGURE 7.19 Reconstructions of the heads of animals in the evolutionary sequence from a plesiomorphic ophiacodontid synapsid *Ophiacodon uniformis* (bottom) of the Late Carboniferous, to a derived sphenacodontid synapsid *Dimetrodon grandis* (second from bottom) of the Early Permian, to the gorgonopsian therapsid *Lycaenops ornatus* (third from bottom) of the Middle Permian, to the derived cynodont therapsid *Thrinaxodon liorhinus* (fourth from bottom) of the Early Triassic, to a true therian mammal (top) in the Late Triassic. *Credit*: From *Evolution: What the Fossils Say and Why It Matters*, by D. R. Prothero, copyright © 2007 Columbia University Press; figure by Carl Buell. Reprinted with permission.

produced gigantism in the insects during this same period of time: a Late Carboniferous atmosphere that contained much more oxygen in it than in the present-day atmosphere (fig. 4.2).

The Terrestrial Ecosystem Established

By the end of the Late Carboniferous the assembly of the structure of the terrestrial ecosystem was complete. The invaders had won—the hostile realm of dry land had been conquered.

Seed plants greened the hills and the highlands of the continents. They eventually penetrated even into the arid deserts of the terrestrial realm. They also now constituted the base of the trophic pyramid, the food source for numerous herbivorous insects and amniotes in the new ecosystem, who in turn constituted the food source for numerous carnivorous insects and amniotes. The spore-bearing plants remained behind, down in the marshes and swamps, and stretched along the banks of the rivers, still tied to the water.

Insect predators flew in the dry air of the skies. Below, insect herbivores attacked the seed plants for food. The long arms race between plants and insect herbivores began—plants evolving ever more complex chemical defenses, trying to poison their attackers, and insects evolving ever more complex chemical mechanisms to digest the plants, eventually enlisting the assistance of both bacterial and fungal commensal species.

Amniote predators hunted in the forests in the hills and highlands. Small carnivores preyed on the numerous insects they found in the plants around them. Large carnivores hunted species of their own kind, the vertebrates. Amniote herbivores ripped and chewed the plants, and the plants began to evolve defenses against these attackers as well—sharp thorns, spines, stinging nettles, and poison. The amphibians remained behind, down in the swamps, streams, and rivers, still laying their eggs in the water like fish.

After 3,530 million years,[83] life had finally left the aquatic world behind—but that exodus had not been easy. Twice, the vertebrate invasion had nearly failed. If it had indeed failed, the fish would have remained in the aquatic world, and the arthropods alone would have ruled the land.

CHAPTER 8

The Legacy of the Devonian Extinctions

The Effect of the End-Frasnian Bottleneck

At the very least, the End-Frasnian Bottleneck delayed the course of vertebrate evolution by some seven million years. It took that long for the harsh climates of the Famennian Gap to ameliorate, and for the Earth to once again become more hospitable to life.

However, the effect of that bottleneck did not merely delay vertebrate evolution; it also altered the direction of vertebrate evolution. Tetrapods suffered a sharp reduction in their previously morphologically diverse forms. In particular, the diversity of sizes seen in Frasnian tetrapods was lost, as only the mid-sized forms survived the bottleneck to re-diversify in the Famennian (fig. 3.4). What diverse Famennian novelties might our cousins in the other Frasnian tetrapod lineages have produced if they had survived the End-Frasnian Bottleneck?

The tetrapodomorph family tree was also sharply pruned. The clade of the elpistostegalian tetrapodomorphs was terminated (table 4.3). These animals had convergently evolved the tetrapod-like traits[1] of a flattened skull with dorsal orbits (fig. 3.1), a sutured dermal intracranial joint, paired frontal bones, and an enlarged endoskeletal shoulder girdle, and they had also lost their dorsal and anal lobe fins. What other tetrapod-like traits would they have continued to evolve in the

Famennian, in parallel with the true tetrapods, if they had survived the End-Frasnian Bottleneck?

In the rivers and lakes, and out in the shallow marine seas of Earth, the great armored fishes (color plates 7 and 9) and the ancient jawless fishes lost half of their diversity. These fishes were dealt a severe blow by the End-Frasnian Bottleneck and were left in a weakened condition in the Famennian. The next bottleneck proved to be fatal.

Out in the oceans, reefs changed forever. The massive Devonian-style skeletal reefs of stromatoporoid sponges, tabulate corals, and rugose corals were destroyed. The Earth has never again seen such huge reef tracts. It took 100 million years before major reef ecosystems again evolved in the seas of the Guadalupian Epoch of the Permian. These reefs were very different, composed instead of calcareous sponges and algae, with contributions to the baffling structure by branching bryozoan colonies and peculiar encrusting brachiopods. Only in the Anisian Age of the Middle Triassic, some 130 million years after the End-Frasnian Bottleneck, would the corals recover and once again become major reef-building organisms. At that time, the first of the scleractinian corals evolved,[2] the major type of corals that are found in modern coral reefs.

The Effect of the End-Famennian Bottleneck

The time delay in vertebrate evolution triggered by the End-Famennian Bottleneck was roughly twice that of the previous bottleneck. It took 14 million years for the harsh glacial–interglacial climatic cycles of the Tournaisian Gap to stabilize and for life on land to flourish once again.

This second Devonian bottleneck also altered the direction of vertebrate evolution, as had the first. The Famennian tetrapods had made major evolutionary advancements for total terrestriality: they had lost their fish-like dentitions and lateral-line systems, and they had abandoned their internal gills for exclusive air-breathing as they became more adapted to life on dry land. In contrast, and as a consequence of the End-Famennian Bottleneck, many of the peculiar tetrapod groups of the Early Carboniferous had abandoned dry land entirely and returned to an aquatic mode of life after the Tournaisian Gap (figs. 7.2 and 7.3).

Others had lost the characteristic synapomorphy itself—the possession of four limbs—and had evolved legless snake-like forms for burrowing underground. Only the whatcheeriids (fig. 7.1) represented an evolutionary advancement in the invasion of land during the Tournaisian Gap. Lauren Cole Sallan and Michael Coates suggest that even the number of fingers on our hands and toes on our feet reflect the effect of the End-Famennian Bottleneck: "Extinctions remove characters from the pool of varying morphologies. For example, digit number is known to be variable among late Famennian tetrapods but stabilizes with a maximum limit of five among all later forms,"[3] that is, after the End-Famennian Bottleneck. If the End-Famennian Bottleneck had not happened, would I be typing this page with seven fingers on each hand?

Even more drastic pruning of the tetrapodomorph family tree occurred in the second Devonian bottleneck. The clade of the tristichopterid tetrapodomorphs was terminated, leaving a large phylogenetic gap between the derived tetrapods and the basal rhizodont and megalichthyid tetrapodomorphs (table 6.3). The rhizodonts did not survive the Carboniferous, and the megalichthyids did not survive the Permian. By the end of the Paleozoic, only the tetrapod clade remained of the once diverse tetrapodomoph lineages.

In the related crossopterygian lobe-finned fishes, the clade of the porolepiform and onychodontid fishes were driven to extinction (table 6.3). The porolepiform crossopterygians had long coexisted with tetrapods, being found in the same environments. They did not survive into the Early Carboniferous. Only the actinistian crossopterygians survive to the present day, as seen in the living coelacanths. The surviving lobe-finned fishes within the sarcopterygian clade, the coelacanths and the lungfishes, are very rare today. As a result, a large phylogenetic gap exists today between limbed vertebrates and finned vertebrates. All of our intermediate cousins are gone, and this is why a modern-day lizard seems to us to be such a radically different animal from a trout.

The clade of the more distantly related great armored fishes (color plates 7 and 9) was totally eliminated after having lost half of its lineages in the End-Frasnian Bottleneck. Similarly, all of the jawless fish lineages between the basal petromyzontiform vertebrates and the gnathostome vertebrates were eliminated, leaving a large phylogenetic gap between modern fishes and their ancient jawless ancestral forms (table 6.3).

Some marine animals, such as the crinoids, or ancient sea lilies, actually benefited from the End-Famennian Bottleneck in fish diversity. These stalked, filter-feeding echinoderms had long suffered heavy predation by the Devonian fishes; once the Devonian fishes were eliminated, the crinoids experienced a blossoming of diversity in a classic example of ecological release.[4] Unlike the majority of animal groups during the Tournaisian Gap, the crinoids experienced their greatest diversity increase in the Paleozoic during this period of time. To echinoderm paleobiologists, the Early Carboniferous has become known as the "Age of Crinoids."[5]

Last, the End-Famennian Bottleneck in vertebrate evolution led to the replacement of the armored fishes and the lobe-finned fishes by the cartilaginous fishes and ray-finned fishes in the Early Carboniferous. In the Late Devonian fossil record, a total of 33 genera of chondrichthyan (cartilaginous) fishes are known, and a total of nine genera of actinopterygian (ray-finned) fishes are known (table 8.1). After the End-Famennian Bottleneck, a total of 96 genera of chondrichthyan fishes have been found in the fossil record of the Early Carboniferous, and a total of 71 genera of actinopterygian fishes (table 8.2). That is, the number of ray-finned fish genera in the Early Carboniferous seas of Earth increased by a factor of eight—almost an entire order of magnitude—and the number of genera of cartilaginous fishes tripled. The simultaneous demise of the placoderm and jawless fishes, and the explosive diversification of the cartilaginous and ray-finned fishes (table 8.2), was not coincidental. The Devonian "Age of Armored Fishes"[6] (color plates 7 and 9) had been replaced by the Carboniferous "Age of Sharks."[7]

What if the Devonian Extinctions Had Never Happened?

Our world is the product of the Devonian extinctions. Only one monophyletic group of four-limbed vertebrates exists on Earth: we, the tetrapods. The cartilaginous fishes and ray-finned fishes fill the seas of Earth; lobe-finned fishes are rare, armored fishes are nonexistent.

TABLE 8.1 Chondrichthyan and actinopterygian fish genera known from the fossil record of the Late Devonian, before the End-Famennian Bottleneck, are listed below.

Vertebrata (vertebrate animals)
– Petromyzontiformes (lampreys; living jawless fishes)
– unnamed clade
– – Pteraspidomorphi
– – unnamed clade
– – – Anaspida
– – – unnamed clade
– – – – Thelodonti
– – – – unnamed clade
– – – – – unnamed clade
– – – – – – Osteostraci
– – – – – – Galeaspida
– – – – – – Pituriaspida †Eifelian
– – – – – Gnathostomata (living jawed vertebrates)
– – – – – – Placodermi (armored fishes)
– – – – – – – Acanthothoraci
– – – – – – – unnamed clade
– – – – – – – – unnamed clade
– – – – – – – – – Rhenanida
– – – – – – – – – Antiarchi
– – – – – – – – unnamed clade
– – – – – – – – – Arthrodira
– – – – – – – – – unnamed clade
– – – – – – – – – – Petalichthyida
– – – – – – – – – – Ptyctodontida
– – – – – – unnamed clade
– – – – – – – Chondrichthyes (living cartilaginous fishes)
– – – – – – – – Late Devonian genera present: *Acmoniodus, Ageolodus, Amelacanthus,*
– – – – – – – – *Cladodus, Cladoselache, Cobelodus, Coronodus, Ctenacanthus, Denaea,*
– – – – – – – – *Diademodus, Dittodus, Eoorodus, Gladbachus, Goodrichthys, Helodus,*
– – – – – – – – *Hercynolepis, Jalodus, Manberodus, Monocladodus, Orodus, Phoebodus,*
– – – – – – – – *Plesioselachus, Prospiraxis, Protacrodus, Psammodus, Sandalodus,*
– – – – – – – – *Siamodus, Stethacanthus, Symmorium, Synthetodus, Tamiobatis, Thrinacodus,*
– – – – – – – – *Tuberospina.*
– – – – – – – unnamed clade
– – – – – – – – Acanthodii
– – – – – – – – Osteichthyes (living bony fishes + descendants)
– – – – – – – – – Actinopterygii (living ray-finned fishes)
– – – – – – – – – – Late Devonian genera present: *Cheirolepis, Gogosardinus, Kentuckia,*
– – – – – – – – – – *Limnomis, Mimia, Moythomasia, Osorioichthys, Stegotrachelus,*
– – – – – – – – – – *Tegeolepis.*
– – – – – – – – – Sarcopterygii (living lobe-finned fishes + descendants)

Source: Phylogenetic data modified from Sepkoski (2002), Benton (2005), Lecointre and Le Guyader (2006), and Sallan and Coates (2010).

† extinct taxa. Where a period is given, taxa went extinct in that specific geologic period.

TABLE 8.2 Chondrichthyan and actinopterygian fish genera known from the fossil record of the Early Carboniferous, after the End-Famennian Bottleneck, are listed below.

Vertebrata (vertebrate animals)
– Petromyzontiformes (lampreys; living jawless fishes)
– unnamed clade
– – Pteraspidomorphi **†Famennian**
– – unnamed clade
– – – Anaspida **†Famennian**
– – – unnamed clade
– – – – Thelodonti **†Frasnian**
– – – – unnamed clade
– – – – – unnamed clade **†Famennian**
– – – – – – Osteostraci **†Famennian**
– – – – – – Galeaspida **†Frasnian**
– – – – – – Pituriaspida †Eifelian
– – – – – Gnathostomata (living jawed vertebrates)
– – – – – – Placodermi **†Famennian** (armored fishes)
– – – – – – – Acanthothoraci **†Frasnian**
– – – – – – – unnamed clade **†Famennian**
– – – – – – – – unnamed clade **†Famennian**
– – – – – – – – – Rhenanida **†Frasnian**
– – – – – – – – – Antiarchi **†Famennian**
– – – – – – – – unnamed clade **†Famennian**
– – – – – – – – – Arthrodira **†Famennian**
– – – – – – – – – unnamed clade **†Famennian**
– – – – – – – – – – Petalichthyida **†Frasnian**
– – – – – – – – – – Ptyctodontida **†Famennian**
– – – – – – unnamed clade
– – – – – – – **Chondrichthyes (living cartilaginous fishes)**
– – – – – – – – **Early Carboniferous genera present:** *Acandylacanthus, Agassizodus,*
– – – – – – – – *Ageleodus, Akmonistion, Anachronistes, Anaclitacanthus, Antliodus,*
– – – – – – – – *Asteroptychius, Belantsea, Bellbonn, Campodus, Chomatodus,*
– – – – – – – – *Chondrenchelys, Cladodus, Coelosteus, Copodus, Cratoselache,*
– – – – – – – – *Ctenacanthus, Ctenopetalus, Cynopodius, Damocles, Debeerius,*
– – – – – – – – *Delphyodontos, Deltoptychius, Denaea, Desmiodus, Dicentrodus, Dicrenodus,*
– – – – – – – – *Diplodocanthus, Diplodoselache, Diplodus, Echinochimaera, Edestus,*
– – – – – – – – *Eucentrurus, Euctenius, Euctenodopsis, Euglossodus, Euphyacanthus,*
– – – – – – – – *Falcatus, Fissodus, Glyphanodus, Goodrichthys, Gregorius, Harpacanthus,*
– – – – – – – – *Harpacodus, Harpagofututor, Helodus, Heteropetalus, Hybocladodus,*
– – – – – – – – *Janassa, Lagarodus, Lambdodus, Leiodus, Lisgodus, Lissodus, Listracanthus,*
– – – – – – – – *Lophosteus, Marracanthus, Mazodus, Mesodmodus, Mesolophodus,*
– – – – – – – – *Moyacanthus, Netsepoye, Onychoselache, Orestiacanthus, Orodus,*
– – – – – – – – *Papillionichthys, Peripristis, Petalodus, Petalorhynchus, Petrodus,*
– – – – – – – – *Phoebodus, Physonemus, Platyxystrodus, Pleuracanthus, Poecilodus,*

TABLE 8.2 (*continued*)

--------- *Polyrhizodus, Pristodus, Psammodus, Psephodus, Rainerichthys, Serratodus,*
--------- *Sikisika, Sphenacanthus, Squatactinus, Squatinactis, Srianta, Stethacanthus,*
--------- *Symmorium, Tamiobatis, Tanaodus, Thrinacoselache, Traquairius,*
--------- *Tristychius, Venustodus, Xenacanthus.*
-------- unnamed clade
--------- Acanthodii
--------- Osteichthyes (living bony fishes + descendants)
---------- **Actinopterygii (living ray-finned fishes)**
----------- **Early Carboniferous genera present:** *Acrolepis, Adroichthys,*
----------- *Aesopichthys, Aestaurichthys, Aetheretmon, Australichthys, Beagiascus,*
----------- *Bendenius, Bishofia, Blairolepis, Borichthys, Brachypareion, Canobius,*
----------- *Cheirodopsis, Chirodus, Coccocephalichthys, Cornuboniscus,*
----------- *Cosmoptychius, Cryphiolepis, Cycloptychius, Cyranorhis, Discoserra,*
----------- *Drydenius, Dwykia, Elonichthys, Eurynotus, Frederichthys, Ganolepis,*
----------- *Gonatodus, Grassator, Guildayichthys, Gyrolepidotus, Holurus, Kalops,*
----------- *Kentuckia, Lineagruan, Mansfieldiscus, Melanecta, Mentzichthys,*
----------- *Mesolepis, Mesonichthys, Mesopoma, Ministrella, Nematoptychius,*
----------- *Novogonatudus, Oxypteriscus, Palaeobergia, Palaeoniscus,*
----------- *Paradrydenius, Paragonatodus, Paramesolepis, Paratarrasius,*
----------- *Phanerosteon, Platysomus, Proceramala, Protamblyptera,*
----------- *Protoeurynotus, Protohaplolepis, Pseudogonatodus, Rhadinichthys,*
----------- *Senekichthys, Soetendalichthys, Strepeoschema, Styracopterus,*
----------- *Sundayichthys, Tarrasius, Wardichthys, Watsonichthys, Wendyichthys,*
----------- *Whiteichthys, Willomoricthys.*
---------- Sarcopterygii (living lobe-finned fishes + descendants)

Source: Phylogenetic data modified from Sepkoski (2002), Benton (2005), Lecointre and Le Guyader (2006), and Sallan and Coates (2010).

† extinct taxa. Where a period is given, taxa went extinct in that specific geologic period. Fish groups that perished in the end-Frasnian extinction are marked **†Frasnian.** Fish groups that perished in the Famennian are marked **†Fammenian.**

These two conditions are the result of the Devonian extinctions—but what if the Devonian extinctions had never happened?

Convergent evolution is a very common phenomenon on Earth.[8] If the elpistostegalian fish had not perished in the end-Frasnian extinction, might tetrapods—vertebrate animals with four limbs—have evolved twice? Would they have evolved convergently? As we saw in chapter 3, for many years it was thought that the tetrapods evolved in the Frasnian from the elpistostegalian lineage of *Panderichthys rhombolepis* to *Elpistostege watsoni* to *Tiktaalik roseae* to *Livoniana multidentata.* It came as quite a surprise when the fossil trackways in Poland proved that tetrapods had already evolved in the early Eifelian, long before the Frasnian. Thus all of the tetrapod-like morphologies seen in the elpistostegalian lineage (fig. 3.1) evolved in parallel, convergently with morphologies independently evolved by the tetrapods.

The idea of two separate groups of animals coexisting and running around on the Earth today, each possessing four limbs but not descending from a single common four-legged ancestor, might at first seem fanciful if not ridiculous, yet that is exactly the situation that we have today with birds and bats. Each group has forelimbs that have been modified into wings, but birds and bats do not stem from a single winged ancestor. Birds independently evolved wings within the theropod dinosaur lineage, and bats independently evolved wings within the placental mammal lineage. Birds are sauropsids; bats are synapsids. You have to go all the way back into the Late Carboniferous to find a common amniote ancestor for these two lineages, as discussed in chapter 7 (table 7.6). In reality, the vertebrate animals convergently evolved wings *three separate times*, not just two.[9] The ancient pterosaurs also independently evolved wings, thus in the Jurassic and Cretaceous skies two groups of flying vertebrates coexisted—pterosaurs and birds—that both had wings but that did not evolve from a single common winged ancestor.

As we saw in chapter 3, four separate lineages of the tetrapodomorphs evolved very similar large-predator morphologies in parallel during the Givetian and Late Devonian: the rhizodonts, the tristichopterids, the elpistostegalians, and the tetrapods. Of these four lineages, only the tetrapods survive today. The elpistostegalians did not survive the End-Frasnian Bottleneck (table 4.3), and the tristichopterids did not survive the End-Famennian Bottleneck. (table 6.3). If those two bottlenecks in vertebrate evolution had not occurred, might further parallel evolution in both the elpistostegalian and tristichopterid lineages have produced animals with four limbs? That is, could the condition of possessing four limbs have evolved independently in three separate vertebrate lineages, just as the condition of possessing forelimbs modified into wings evolved independently in the three separate vertebrate lineages of pterosaurs, birds, and bats?

If having four legs was a convergent trait instead of a synapomorphy, then obviously we would have to invent separate names for each of the convergent vertebrate lineages that possessed four limbs. Tetrapodo-tristichopteridae, Tetrapodo-elpistostegalia, and Tetrapodo-tetrapoda would be a bit cumbersome; we do not use Ptero-dinosauria and Ptero-mammalia for the birds and bats that convergently possess *ptera* (wings)—although we do say Pterosauria for the reptiles with wings! In a world where the Devonian extinctions had never occurred, perhaps we, with our characteristic modesty, might have called ourselves the Eutetrapoda, the "true four-limbed vertebrates." Our cousins the elpistostegalian tetrapods we might have called the Paratetrapoda, the "near, or similar four-limbed vertebrates," and our more distant tristichopterid tetrapod relatives the Pseudotetrapoda, the "false four-limbed vertebrates."

What if the lobe-finned fishes (fig. 3.1) were common today, and the sharks and ray-finned fishes were rare? There would no longer be a large morphological gap between the tetrapods, the limbed vertebrates, and the fishes. The idea of a "fish with legs" (fig. 3.1) would be much less unusual than the much more bizarre development of modern-day frogs, where what hatches out of a frog egg does not look at all like a frog. A juvenile frog, a tadpole, looks like a fish. It has no legs at all. At least the lobe-finned fishes, with their fleshy and stumpy lobed fins, look much more like a limbed vertebrate than a tadpole does. The idea that we evolved from fish might not seem so strange to people if every time they bought a fish at the supermarket that fish possessed four stumpy limb-like fins, complete with internal bones.

What if air-breathing, lobe-finned lungfishes were numerous today, instead of rare? Only three families of lungfish exist in the modern world, living in widely separated geographic regions of the present-day Earth. These families are the Neoceratodontidae in Australia, the Lepidosirenidae in South America, and the Protopteridae in Africa. Curiously, the ray-finned fishes also convergently evolved the capability of locomoting on land by crawling with their pectoral fins, although it is a rare trait in this group. The three groups of land-crawling, air-breathing ray-finned fishes are the mudskippers,[10] the climbing perch,[11] and the walking catfish (*Clarias batrachus*).[12] As their name suggests, the climbing perch can actually climb up into the branches of shrubs in search of food! The walking catfish (*Clarias batrachus*) is the most impressive of the actinopterygian fishes that have convergently evolved the sarcopterygian-like ability to breathe air and locomote on land using their pectoral fins. This fish is over 300 millimeters (about a foot) long and is native to Asia and Polynesia. It has been introduced into Florida, where it is now an invasive species pest. Walking catfish will leave streams and rivers after rainfalls, when the land is wet, and walk considerable distances to ponds stocked with farmed fish, which they then proceed to eat with gusto. Fish farmers in Florida have actually been forced to build low fences around their fish ponds to keep the walking catfish out! The walking catfish is such a pest that the fish farmers of Florida have successfully lobbied to have the possession of these fishes banned by law.

What if walking, air-breathing fish were not rare curiosities like the ray-finned mudskippers, climbing perch, and walking catfish? What if the Devonian extinctions had not brutally pruned the branches of the family tree of the lobe-finned lungfishes, and these fishes could be found today in most rivers and streams? It is not just the lobe-finned lungfishes—in the Devonian, multiple fish lineages were convergently evolving enlarged and advanced lung systems to breathe more air. For example, fossil species of the placoderm genus *Bothriolepis* (color plate 9) are frequently found with our ancestors, the tetrapodomorph fishes, in the Late Devonian (as discussed in chapters 3 and 5). We coexisted with one another, and these armored fishes were subject to the same environmental pressures in anoxic river waters to develop more efficient air breathing as the tetrapodomorph fishes were, particularly during the Frasnian with its low oxygen content in the atmosphere itself, as discussed in chapter 4. And, as one might predict, some fossils of *Bothriolepis* species show spatial evidence inside their dermal armor of the convergent evolution of lungs.[13] What if it were not at all unusual to go down to a river and to see several different types of fishes crawling about on the river banks and breathing air? If, like common pet turtles in aquaria, modern-day children kept air-breathing fish as pets at home, would their parents still be skeptical that land vertebrates are the descendants of fish?

What if the great armored-fish predators (color plate 7) had never gone extinct? Might the seas of Earth still be populated by giant dinichthyids and titanichthyids, the "terrible fishes" and "titanic fishes," that were seven meters (23 feet) long, with massive bony head shields and jaws like a snapping turtle? Would our present-day toothy predators like the great white shark have ever evolved if these placoderm predators had survived the Devonian bottlenecks in vertebrate evolution? Would movie producers instead be producing horror films of people being devoured by terrible armored fishes instead of great white sharks?

The Next Invasion . . .

At the close of chapter 7, we noted that it took 3,530 million years for life to finally leave the aquatic world behind and to successfully invade the terrestrial world. After another 300 million years of successful life on land, in the year 1969, individuals of the synapsid amniote species *Homo sapiens* left the terrestrial world itself behind and traveled across the sea of empty space to the Moon.

The next goal for life is in sight: the planet Mars. Mars is the only planet in our solar system that human beings could colonize, the only planet on which we could flourish. Our sister planet sunwards, Venus, is similar to the Earth in size and surface gravity, but there all similarities between the two planets end. The thick carbon dioxide atmosphere of Venus is about 93 times as dense as the atmosphere of the Earth at sea level.[14] The greenhouse effect of such a dense carbon dioxide atmosphere, plus the fact that Venus is closer to the heat of the sun than Earth, combine to make the surface of Venus the hottest in the entire solar system—even hotter than on the planet Mercury, which is much closer to the sun than Venus. Under the thick gloom of the perpetual clouds, the average surface temperature on the planet is 464°C (867°F), temperatures at which the metal lead will melt and trickle downhill like water.[15] We never could live, much less flourish, under the crushing pressure and hellish temperatures present on the surface of Venus.

Closer to home, we could establish research stations on the Moon, but we could not establish flourishing colonies of humans. The Moon is too alien, too hostile, too small. A single "day" on the Moon lasts for almost two Earth weeks, during which the land is exposed constantly to the unshielded, blowtorch-like intensity of radiation from the sun.

Likewise, a single "night" on the Moon lasts for over 13 Earth-nights, during which the land is exposed to the intense cold of outer space and with the only light coming from the stars, which do not twinkle but shine like points of ice in the almost total vacuum. The Moon is in gravitational lock with Earth, such that one side is perpetually facing our planet, producing these long day and night phases as the Moon orbits the Earth relative to the sun. Thus only the polar regions of the Moon receive constant sunlight, and in these regions we could grow agricultural crops in sealed greenhouses. Our robot probes to the Moon have proved that considerable amounts of water also exist frozen in the bottom of craters in the polar regions, bottoms that are never exposed to light from the sun. In the polar regions, we could build permanent research stations, but not colonies. Because of its small size, gravity is much lower on the surface of the Moon—only 16 percent of that of the Earth. In the last century, when humans traveled to the Moon, our lunar explorers discovered that they bounced off the ground with each step. Prolonged stays in such near-weightless conditions would create serious health problems: our bones would become brittle, and our muscles would atrophy and fail. Thus a research station on the Moon might be an interesting, truly out-of-this-world place to visit and work, but we could not live there permanently. We could not have children there, and watch them grow and prosper as conquerors of a new world.

In contrast to Venus or the Moon, Mars is much more like Earth. Gravity on the surface of Mars is 38 percent that of the Earth. Water is very abundant and widespread on Mars, from the vast polar ice caps to frozen groundwater in the soil, even in the low latitudes of the planet. Mars has an atmosphere of carbon dioxide, which is perfect plant food for growing our crops in sealed greenhouses. We could even use the abundant carbon dioxide to make cheap rocket fuel for shuttle craft to travel from colony to colony on the planet.

Esthetically, Mars would not seem so alien to us. Unlike the perpetually jet-black skies and blindingly bright ash-colored landscapes of the barren Moon, the sky of Mars is orangey-beige, and its desert landscape of rocks and sand dunes is a riot of reds, oranges, chocolate browns, and blacks. Unlike the approximately two weeks that each Moon day lasts, the length of the day on Mars is almost exactly the same as that on

Earth: 24 hours and 39 minutes, as opposed to 24 hours. Each day, we could watch beautiful, blue-tinted Martian sunrises and sunsets (color plate 12). The axial tilt of Mars is similar to that of the Earth as well: 25° as opposed to 23°. Thus Mars has four seasons, just as the Earth does, and we could mark the passage of time through winter, spring, summer, and fall; each season lasts twice as long on Mars, however, as the Martian year is almost twice as long as an Earth year. Our robot rovers on Mars have photographed cirrus clouds moving across the skies of Mars, just like the high-altitude, herring-bone ice clouds we see on Earth, giving us a feel for what it would be like to actually be there ourselves. These same robot probes have photographed Martian dust devils, just like those we see in our desert regions of Earth, tracking across the rocky landscapes. Although Mars is a smaller planet than the Earth, it presents the explorer with even grander vistas: Mars has the tallest mountains in the solar system, and the largest canyons.[16]

In chapter 1, we saw that four particularly serious problems had to be solved before life could leave the aquatic world and invade the hostile realm of the terrestrial world: dehydration, gravity, temperature fluctuations, and radiation poisoning. Curiously, invading the Martian world will involve solving very similar problems: Mars is an arid desert world (in that water at its surface is not liquid), its gravity is different from the Earth, it is a colder world than the Earth, and the radiation flux at its surface is more like that of the ancient Earth before the plants created the protective ozone layer of the Earth's atmosphere.

Life succeeded in invading the hostile terrestrial world of the Earth. It did not fail, even given the serious setbacks of the Devonian extinctions. Will a member of one clade of these conquerors, the synapsid amniote *Homo sapiens*, succeed in the invasion of the hostile Martian world, be victorious in bringing plant and animal life to our sister planet? Or will it fail? Time will tell . . .

Notes

Chapter 1: The Evolution of Life on Land

1. The "Ma" in table 1.1 is an abbreviation for the Latin word *mega-annum*, which translates as "one million years" or 10^6 years in English.

2. The Swedish inventor Alfred Nobel established no prize in evolutionary biology, and in general the Nobel Prize was to be awarded for "practical" results, thus favoring empiricists over theoreticians. To compensate for this glaring bias, the Swedish industrialist Holger Crafoord and his wife Anna-Greta established the Crafoord Prize, which can be awarded to evolutionary and theoretical biologists.

3. Hence the Crafoord Prize for Carl Woese.

4. See discussion in Lecointre and Le Guyader (2006).

5. An α-proteobacterium; Ibid.

6. Convergent evolution has occurred frequently in the history of life; see discussion in McGhee (2011).

7. See discussion in Mojzsis et al. (1996), Buick (2001).

8. $^{12}CO_2 + H_2O \rightarrow {}^{12}CH_2O + O_2\uparrow$.

9. Proteobacteria: the purple non-sulfur bacteria are α-proteobacteria, and the purple sulfur bacteria are γ-proteobacteria. Photosynthetic anaerobic sulfur bacteria use hydrogen sulfide rather than hydrogen hydroxide (water) in photosynthesis ($^{12}CO_2 + 2H_2S \rightarrow {}^{12}CH_2O + H_2O + 2S$) and produce sulfur rather than oxygen as a waste product.

10. The green sulfur bacteria and the green non-sulfur bacteria.

11. Phylogenetic lineage Bacteria: Firmicutes: Clostridia: Clostridiales: Heliobacteriaceae.

12. See discussion in Xiong et al. (2000).

13. See discussion in Dauphas et al. (2004).

14. See discussion in Schopf (1993), Hofmann et al. (1999), Lecointre and Le Guyader (2006).

15. See discussion in McKay et al. (1996).

16. See discussion in Hartmann (2003:262–265).

17. See "A Recipe Book of Life" in chapter 4 of Ward (2005).

18. See discussion in Mojzsis et al. (1996).

19. See discussion in Hartmann (2003:262–265).

20. See discussion in Schopf (1993), Hofmann et al. (1999).

21. Thus the oldest cyanobacteria are cited as 3.4 billion years old in the phylogenetic classification of life in Lecointre and Le Guyader (2006:69).

22. See discussion in Brocks et al. (1999).

23. 0.2 percent to 1.0 percent of the atmosphere's present 21 percent level of oxygen; see discussion in Lane (2002).

24. The atmosphere is now almost 0.039 percent carbon dioxide and steadily rising, due to the burning of fossil hydrocarbons.

25. See discussion in Lane (2002).

26. $CH_4 + 2O_2 \rightarrow CO_2 + 2H_2O$.

27. 5 percent to 18 percent of the atmosphere's present 21 percent level of oxygen; see discussion in Lane (2002).

28. See discussion in Lecointre and Le Guyader (2006).

29. See discussion in Prave (2002).

30. See discussion in Labandeira (2005).

31. Ibid.

32. See discussion in Yuan et al. (2005).

33. Ibid.

34. See discussion in Redecker et al. (2000).

35. Prasinophyte green-algal fossils similar to *Pterosperma* have been dated to 1.2 billion years old; see discussion in Lecointre and Le Guyader (2006).

36. See figure 3 in Erwin et al. (2011).

37. See discussion in Erwin (2001).

38. *Phaneros*, φανερός, is the Greek word for visible, and *zoion*, ζωον, is the Greek word for animal.

39. See discussion in Lutz and Rhoads (1977), Babarro and De Zwaan (2008).

40. See discussion in Berner (2006). We will consider these models in more detail in chapter 3.

41. $O + O_2 \rightarrow O_3$.

42. See discussion in Labandeira (2005), Strother et al. (2010), Versteegh and Riboulleau (2010).

43. See discussion in Labandeira (2005).

44. See discussion in MacNaughton et al. (2002).
45. See discussion in Grinspoon (2003:121).
46. See discussion in Kessler and Müller (1988).
47. See discussion in Cowen (2000).
48. See discussion in Grinspoon (2003:146).
49. See discussion in McKay (1996a, 1996b), Zubrin and Wagner (2011:342–344).

Chapter 2: The Plants Establish a Beachhead

1. See discussion in McGhee (2011).
2. See discussion in Lecointre and Le Guyader (2006).
3. See discussion in Labandeira (2005).
4. The Late Carboniferous in table 2.2 is sometimes referred to as the Pennsylvanian, particularly in North America, and the Early Carboniferous is sometimes referred to as the Mississippian.
5. Late Arenig Age in older timescales.
6. See discussion in Rubinstein et al. (2010), Wellman (2010).
7. See discussion in Wellman et al. (2003), Wellman (2010).
8. See discussion in Rubinstein et al. (2010), Wellman (2010).
9. Caradoc Age in older timescales.
10. See discussion in Wellman et al. (2003).
11. See discussion in Wellman (2003, 2010).
12. See discussion in Lecointre and Le Guyader (2006).
13. Ashgill Age in older timescales.
14. See discussion in Lecointre and Le Guyader (2006).
15. See discussion in Steemans et al. (2010).
16. See discussion in Redecker et al. (2000).
17. See discussion in Taylor and Taylor (1993).
18. See discussion in Redecker et al. (2000).
19. See discussion in Taylor et al. (1995).
20. See discussion in Edwards and Wellman (2001), Steemans et al. (2010).
21. See discussion in Raymond et al. (2006).
22. See discussion in Darwin (1859, Chapter IX).
23. See discussion in Edwards and Wellman (2001).
24. See discussion in Berry and Fairon-Demaret (2001).
25. See discussion in Taylor et al. (1995), Taylor and Osborn (1996).
26. See discussion in Edwards and Wellman (2001), Steemans et al. (2010).
27. See discussion in Edwards and Wellman (2001).
28. Ibid.
29. See discussion in Kenrick and Crane (1997a, 1997b).

30. See discussion in McGhee (2011).

31. See discussion in Stein et al. (2007).

32. See discussion in Stein et al. (2012).

33. See discussion in Meyer-Berthaud et al. (2010).

34. See discussion in Stein et al. (2012).

35. See discussion in McGhee (2011).

36. See discussion in Gerrienne et al. (2010).

37. The word "Frasnian" is pronounced as if it were spelled "Franian"; that is, the "s" is silent. The pronunciation is French, and comes from the *Assise de Frasnes* boundary strata in Belgium.

38. In the late **VCo** spore zone, or the Fa2c chronozone of older timescales. The Famennian spore zonation timescale will be discussed in detail in chapter 5; see tables 5.1 and 5.2.

39. See discussion in Algeo et al. (2001).

40. See discussion in Prestianni and Gerrienne (2010).

41. Ibid.

42. See discussion in DiMichele and Hook (1992).

Chapter 3: The First Animal Invasion

1. The phylogenetic status of the Placozoa, of which there is only one species known, is still uncertain; they may be simplified eumetazoans, i.e., their simplicity of form may be a secondary characteristic, not primary. See discussion in Lecointre and Le Guyader (2006).

2. See discussion in Lecointre and Le Guyader (2006).

3. See discussion in Pritchard et al. (1993).

4. In addition, many arachnids have evolved tiny book lungs to assist their tracheal breathing.

5. See discussion in McGhee (2011).

6. Caradoc Age in older timescales.

7. See discussion in Draganits et al. (2001).

8. See discussion in Shear and Edgecombe (2010).

9. See discussion in Lecointre and Le Guyader (2006).

10. See discussion in Dunlop (2010).

11. See discussion in Wilson and Anderson (2004).

12. See discussion in Wilson (2006).

13. See discussion in Shear and Edgecombe (2010) and Shear and Selden (2001), respectively.

14. See discussion in Dunlop (2010).

15. See discussion in Lecointre and Le Guyader (2006).

16. See discussion in Dunlop (2010).

17. See discussion in Shear and Selden (2001) and Engel and Grimaldi (2004), respectively.

18. This insect species has been disputed, however. It is undoubtedly a bristletail, but it may be a recent contaminant and thus not of Emsian age; see discussion in Shear and Selden (2001, 34).

19. See discussion in Shear and Edgecombe (2010).

20. See discussion in Dunlop (2010).

21. See discussion in Shear and Edgecombe (2010).

22. See discussion in Dunlop (2010).

23. See discussion in Shear and Selden (2001).

24. Ibid.

25. Ibid.

26. See discussion in Brainerd (1994) Duncker (2004).

27. See discussion in Brainerd (1994).

28. Some fish, however, have convergently evolved both internal fertilization and viviparity; that is, they give live birth. See discussion in McGhee (2011).

29. See discussion in Lecointre and Le Guyader (2006).

30. They are known only from fossil scales, which appear to be acanthodian; see discussion in Lecointre and Le Guyader (2006).

31. See discussion in Zhu et al. (2009).

32. See discussion in Shubin et al. (1997), Clack (2009).

33. See discussion in Zhu et al. (2009).

34. See discussion in Lecointre and Le Guyader (2006).

35. Clack (2002, 2012) and George and Blieck (2011) consider the lungfishes to be more closely related to the tetrapodomorphs than the crossopterygians. Benton (2005) considers the crossopterygians to be more related to the tetrapodomorphs and the lungfishes to be basal sarcopterygians. Zhu et al. (2009) considers the crossopterygians to be basal (and paraphyletic) and the lungfishes to be more derived than the basal tetrapodomorphs.

36. See discussion in Niedźwiedzki et al. (2010).

37. See discussion in Clack (2002, 2012).

38. See discussion in Niedźwiedzki et al. (2010).

39. Ibid.

40. See discussion in Carroll (2009, 44).

41. See discussion in Janvier and Clément (2010, 41).

42. See discussion in Long et al. (2006, 199 and 201, respectively). For an opposing point of view, see discussion in Snitting (2008).

43. See discussion in Long et al. (2006, 199).

44. Ibid. (201).

45. See discussion in Janvier and Clément (2010, 41).

46. See discussion in Ahlberg and Johanson (1998).

47. See discussion in Zhu et al. (2009).
48. See discussion in McGhee (2011).
49. See discussion in Ahlberg and Johanson (1998, 792–794).
50. See discussion in Blieck et al. (2007).
51. Ibid.
52. Ibid.
53. Ibid.
54. Ibid.
55. See discussion in Ahlberg et al. (2008), Clack et al. (2012).
56. See discussion in Ahlberg et al. (2008, 1203).
57. See discussion in Ahlberg et al. (2008), Clack et al. (2012).
58. See discussion in Ahlberg (1995).
59. See figure 4 in Niedźwiedzki et al. (2010).
60. See discussion in Stössel (1995).
61. See discussion in King et al. (2011).
62. See discussion in Ahlberg (1995, 1998).
63. Ibid.
64. See discussion in Ahlberg (1991, 1995), Clack (2002, 2012).
65. See discussion in Ahlberg (1995).
66. See discussion in Zhu et al. (2002).
67. See discussion in Daeschler (2000).
68. See discussion in Ahlberg et al. (2008).
69. See discussion in Campbell and Bell (1977).
70. See discussion in Niedźwiedzki et al. (2010).
71. See discussion in Stössel (1995).
72. The discovery of a Scottish *Holoptychius* fossil jaw, with a length of 470 millimeters (18.5 inches), is reported by Hugh Miller in his classic Devonian book, *The Old Red Sandstone* (1858, 254–255).
73. See discussion in Ahlberg (1998).
74. See discussion in Zhu et al. (2002).
75. See discussion in Young (2010).
76. See discussion in Campbell and Bell (1977).
77. See discussion in Young (2010).
78. Daeschler, 2000; George and Blieck, 2011; for an analysis of the osmotic tolerances of the early tetrapods, see discussion in Laurin and Soler-Gijón (2010).

Chapter 4: The First Catastrophe and Retreat

1. The end-Frasnian biodiversity crisis is also referred to as the "Late" Devonian biodiversity crisis because it took place *within* the Late Devonian Epoch; see discussion in McGhee (1996).

2. Curiously, this similarity in proportional diversity loss in the two fish groups extends even to the species level—both groups lost a bit more than half of their species—even though the placoderms were an order of magnitude more diverse than the acanthodians. See discussion in McGhee (1996, 123–125).

3. See discussion in Copper (1994).

4. For an extensive list of the victims of the end-Frasnian extinction, see discussion in McGhee (1996).

5. Ibid.; see discussion in Bond (2006).

6. See discussion in McGhee (1996).

7. See discussion in Schindler (1990, 1993), McGhee (1996, 2001a).

8. Upper *gigas* Zone in older zonal timescales.

9. Uppermost *gigas* Zone in older zonal timescales. See discussion in Streel et al. (2000, fig. 9).

10. In the Sonyea Group, the very youngest part of which still falls in the *punctata* conodont zone. See discussion in Raymond and Metz (1995).

11. In the West Falls Group; see discussion in Raymond and Metz (1995). The oldest strata containing the minimum diversity *torquata-gracilis* spore assemblage in the West Falls Group is the lower Hanover Shale, which dates to the Late *rhenana* conodont zone; see discussion in Streel (2000, fig. 12).

12. Clack (2012, 112).

13. See discussion in McGhee (1996, 86–88).

14. Chronozones Fa1a, Fa1b, and Fa2a in older timescales.

15. See discussion in Streel et al. (2000, 128), Blieck et al. (2010, fig. 2).

16. See discussion in Poty (1999, 11).

17. See discussion in Averbuch et al. (2005).

18. Ibid.

19. Ibid.

20. See discussion in Raymo and Ruddiman (1992).

21. A maximum measured value of $^{87}Sr/^{86}Sr = 0.070842$ is reported in this interval by Averbuch et al. (2005).

22. See discussion in John et al. (2010).

23. See discussion in Courtillot and Renne (2003), Courtillot et al. (2010).

24. See discussion in Racki (2005).

25. See discussion in Scott and Glasspool (2006, 10863).

26. Ibid. (10862, 10864).

27. See discussion in Cressler (2001). The strata are also dated to the **VH** spore zone (see table 5.2 in the next chapter) and the Fa2c chronozone of older timescales. The charcoal is found in the same Red Hill outcrop where two Famennian tetrapod species are found, as will be discussed in chapter 5.

28. See discussion in Scott and Glasspool (2006, fig. 1), Marynowski and Filipiak (2007), Marynowski et al. (2010).

29. See discussion in Scott and Glasspool (2006).

30. Again, assuming a Famennian conodont zone represents approximately 0.7 million years.

31. See discussion in Berner et al. (2003), Berner (2006).

32. See discussion in Scott and Glasspool (2006, 10861).

33. Ibid. (10862).

34. Ibid. (10863).

35. See discussion in Algeo and Ingall (2007).

36. See discussion in Raymond and Metz (1995).

37. See discussion in DiMichele and Hook (1992).

38. Ibid.

39. See discussion in Brezinski et al. (2010).

40. See discussion in Hartkopf-Fröder et al. (2007), Marynowski et al. (2008).

41. See discussion in Schieber (2009).

42. See discussion in Marynowski et al. (2010).

43. See discussion in John et al. (2010).

44. See discussion in Marynowski et al. (2008).

45. See discussion in Rimmer et al. (2004).

46. See discussion in McGhee et al. (1986), McGhee (1996, 218–224).

47. See discussion in McGhee (1996, fig. 7.11).

48. See discussion in Stein et al. (2007).

49. See discussion in Algeo et al. (2001).

50. See discussion in Goddéris and Joachimski (2004).

51. See discussion in Murphy et al. (2000), Sageman et al. (2003).

52. See discussion in Racki and Wignall (2001).

53. See discussion in Murphy et al. (2001).

54. See discussion in Riquier et al. (2006).

55. See discussion in Bond and Wignall (2005).

56. See discussion in Riquier et al. (2006).

57. Paleopedology is the scientific study of fossil soil horizons, not fossil children. I am not making this up!

58. See discussion in Retallack et al. (2009).

59. Ibid. (fig. 1D).

60. See discussion in Streel et al. (2000).

61. Ibid. (147); see also Raymond and Metz (1995, fig. 2).

62. See discussion in Streel et al. (2000, 128).

63. Ibid. (148).

64. Four new species of tetrapods appeared in the fossil record at the same time as the land plants recovered their diversity, as will be discussed in chapter 5.

65. See discussion in McGhee (1996, 115–121).

66. Ibid. (121–123).

67. See discussion in Averbuch et al. (2005).

68. See discussion in Brand (1989).

69. See discussion in Veizer et al. (1999).

70. See discussion in Joachimski and Buggisch (2002, fig. 2).

71. Ibid.

72. See discussion in Joachimski et al. (2004, 2009).

73. Michael Joachimski personal communication, 10 July 2012.

74. I have plotted only the average sea surface temperature values in the trend line in the figure, rather than plotting the huge cloud of data points given in the original figure in Joachimski et al. (2009, fig. 7).

75. See discussion in Joachimski et al. (2009).

76. See discussion in Gong and Xu (2003), Joachimski and Buggisch (2003); see also Racki (2005). Joachimski et al. (2009) argue that the distribution of microbial reefs in time is paleobiological support for the conodont oxygen isotope temperature curve: microbial reefs were common in the Early Devonian Pragian and the Late Devonian Famennian, both interpreted as times of very warm sea-surface temperatures (low $\delta^{18}O$ ratio values in conodont skeletal apatite). However, this argument is circular because it is the interpretation of the low $\delta^{18}O$ ratios as periods of hot temperatures that is in question; that is, the data simply demonstrate a correlation between microbial reef diversity and low $\delta^{18}O$ ratios in the fossil record. Independent biological evidence of the preference of microbial reefs for hot-water conditions is needed to support the argument that the correlated low $\delta^{18}O$ ratios do indeed represent hot-water conditions rather than sea-water salinity conditions.

77. See the discussion in Racki (2005).

78. Joachimski et al. (2004, fig.7) assumed $\delta^{18}O = -1‰$ VSMOW for Devonian sea water.

79. See discussion in Johnson et al. (1985).

80. See discussion in Filer (2002).

81. This interpretation is not universally accepted. For example, Sandberg et al. (2002) propose that a major sea-level drop, not rise, occurred in the Upper Kellwasser Horizon.

82. See discussion in Chen and Tucker (2004, fig. 3).

83. To make matters even more confusing, Chen and Tucker (2003, fig. 12) explicitly show the major Chinese sea-level fall occurring *before* the sea-level rise that corresponds to the Upper Kellwasser Horizon in the sea-level curve of Johnson et al. (1985).

84. See discussion in Bond and Wignall (2008).

85. Ibid.

86. See discussion in Hallam and Wignall (1997).

87. See discussion in McGhee (1984, 1988).

88. See discussion in Bambach et al. (2004).

89. See discussion in Scheckler (1986), Raymond and Metz (1995).

90. See discussion in Stigall (2010, 1).

91. For an extensive discussion of the numerous other contenders for the end-Frasnian kill mechanism, see discussion in McGhee (1996).

92. See discussion in Johnson et al. (1985). This attribution is made by Bond and Wignall (2008, 107). The English paleontologist Michael House also noted a correlation between the occurrence of black shales and extinction events in the fossil record in 1985 (House, 1985).
93. See discussion in Scott and Glasspool (2006).
94. See discussion in Clack (2007, 519).
95. Ibid.
96. See discussion in Kump et al. (2005).
97. See discussion in Harfoot et al. (2008).
98. Ibid. (251).
99. See discussion in Copper (1977).
100. See discussion in Alvarez et al. (1980), Bailey et al. (1994), Toon et al. (1994).
101. See discussion in Alvarez et al. (1980).
102. See discussion in McGhee (1981).
103. See discussion in McGhee, 1982. See summaries in McGhee (1996, 2001b, 2005).
104. See discussion in McGhee (1996).
105. See discussion in Reimold et al. (2005).
106. See discussion in Kaufmann et al. (2004).
107. See discussion in Gradstein et al. (2004).
108. See discussion in McGhee et al. (1984).
109. See discussion in McGhee et al. (1986), McGhee (1996).
110. See discussion in Courtillot (1999), Racki et al. (2002).
111. See discussion in Courtillot (1999), Alvarez (2003), Morgan et al. (2004), McGhee (2005).
112. Algeo et al. (1995) refer to the model as the innocuous sounding "Devonian plant hypothesis." Other sources have been more explicit: in discussing the model in the August 1995 issue of the popular science magazine *Earth*, Ruth Flanagan refers to land plants as "mass murderers of the Devonian."
113. $CO_2 + H_2O \rightarrow CH_2O + O_2$
114. $CO_2 + CaSiO_3 \rightarrow CaCO_3 + SiO_2$
115. See discussion in Algeo et al. (2001, 233).
116. See discussion in Hallam and Wignall (1997, 91).
117. See discussion in Averbuch et al. (2005).
118. See note 114.
119. See discussion in Raymo and Ruddiman (1992).
120. See note 113.
121. See discussion in Racki (2005).
122. See discussion in Raymo and Ruddiman (1992).
123. See discussion in Kent and Muttoni (2008).
124. See discussion in Irving (2008).
125. See discussion in Averbuch et al. (2005).

126. See discussion in McGhee (1982).

127. See discussion in Kaufmann et al. (2004).

128. See discussion in Gradstein et al. (2004).

129. Maximum age for the Lower Kellwasser Horizon is 377.2 + 1.7 = 378.9 million years; minimum age for the Frasnian/Famennian boundary is 374.5 – 2.6 = 371.9 million years; thus the maximum overlap in the ± age uncertainty of the two dates is 378.9 – 371.9 = 7.0 million years.

130. Minimum age for the Lower Kellwasser Horizon is 377.2 – 1.7 = 375.5 million years; maximum age for the Frasnian/Famennian boundary is 374.5 + 2.6 = 377.1 million years; thus the minimum overlap in the ± age uncertainty of the two dates is 377.1 – 375.5 = 1.6 million years.

131. See discussion in McGhee (1997).

132. See discussion in Stigall (2010, 4); see also the research review in Kerr (2012).

133. See discussion in McKinney and Lockwood (1999, 450).

134. See discussion in Scott and Glasspool (2006, 10863).

135. See discussion in DiMichele and Hook (1992, 218).

136. Ibid. (232).

Chapter 5: The Second Animal Invasion

1. See discussion in Wade (2006).

2. See discussion in Clack (2002, 137).

3. See discussion in Ahlberg (1995, 424).

4. See discussion in Clack (2012, 184).

5. See discussion in McGhee (1996, 113–115).

6. See discussion in Blieck et al. (2007).

7. See discussion in Clack et al. (2012).

8. Chronozone Fa2b of older timescales. See discussion and literatured cited in Blieck et al. (2007).

9. See discussion in Blieck et al. (2007).

10. See discussion in Blieck et al. (2007); within the Fa2c chronozone of older timescales, Daeschler (2000).

11. See discussion in Blieck et al. (2007); the Strunian Age or chronozone Fa2d of older timescales.

12. See discussion in Clack et al. (2012).

13. The shoulder girdle, in particular, was more derived than that of *Ichthyostega*; see discussion in Clack (2012, 180).

14. See discussion in Lebedev (2004).

15. See discussion in Ahlberg et al. (2008, 1203).

16. For a discussion of convergence due to reverse evolution, see discussion in McGhee (2011).

17. See discussion in Blom (2005).
18. See discussion in Clément et al. (2004).
19. See discussion in Blieck et al. (2007).
20. See discussion in Clément et al. (2004).
21. See discussion in Lebedev (2004).
22. See discussion in Clack (2002, 2006), Benton (2005).
23. See discussion in Clack et al. (2012).
24. See discussion in Blom (2005).
25. See discussion in Clack (2002, 2006).
26. See discussion in Daeschler et al. (1994).
27. See discussion in Lebedev and Clack (1993), Clack (2002).
28. See discussion in Ahlberg (1995).
29. See discussion in Lebedev (2004).
30. See discussion in Blom et al. (2007).
31. Ibid.
32. See discussion in Lebedev (2004).
33. See discussion in Clack (2002), Astin et al. (2010).
34. See discussion in Clément et al. (2004).
35. See discussion in Daeschler et al. (2009), Cressler et al. (2010).
36. For an analysis of the osmotic tolerances of the early tetrapods, see Laurin and Soler-Gijón (2010).
37. See discussion in Clack (2002).
38. See discussion in Lebedev and Clack (1993).
39. See discussion in Daeschler et al. (2009), Cressler et al. (2010).

Chapter 6: The Second Catastrophe and Retreat

1. The end-Famennian biodiversity crisis is also referred to as the "end-Devonian" biodiversity crisis because it "ended" the Devonian Epoch of geologic time.
2. A paraphyletic taxon; gymnosperms = non-angiosperm spermatophytes. *Pitus* was a lyginopterid spermatophyte, see table 2.1.
3. A paraphyletic taxon; pteridosperms = non-core seed plants.
4. See discussion in DiMichele and Hook (1992, 225).
5. The family Prionoceratidae; see House (2002, 11). The two genera were *Mimimitoceras* and *Imitoceras*; see Walliser (1996, 240).
6. See discussion in Hallam and Wignall (1997).
7. See discussion in Streel et al. (2000), Strother et al. (2010).
8. See discussion in Grahn and Paris (2011), Wei et al. (2012).
9. See discussion in Poty (1999).
10. See discussion in Hallam and Wignall (1997).

11. See discussion in Raymond and Metz (1995).

12. See discussion in Bambach et al. (2004, fig. 6).

13. See discussion in Streel et al. (2000).

14. See discussion in Streel (2009), Blieck et al. (2010).

15. See discussion in Coates and Clack (1995).

16. See discussion in Clack (2012, 266).

17. See discussion in DiMichele and Hook (1992).

18. For example, Famennian-style microbial reef mounds in the marine realm were decimated in the end-Famennian crisis, and only began to recover in the late Tournaisian; see Webb (2002).

19. See discussion in Isaacson et al. (2008).

20. Assuming a Famennian conodont zone represents about 0.7 million years; see chapter 4.

21. See discussion in Isaacson et al. (2008).

22. See discussion in Brezinski et al. (2010).

23. Ibid.

24. See discussion in Isaacson et al. (2008).

25. See discussion in Brezinski et al. (2010).

26. See discussion in Retallack et al. (2009).

27. Assuming a duration of 0.7 million years for a conodont zone; see chapter 4. The base of the **VH** spore zone is seven conodont zones after the Famennian Gap, and five conodont zones before the base of the Early Carboniferous; see table 6.2.

28. See discussion in Retallack et al. (2009, 1143).

29. See discussion in Brezinski et al. (2009, 322).

30. Ibid.

31. See discussion in Falcon-Lang (2000).

32. See discussion in Brezinski et al. (2009).

33. See discussion in Scott and Glasspool (2006, fig. 1), Marynowski and Filipiak (2007), Marynowski et al. (2010).

34. See discussions in Daeschler et al. (2009), Cressler et al. (2010).

35. See discussion in House (2002).

36. See discussion in Kaiser et al. (2006).

37. See discussion in Kaiser et al. (2008).

38. See discussion in Retallack et al. (2009).

39. See discussion in Isbell et al. (2003).

40. Caputo et al. (2008) date these strata to the **BP** and **PC** spore zones, which are time equivalent to the Lower *crenulata* to *isosticha* conodont zonal interval; see table 7.2.

41. See discussion in Saltzman et al. (2000, 347).

42. See discussion in Kammer and Matchen (2008).

43. Caputo et al. (2008) date these strata to the **TC** and **NM** spore zones. The **NM** spore zone can be subdivided into the **DP** and **ME** spore zones, thus the **TC-NM**

zonal interval is time equivalent to the Upper *commutata*–Middle *bilineatus* conodont zonal interval; see table 7.2.

44. See discussion in Gulbranson et al. (2010). The mid-point of the Visean, which spans in geologic time from 345.3 million to 328.3 million years ago, is 336.8 million years.

45. See discussion in Mii et al. (1999).

46. See discussion in Frank et al. (2008).

47. See discussion in Algeo et al. (1995).

48. Atmospheric CO_2 downdraw by terrestrial photosynthesis is also hypothesized to have had a negative effect on the marine photosynthetic arcritarchs during this same interval of time; see Strother et al. (2010).

49. For detailed discussion of the Late Paleozoic Ice Age, see Fielding et al. (2008).

50. See discussion in McGhee et al. (2012, 2013).

51. The last two conodont zones; assuming an average of 0.7 million years per Famennian conodont zone, two zones would represent the time span of 1.4 million years.

52. The Lower *crenulata* to *isosticha* zonal interval spans three conodont zones. Assuming an average duration of 1.39 million years per Tournaisian conodont zone, three zones would represent the time span of 4.17 million years.

53. See discussion in Zachos et al. (2001).

54. See discussion in McGhee (2001b).

55. See discussion in Zachos et al. (2001).

56. See discussion in Sepkoski (1986), Lewis et al. (2008).

57. See discussion in Lewis et al. (2008).

58. See discussion in Hayward (2002).

59. See discussion in Isbell et al. (2003).

60. See discussion in Sepkoski (1986), Stanley (2007), McGhee et al. (2012).

61. Joachimski and Buggisch (2002) report twin +3‰ $\delta^{13}C$ excursions in carbonate carbon for the twin Kellwasser horizons, in contrast to the +0.8‰ $\delta^{13}C$ excursion reported for the Oi-1 glacial pulse reported by Zachos et al. (2001).

62. See discussion in Sepkoski (1986).

63. Ibid.

64. Kaiser et al. (2006) report a +1.2‰ $\delta^{13}C$ excursion in carbonate carbon for the Hangenberg Black Shale, in contrast to the +0.8‰ $\delta^{13}C$ excursion reported for the Mi-1 glacial pulse reported by Zachos et al. (2001).

65. That is, Frasnian atmospheric carbon dioxide levels were higher by a factor of four to five than the pre-industrial level of 0.03 percent, and Famennian carbon dioxide levels were higher by a factor of two to three. See Berner (2006, fig. 18).

66. Due to post–Industrial Revolution burning of fossil fuels by humans, the carbon dioxide level in the Earth's atmosphere has risen to almost 0.039 percent.

67. That is, an Early Oligocene carbon dioxide level that was a factor of two to three higher than the pre-industrial level of 0.03 percent; see Berner (2006, fig. 18).

68. See discussion in Pagani et al. (2005).

69. See discussions in Stanley (2007), Raup and Sepkoski (1982).

70. See discussion in Stanley (2007).

71. See discussion in Raymo and Ruddiman (1992), Zachos et al. (2001), Barrett (2003), DeConto and Pollard (2003), Averbuch et al. (2005), Pagani et al. (2005), Berner (2006), Irving (2008), Kent and Muttoni (2008).

Chapter 7: Victory at Last

1. See discussion in McGhee (1996, 113–115).

2. See discussion in Carroll (2009, 60).

3. See discussion in Clack (2012, 267).

4. See discussion in Sallan and Coates (2010, 10131).

5. See discussion in Newell (1967).

6. Newly discovered Scottish and Russian fossil localities may yield new Tournaisian tetrapod species; if so, these still will be from southeast Laurussia. See discussion in Clack (2012, 183, 290).

7. The five species that are found outside of the southeastern region of Laurussia are *Greererpeton burkemorani* from West Virginia, *Whatcheeria deltae* and *Sigournea multidentata* from Iowa, and *Antlerpeton clarkii* from Nevada, all in North America, and *Ossinodus pueri* from Australia.

8. See discussion in Clack and Finney (2005).

9. Tentative evidence from bone fragments suggest that a whatcheeriid-like tetrapod may have been present earlier in Pennsylvania, central Laurussia, in the **VH** spore zone of the Famennian; see discussion in Daeschler et al. (2009).

10. See discussion in Clack (2012).

11. The Middle Paddock site, early Holkerian Age in the British regional geologic timescale. See discussion in Carroll (2009, fig. 4.1).

12. See discussion in Coates et al. (2008).

13. See discussion in Clack (2002), Coates et al. (2008). These species are *Eucritta melanolimnetes, Balanerpeton woodi, Ophidererpeton kirtonense, Eldeceeon rolfei, Silvanerpeton miripedes*, and *Westlothiana lizziae*.

14. Early Brigantian Age, in the regional geologic timescale in Scotland.

15. Late Holkerian Age, in the regional geologic timescale in Scotland.

16. See discussion in Benton (2005), Ward et al. (2006).

17. See discussion in Ward et al. (2006), Sallan and Coates (2010).

18. See discussion in Benton (2005), Ward et al. (2006), Sallan and Coates (2010).

19. See discussion in Shear and Selden (2001, 36).

20. See discussion in Clack (2012, 299–304).

21. See discussion in Clack (2002), Clack and Finney (2005), Benton (2005).

22. See discussion in Clack (2002).

23. Ibid.

24. See discussion in Clack (1998, 66).

25. For the serious-minded editors of the journal *Nature*, Clack stated that the etymology of the species name "from the black lagoon" refers to the paleoenvironment of the locality at East Kirkton, Scotland, where the first fossil was found (Clack, 1998, 66.). However, the name "the true creature from the black lagoon" is entirely too similar to the title of the classic science fiction horror movie, *The Creature from the Black Lagoon*, to be a coincidence. In that movie, the amphibious monster, a supposed living fossil that survived unchanged from the Devonian, is itself a peculiar mélange of human, reptile, and fish crown-group traits.

26. See discussion in Clack (2002).

27. See discussion in Clack (2012, 411).

28. See discussion in Clack (2012, 308).

29. See discussion in Clack (2002, 273–277), Benton (2005, 103–104).

30. That is, the lepospondyls are extinct if they are not the ancestors of the lissamphibians—if they are the ancestors of the lissamphibians, then the batrachomorphs are extinct.

31. See discussion in Sallan and Coates (2010).

32. See discussion in Clack (2002).

33. See discussion in Benton (2005).

34. Middle Holkerian Age, in the regional geologic timescale in Scotland.

35. See discussion in Clack (2002), Sallan and Coates (2010).

36. See discussion in Clack (2002, 291–300).

37. See discussion in Carroll (2009).

38. See discussion in Clack (2002, 199).

39. See discussion in Carroll (2009, 74).

40. See discussion in Clack (2012, 277).

41. See discussion in Labandeira (2005, 260).

42. See discussion in Cowen (2000, 144–146).

43. For a discussion of the convergent evolution of flight, see McGhee (2011).

44. Some, however, question whether this species is actually a dipluran; see discussion in Grimaldi and Engel (2005, 118).

45. These species are *Arthropleura armata, Euphoberia armigera, Pleurojulus biornatus, P. levis, Hexecontasoma carinatum, Mazoscolopendra richardsoni, Protosolpuga carbonaria*, and *Geralinura carbonaria*.

46. See discussion in Kraus and Brauckmann (2003).

47. See discussion in Wilson (2006).

48. See discussion in Wilson and Hannibal (2005).

49. Named after Quaker paleontologist Eugene (Gene) Richardson, who spent his life studying the fauna of the Mazon Creek *Lagerstätte* at the Field Museum of Natural History in Chicago, and who befriended many amateur paleontologists who collected the Mazon Creek ironstones.

50. See discussion in Shear and Edgecombe (2010).

51. See discussion in Dunlop (2010).

52. See discussion in Grimaldi and Engel (2005).

53. See discussion in Wootton et al. (1998, 749).

54. See discussion in Prokop et al. (2005).

55. All from strata dated to the mid-Serpukhovian, or late Namurian-A age of older timescales. See discussion in Carroll (2009, fig. 4.1).

56. See discussion in Grimaldi and Engel (2005).

57. Ibid.

58. Ibid.

59. Ibid.

60. See discussion in Berner et al. (2007).

61. See discussion in Prokop et al. (2005).

62. See discussion in Dudley (1998), Lane (2002).

63. See discussion in Falkowski et al. (2005).

64. See discussion in Grimaldi and Engel (2005, 178).

65. See discussion in Ward et al. (2006).

66. For a discussion of the phenomenon of ecological convergence, see McGhee (2011).

67. See discussion in Benton (2005).

68. See discussion in Voigt and Ganzelewski (2010).

69. See discussion in Benton (2005).

70. Westphalian-A age in older timescales; see discussion in Carroll (2009, fig. 4.1).

71. See discussion in Cowen (2000, 145).

72. See discussion in Benton (2005).

73 Ibid., see also Ward et al. (2006).

74. See discussion in Benton (2005).

75. See discussion in Ward et al. (2006).

76. See discussion in Benton (2005).

77. Ibid.

78. See discussion in Ward et al. (2006).

79. See discussion in Benton (2005).

80. See discussion in Ward et al. (2006).

81. Ibid.

82. See discussion in Benton (2005).

83. That is 3,830 Ma, age of the oldest geochemical evidence for life on Earth, minus 300 Ma, age of the latest Gzhelian.

Chapter 8: The Legacy of the Devonian Extinctions

1. See discussion in Coates et al. (2008).
2. See discussion in Flügel and Stanley (1984), Fois and Gaetani (1984).
3. See discussion in Sallan and Coates (2010, 10135).
4. See discussion in Sallan et al. (2011).
5. See discussion in Kammer and Ausich (2006).
6. See discussion in Young (2010).
7. See discussion in Benton (2005).
8. See discussion in McGhee (2011).
9. Ibid.
10. Nine genera of these fish exist today in the phylogenetic lineage Actinopterygii: Perciformes: Gobiidae: Oxudercinae.
11. Four genera of these fish exist today in the phylogenetic lineage Actinopterygii: Perciformes: Anabantidae.
12. Phylogenetic lineage Actinopterygii: Siluriformes: Clariidae.
13. See discussion in Benton (2005), Clack (2007).
14. A density of 9,300 kilopascals as opposed to 101 kilopascals on Earth.
15. Lead melts at 327°C.
16. See discussion in Croswell (2003).

References

Ahlberg, P. E. 1991. Tetrapod or near-tetrapod fossils from the Upper Devonian of Scotland. *Nature* 354: 298–301.

——. 1995. *Elginerpeton pancheni* and the earliest tetrapod clade. *Nature* 373: 420–425.

——. 1998. Postcranial stem tetrapod remains from the Devonian of Scat Craig, Morayshire, Scotland. *Zoological Journal of the Linnean Society* 122: 99–141.

Ahlberg, P. E., J. A. Clack, E. Lukševičs, H. Blom, and I. Zupinš. 2008. *Ventastega curonica* and the origin of tetrapod morphology. *Nature* 453: 1199–1204.

Ahlberg, P. E. and Z. Johanson. 1998. Osteolepiforms and the ancestry of tetrapods. *Nature* 395: 792–794.

Algeo, T. J., R. A. Berner, J. B. Maynard, and S. E. Scheckler. 1995. Late Devonian oceanic anoxic events and biotic crisis: "Rooted" in the evolution of vascular land plants? *GSA Today* 5: 63–66.

Algeo, T. J. and E. Ingall. 2007. Sedimentary C_{org}: P ratios, paleocean ventilation, and Phanerozoic atmospheric pO_2. *Palaeogeography, Palaeoclimatology, Palaeoecology* 256: 130–155.

Algeo, T. J., S. E. Scheckler, and J. B. Maynard. 2001. Effects of the Middle to Late Devonian spread of vascular land plants on weathering regimes, marine biotas, and global climate. In *Plants Invade the Land*. ed. P. G. Gensel and D. Edwards. New York: Columbia University Press, pp. 213–236.

Alvarez, L. W., W. Alvarez, F. Asaro, and H. V. Michel. 1980. Extraterrestrial cause for the Cretaceous–Tertiary extinction. *Science* 208: 1095–1108.

Alvarez, W. 2003. Comparing the evidence relevant to impact and flood basalt at times of major mass extinctions. *Astrobiology* 3: 153–161.

Andrews, H. N. 1960. Notes on Belgian specimens of *Sporogonites. Palaeobotanist* 7: 85–89.

Astin, T. R., J. E. A. Marshall, H. Blom, and C. M. Berry. 2010. The sedimentary environment of the Late Devonian East Greenland tetrapods. In *The Terrestrialization Process: Modelling Complex Interactions at the Biosphere-Geosphere Interface.* ed. M. Vecoli, G. Clément, and B. Meyer-Berthaud. Geological Society, London, Special Publications 339: 93–109.

Averbuch, O., N. Tribovillard, X. Devleeschouwer, L. Riquier, B. Mistiaen, and B. van Vliet–Lanoe. 2005. Mountain building–enhanced continental weathering and organic carbon burial as major causes for climatic cooling at the Frasnian–Famennian boundary (c. 376 Ma)? *Terra Nova* 17: 25–34.

Babarro, J. M. F. and A. De Zwaan. 2008. Anaerobic survival potential of four bivalves from different habitats: A comparative survey. *Comparative Biochemistry and Physiology, Part A* 151: 108–113.

Bailey, M. E., S. V. M. Clube, G. Hahn, W. M. Napier, and G. B. Valsechi. 1994. Hazards due to giant comets: Climate and short-term catastrophism. In *Hazards due to Comets and Asteroids.* ed. T. Gehrels. Tucson, Ariz.: University of Arizona Press, pp. 479–536.

Bambach, R. K., A. H. Knoll, and S. C. Wang. 2004. Origination, extinction, and mass depletions of marine diversity. *Paleobiology* 30: 522–542.

Barrett, P. 2003. Cooling a continent. *Nature* 421: 221–223.

Benton, M. J. 2005. *Vertebrate Palaeontology, Third Edition.* Oxford, England: Blackwell Publishing.

Berner, R. A. 2006. GEOCARBSULF: A combined model for Phanerozoic atmosphere O_2 and CO_2. *Geochimica et Cosmochimica Acta* 70: 5653–5664.

Berner, R. A., D. J. Beerling, R. Dudley, J. M. Robinson, and R. A. Wildman. 2003. Phanerozoic atmospheric oxygen. *Annual Review of Earth and Planetary Sciences* 31: 105–134.

Berner, R. A., J. M. VandenBrooks, and P. D. Ward. 2007. Oxygen and evolution. *Science* 316: 557–558.

Berry, C. M. and M. Fairon-Demaret. 2001. The Middle Devonian flora revisited. In *Plants Invade the Land.* ed. P. G. Gensel and D. Edwards. New York: Columbia University Press, pp. 120–139.

Blieck, A., G. Clément, H. Blom, H. Lelievre, E. Luksevics, M. Streel, J. Thorez, and G. C. Young. 2007. The biostratigraphical and palaeogeographical framework of the earliest diversification of tetrapods (Late Devonian). In *Devonian Events and Correlations.* ed. R. T. Becker and W. T. Kirchgasser. Geological Society, London, Special Publications 278: 219–235.

Blieck, A., G. Clément, and M. Streel. 2010. The biostratigraphical distribution of earliest tetrapods (Late Devonian): A revised version with comments on biodiversification. In *The Terrestrialization Process: Modelling Complex Interactions at the Biosphere–Geosphere Interface.* ed. M. Vecoli, G. Clément, and

B. Meyer–Berthaud. Geological Society, London, Special Publications 339: 129–138.

Blom, H. 2005. Taxonomic revision of the Late Devonian tetrapod *Ichthyostega* from East Greenland. *Palaeontology* 48(1): 111–134.

Blom, H., J. A. Clack, P. E. Ahlberg, and M. Friedman. 2007. Devonian vertebrates from East Greenland: A review of faunal composition and distribution. *Geodiversitas* 29(1): 119–141

Bond, D. 2006. The fate of the homoctenids (*Tentaculitoidea*) during the Frasnian–Famennian mass extinction (Late Devonian). *Geobiology* 4: 167–177.

Bond, D. P. G. and P. B. Wignall. 2005. Evidence for Late Devonian (Kellwasser) anoxic events in the Great Basin, western United States. In *Understanding Late Devonian and Permian–Triassic Biotic and Climatic Events*. ed. D. J. Over, J. R. Morrow, and P. B. Wignall. Amsterdam, Netherlands: Elsevier B.V., pp. 225–262.

——. 2008. The role of sea-level change and marine anoxia in the Frasnian–Famennian (Late Devonian) mass extinction. *Palaeogeography, Palaeoclimatology, Palaeoecology* 263: 107–118.

Brainerd, E. L. 1994. The evolution of lung-gill bimodal breathing and the homology of vertebrate respiratory pumps. *American Zoologist* 34: 289–299.

Brand, U. 1989. Global climatic changes during the Devonian–Mississippian: Stable isotope biogeochemistry of brachiopods. *Palaeogeography, Palaeoclimatology, Palaeoecology (Global and Planetary Change Section)* 75: 311–329.

Brezinski, D. K., C. B. Cecil, and V. W. Skema. 2010. Late Devonian glacigenic and associated facies from the central Appalachian Basin, eastern United States. *Geological Society of America Bulletin* 122(1/2): 265–281.

Brezinski, D. K., C. B. Cecil, V. W. Skema, and C. A. Kertis. 2009. Evidence for long-term climate change in Upper Devonian strata of the central Appalachians. *Palaeogeography, Palaeoclimatology, Palaeoecology* 284: 315–325.

Brocks, J. J., G. A. Logan, R. Buick, and R. E. Summons. 1999. Archaen molecular fossils and the early rise of eukaryotes. *Science* 285: 1033–1036.

Buick, R. 2001. Life in the Archaean. In *Palaeobiology II*, ed. D. E. G. Briggs and P. R. Crowther. Oxford, England: Blackwell Science, pp. 13–21.

Campbell, K. S. W. and M. W. Bell. 1977. A primitive amphibian from the Late Devonian of New South Wales. *Alcheringa* 1: 369–381.

Caputo, M. V., J. H. G. Melo, M. Streel, and J. L. Isbell. 2008. Late Devonian and Early Carboniferous glacial records of South America. In *Resolving the Late Paleozoic Ice Age in Time and Space*. ed. C. R. Fielding, T. D. Frank, and J. L. Isbell. Boulder, Colo.: Geological Society of America Special Paper 441, pp. 161–173.

Carroll, R. 2009. *The Rise of Amphibians: 365 Million Years of Evolution*. Baltimore, Md.: The Johns Hopkins University Press.

Chen, D. and M. E. Tucker. 2003. The Frasnian–Famennian mass extinction: Insights from high–resolution sequence stratigraphy and cyclostratigraphy in South China. *Palaeogeography, Palaeoclimatology, Palaeoecology* 193: 87–111.

——. 2004. Palaeokarst and its implication for the extinction event at the Frasnian–Famennian boundary (Guilin, South China). *Journal of the Geological Society, London* 161: 895–898.

Clack, J. A. 1998. A new Early Carboniferous tetrapod with a *mélange* of crown-group characters. *Nature* 394: 66–69.

——. 2002. *Gaining Ground: The Origin and Evolution of Tetrapods*. Bloomington, Ind.: Indiana University Press.

——. 2006. The emergence of early tetrapods. *Palaeogeography, Palaeoclimatology, Palaeoecology* 232: 167–189.

——. 2007. Devonian climate change, breathing, and the origin of the tetrapod stem group. *Integrative and Comparative Biology* 47(4): 510–523.

——. 2009. The fin to limb transition: New data, interpretations, and hypotheses from paleontology and developmental biology. *Annual Review of Earth and Planetary Sciences* 37: 163–179.

——. 2012. *Gaining Ground: The Origin and Evolution of Tetrapods, Second Edition*. Bloomington, Ind.: Indiana University Press.

Clack, J. A., P. E. Ahlberg, H. Blom, and S. M. Finney. 2012. A new genus of Devonian tetrapod from north-east Greenland, with new information on the lower jaw of *Ichthyostega*. *Palaeontology* 55(1): 73–86.

Clack, J. A. and S. M. Finney. 2005. *Pederpes finneyae*, an articulated tetrapod from the Tournaisian of western Scotland. *Journal of Systematic Palaeontology* 2(4): 311–346.

Clément, G., P. E. Ahlberg, A. Blieck, H. Blom, J. A. Clack, E. Potyll, J. Thorezil, and P. Janvier. 2004. Devonian tetrapod from western Europe. *Nature* 427: 412–413.

Coates, M. I. and J. A. Clack. 1995. Romer's gap: Tetrapod origins and terrestriality. *Bulletin du Muséum National d'Histoire Naturelle, Paris* 17: 373–388.

Coates, M. I., M. Ruta, and M. Friedman. 2008. Ever since Owen: Changing perspectives on the early evolution of tetrapods. *Annual Review of Ecology, Evolution, and Systematics* 39: 571–592.

Copper, P. 1977. Paleolatitudes in the Devonian of Brazil and the Frasnian–Famennian mass extinction. *Palaeogeography, Palaeoclimatology, Palaeoecology* 21: 165–207.

——. 1994. Ancient reef ecosystem expansion and collapse. *Coral Reefs* 13: 3–11.

Courtillot, V. E. 1999. *Evolutionary Catastrophes: The Science of Mass Extinction*. Cambridge, England: Cambridge University Press.

Courtillot, V. E., V. A. Kravchinksy, X. Quidelleur, P. R. Renne, and D. P. Gladkochub. 2010. Preliminary dating of the Viluy traps (Eastern Siberia): Eruption at the time of Late Devonian extinction events? *Earth and Planetary Science Letters* 300: 239–245.

Courtillot, V. E. and P. R. Renne. 2003. On the ages of flood basalt events. *Comptes Rendus Geoscience* 335: 113–140.

Cowen, R. 2000. *History of Life* (3rd edition). Oxford, England: Blackwell Science.

Cressler, W. L. 2001. Evidence of earliest known wildfires. *Palaios* 16(2): 171–174.

Cressler, W. L., E. B. Daeschler, R. Slingerland, and D. A. Peterson. 2010. Terrestrialization in the Late Devonian: A palaeoecological overview of the Red Hill site, Pennsylvania, USA. In *The Terrestrialization Process: Modelling Complex Interactions at the Biosphere–Geosphere Interface*. ed. M. Vecoli, G. Clément, and B. Meyer-Berthaud. Geological Society, London, Special Publications 339: 111–128.

Croswell, K. 2003. *Magnificent Mars*. New York: Free Press.

Daeschler, E. B. 2000. Early tetrapod jaws from the Late Devonian of Pennsylvania, USA. *Journal of Paleontology* 74(2): 301–308.

Daeschler, E. B., J. A. Clack, and N. H. Shubin. 2009. Late Devonian tetrapod remains from Red Hill, Pennsylvania, USA: How much diversity? *Acta Zoologica* 90(Supplement 1): 306–317.

Daeschler, E. B., N. H. Shubin, K. S. Thomson, and W. W. Amaral. 1994. A Devonian tetrapod from North America. *Science* 265: 639–642.

Darwin, C. 1859. *On the Origin of Species by Means of Natural Selection, or the Preservation of Favoured Races in the Struggle for Life*. London, England: John Murray.

Dauphas, N., M. van Zuilen, M. Wadhwa, A. M. Davis, B. Marty, and P. E. Janney. 2004. Clues from Fe isotope variations on the origin of Early Archean BIFs from Greenland. *Science* 306: 2077–2080.

DeConto, R. M. and D. Pollard. 2003. Rapid Cenozoic glaciation of Antarctica induced by declining atmospheric CO_2. *Nature* 421: 245–249.

DiMichele, W. A. and R. W. Hook. 1992. Paleozoic terrestrial ecosystems. In *Terrestrial Ecosystems through Time*. ed. A. K. Behrensmeyer, J. D. Damuth, W. A. DiMichele, R. Potts, H.–D. Sues, and S. L. Wing. Chicago, Ill.: University of Chicago Press, pp. 205–325.

Donoghue, M. J. 2005. Key innovations, convergence, and success: Macroevolutionary lessons from plant phylogeny. In *Macroevolution: Diversity, Disparity, Contingency*. ed. E. S. Vrba and N. Eldredge. *Paleobiology* 31(2) Supplement, pp. 77–93.

Draganits, E., S. J. Braddy, and D. E. G. Griggs. 2001. A Gondwanan coastal arthropod ichnofauna from the Muth Formation (Lower Devonian, Northern India): Paleoenvironment and tracemaker behavior. *Palaios* 16(2): 126–147.

Dudley, R. 1998. Atmospheric oxygen, giant Paleozoic insects and the evolution of aerial locomotor performance. *Journal of Experimental Biology* 201: 1043–1050.

Duncker, H.–R. 2004. Vertebrate lungs: Structure, topography and mechanics; a comparative perspective of the progressive integration of respiratory system, locomotor apparatus and ontogenetic development. *Respiratory Physiology and Neurobiology* 114: 111–124.

Dunlop, J. A. 2010. Geological history and phylogeny of Chelicerata. *Arthropod Structure and Development* 39: 124–142.

Edwards, D. 1970. Fertile Rhyniophytina from the Lower Devonian of Britain. *Palaeontology* 13: 451–461.

Edwards, D. S. 1980. Evidence for the sporophytic status of the Lower Devonian plant *Rhynia gwynne–vaughanii* Kidston and Lang. *Review of Palaeobotany and Palynology* 29: 177–188.

Edwards, D. and C. Wellman. 2001. Embryophytes on land: The Ordovician to Lochkovian (Lower Devonian) record, in *Plants Invade the Land.* ed., P. G. Gensel and D. Edwards. Columbia University Press, New York, 3–28.

Engel, M. S. and D. A. Grimaldi. 2004. New light shed on the oldest insect. *Nature* 427: 627–630.

Erwin, D. H. 2001. Metazoan origins and early evolution. In *Palaeobiology II.* ed. D. E. G. Briggs and P. R. Crowther. Oxford, England: Blackwell Science, pp. 25–31.

Erwin, D. H., M. Laflamme, S. M. Tweedt, E. A. Sperling, D. Pisani, and K. J. Peterson. 2011. The Cambrian conundrum: Early divergence and later ecological success in the early history of animals. *Science* 334: 1091–1097.

Falcon-Lang, H. J. 2000. Fire ecology of the Carboniferous tropical zone. *Palaeogeography, Palaeoclimatology, Palaeoecology* 164: 339–355.

Falkowski, P., M. Katz, A. Milligan, K. Fennel, B. Cramer, M. P. Aubry, R. A. Berner, and W. M. Zapol. 2005. The rise of atmospheric oxygen levels over the past 205 million years and the evolution of large placental mammals. *Science* 309: 2202–2204.

Fielding, C. R., T. D. Frank, and J. L. Isbell. 2008. *Resolving the Late Paleozoic Ice Age in Time and Space.* Boulder, Colorado: Geological Society of America Special Paper 441: 1–354.

Filer, J. K. 2002. Late Frasnian sedimentation cycles in the Appalachian basin: Possible evidence for high frequency eustatic sea-level changes. *Sedimentary Geology* 154: 31–52.

Filipiak, P. 2004. Miospore stratigraphy of Upper Famennian and Lower Carboniferous deposits of the Holy Cross Mountains (central Poland). *Review of Palaeobotany and Palynology* 128: 291–322.

Flügel, E. and G. D. Stanley. 1984. Re–organization, development and evolution of post-Permian reefs and reef-organisms. *Paleontographica Americana* 54: 177–186.

Fois, E. and M. Gaetani. 1984. The recovery of reef-building communities and the role of cnidarians in carbonate sequences of the Middle Triassic (Anisian) in the Italian Dolomites. *Paleontographica Americana* 54: 191–200.

Frank, T. D., L. P. Birgenheier, I. P. Montañez, C. R. Fielding, and M. C. Rygel. 2008. Late Paleozoic climate dynamics revealed by comparison of ice-proximal stratigraphic and ice-distal isotopic records. In *Resolving the Late Paleozoic Ice Age in Time and Space.* ed. C. R. Fielding, T. D. Frank, and J. L. Isbell. Boulder, Colorado: Geological Society of America Special Paper 441, pp. 331–342.

Friedman, M. and L. C. Sallan. 2012. Five hundred years of extinction and recovery: A Phanerozoic survey of large-scale diversity patterns in fishes. *Palaeontology* 55: 707–742.

George, D. and A. Blieck. 2011. Rise of the earliest tetrapods: An Early Devonian origin from marine environment. *PLoS ONE* 6(7): e22136. doi: 10.1371/journal.pone.0022136

Gerrienne, P., B. Meyer–Berthaud, H. Lardeux, and S. Régnault. 2010. First record of *Rellimia* Leclerq & Bonamo (Aneurophytales) from Gondwana, with comments on the earliest lignophytes. In *The Terrestrialization Process: Modelling Complex Interactions at the Biosphere-Geosphere Interface*. ed. M. Vecoli, G. Clément, and B. Meyer-Berthaud. Geological Society, London, Special Publications 339: 81–92.

Goddéris, Y. and M. M. Joachimski. 2004. Global change in the Late Devonian: Modelling the Frasnian–Famennian short–term carbon isotope excursions. *Palaeogeography, Palaeoclimatology, Palaeoecology* 202: 309–329.

Gong, Y.–M. and R. Xu. 2003. Conodont apatite $\delta^{18}O$ signatures indicate climatic cooling as a trigger of the Late Devonian mass extinction: Comment. *Geology* 31(4): 383.

Gould, S. J. 1989. *Wonderful Life: The Burgess Shale and the Nature of History*. New York: W. W. Norton & Company.

Gradstein, F., J. Ogg, and A. Smith. 2004. *A Geologic Time Scale 2004*. Cambridge, England: Cambridge University Press.

Grahn, Y. and F. Paris. 2011. Emergence, biodiversification and extinction of the chitinozoan group. *Geological Magazine* 148(2): 226–236.

Grimaldi, D. and M. S. Engel. 2005. *Evolution of the Insects*. Cambridge, England: Cambridge University Press.

Grinspoon, D. 2003. *Lonely Planets: The Natural Philosophy of Alien Life*. New York: HarperCollins Publishers.

Gulbranson, E. L., I. P. Montañez, M. D. Schmitz, C. O. Limarino, J. L. Isbell, S. A. Marenssi, and J. L. Crowley. 2010. High-precision U-Pb calibration of Carboniferous glaciation and climate history, Paganzo Group, NW Argentina. *Geological Society of America Bulletin* 122(9/10): 1480–1498.

Hallam, A. and P. B. Wignall. 1997. *Mass Extinctions and their Aftermath*. Oxford, England: Oxford University Press.

Harfoot, M. B., J. A. Pyle, and D. J. Beerling. 2008. End-Permian ozone shield unaffected by oceanic hydrogen sulphide and methane releases. *Nature Geoscience* 1(4): 247–252.

Hartkopf-Fröder, C., M. Kloppisch, U. Mann, P. Neumann-Mahlkau, R. G. Schaefer, and H. Wilkes. 2007. The end-Frasnian mass extinction in the Eifel Mountains, Germany: New insights from organic matter composition and preservation. *Geological Society of London Special Publications* 278: 173–196.

Hartmann, W. K. 2003. *A Traveler's Guide to Mars*. New York: Workman Publishing.

Hayward, B. W. 2002. Late Pliocene to Middle Pleistocene extinctions of deep-sea benthic foraminifera (*Stilostomella* extinction) in the southwest Pacific. *Journal of Foraminiferal Research* 32(3): 274–307.

Hofmann, H. J., K. Grey, A. H. Hickman, and R. I. Thorpe. 1999. Origin of 3.45 Ga coniform stromatolites in Warrawoona Group, Western Australia. *Geological Society of America Bulletin* 111(8): 1256–1262.

House, M. R. 1985. Correlation of mid-Palaeozoic ammonoid evolutionary events with global sedimentary perturbations. *Nature* 313: 17–22.

——. 2002. Strength, timing, setting and cause of mid-Palaeozoic extinctions. *Palaeogeography, Palaeoclimatology, Palaeoecology* 181: 5–25.

Irving, E. 2008. Why Earth became so hot 50 million years ago and why it then cooled. *Proceedings of the National Academy of Sciences USA* 105(42): 16061–16062.

Isaacson, P. E., E. Díaz-Martínez, G. W. Grader, J. Kalvoda, O. Babek, and F. X. Devuyst. 2008. Late Devonian–earliest Mississippian glaciation in Gondwanaland and its biogeographic consequences. *Palaeogeography, Palaeoclimatology, Palaeoecology* 268: 126–142.

Isbell, J. L., M. F. Miller, K. L. Wolfe, and P. A. Lenaker. 2003. Timing of late Paleozoic glaciation in Gondwana: Was glaciation responsible for the development of Northern Hemisphere cyclothems? In *Extreme Depositional Environments: Mega End Members in Geologic Time*. ed. M. A. Chan and A. W. Archer. Boulder, Colo.: Geological Society of America Special Paper 370, pp. 5–24.

Janvier, P. and G. Clément. 2010. Muddy tetrapod origins. *Nature* 463: 40–41.

Joachimski, M. M., S. Breisig, W. Buggisch, J. A. Talent, R. Mawson, M. Gereke, J. R. Morrow, J. Day, and K. Weddige. 2009. Devonian climate and reef evolution: Insights from oxygen isotopes in apatite. *Earth and Planetary Science Letters* 284: 599–609.

Joachimski, M. M. and W. Buggisch. 2002. Conodont apatite $\delta^{18}O$ signatures indicate climatic cooling as a trigger of the Late Devonian mass extinction. *Geology* 30(8): 711–714.

——. 2003. Conodont apatite $\delta^{18}O$ signatures indicate climatic cooling as a trigger of the Late Devonian mass extinction: Reply. *Geology* 31(4): 384.

Joachimski, M. M., R. van Gelden, S. Breisig, W. Buggisch, and J. Day. 2004. Oxygen isotope evolution of biogenic calcite and apatite during the Middle and Late Devonian. *International Journal of Earth Science* 93: 542–553.

John, E. H., P. B. Wignall, R. J. Newton, and S. H. Bottrell. 2010. $\delta^{34}S_{CAS}$ and $\delta^{18}O_{CAS}$ records during the Frasnian–Famennian (Late Devonian) transition and their bearing on mass extinction models. *Chemical Geology* 275: 221–234.

Johnson, J. G., G. Klapper, and C. A. Sandberg. 1985. Devonian eustatic fluctuations in Euramerica. *Geological Society of America Bulletin* 96: 567–587.

Kaiser, S. I., T. Steuber, and R. T. Becker. 2008. Environmental change during the late Frasnian and early Tournaisian (Late Devonian-Early Carboniferous):

Implications from stable isotopes and conodont biofacies in southern Europe. *Geological Journal* 43: 241–260.

Kaiser, S. I., T. Steuber, R. T. Becker, and M. M. Joachimski. 2006. Geochemical evidence for major environmental change at the Devonian-Carboniferous boundary in the Carnic Alps and the Rhenish Massif. *Palaeogeography, Palaeoclimatology, Palaeoecology* 240: 146–160.

Kammer, T. W. and W. I. Ausich. 2006. The "Age of Crinoids": A Mississippian biodiversity spike coincident with widespread carbonate ramps. *Palaios* 21: 238–248.

Kammer, T. W. and D. L. Matchen. 2008. Evidence for eustasy at the Kinderhookian–Osagean (Mississippian) boundary in the United States: Response to late Tournaisian glaciation? In *Resolving the Late Paleozoic Ice Age in Time and Space*. ed. C. R. Fielding, T. D. Frank, and J. L. Isbell. Boulder, Colo.: Geological Society of America Special Paper 441, pp. 261–274.

Kaufmann, B., E. Trapp, and K. Mezger. 2004. The numerical age of the Upper Frasnian (Upper Devonian) Kellwasser horizons: A new U-Pb zircon date from Steinbruch Schmidt (Kellerwald, Germany). *Journal of Geology* 112: 495–501.

Kenrick, P. and P. R. Crane. 1997a. The origin and early evolution of plants on land. *Nature* 389: 33–39.

——. 1997b. *The Origin and Early Diversification of Land Plants: A Cladistic Study*. Washington, D.C.: Smithsonian Institution Press.

Kent, D. V. and G. Muttoni. 2008. Equatorial convergence of India and early Cenozoic climate trends. *Proceedings of the National Academy of Sciences USA* 105(42): 16065–16070.

Kerr, R. A. 2012. More than one way for invaders to wreak havoc. *Science* 335: 646.

Kessler, W. and G. Müller. 1988. Minor and trace-element data from iron oxides from iron-formations of the Iron Quadrangle, Minas Gerais, Brazil. *Mineralogy and Petrology* 39: 245–250.

King, H. M., N. H. Shubin, M. I. Coates, and M. E. Hale. 2011. Behavioral evidence for the evolution of walking and bounding before terrestriality in sarcopterygian fishes. *Proceedings of the National Academy of Sciences USA* 108(52): 21146–21151.

Kraus, O. and C. Brauckmann. 2003. Fossil giants and surviving dwarfs: Arthropleurida and Pselaphognatha (Atelocerata, Diplopoda); characters, phylogenetic relationships and construction. *Verhandlungen des naturwissenschaftlichen Vereins in Hamburg* 40: 5–50.

Kump, L. R., A. Pavlov, and M. A. Arthur. 2005. Massive release of hydrogen sulfide to the surface ocean and atmosphere during intervals of oceanic anoxia. *Geology* 33(5): 397–400.

Labandeira, C. C. 2005. Invasion of the continents: Cyanobacterial crusts to tree-inhabiting arthropods. *Trends in Ecology and Evolution* 20(5): 253–262.

Lane, N. 2002. *Oxygen: The Molecule that Made the World*. Oxford, England: Oxford University Press.

Laurin, M. and R. Soler-Gijón. 2010. Osmotic tolerance and habitat of early stegocephalians: Indirect evidence from parsimony, taphonomy, palaeobiogeography, physiology, and morphology. In *The Terrestrialization Process: Modelling Complex Interactions at the Biosphere-Geosphere Interface*. ed. M. Vecoli, G. Clément, and B. Meyer-Berthaud. Geological Society, London, Special Publications 339: 151–179.

Lebedev, O. A. 2004. A new tetrapod *Jakubsonia livnensis* from the early Famennian (Devonian) of Russia and palaeoecological remarks on the Late Devonian tetrapod habitats. *Acta Universitatis Latviensis, Series Earth and Environment Sciences* 679: 79–98.

Lebedev, O. A. and J. A. Clack. 1993. Upper Devonian tetrapods from Andreyevka, Tula, Russia. *Palaeontology* 36(6): 721–734.

Lecointre, G. and H. Le Guyader. 2006. *The Tree of Life: A Phylogenetic Classification*. Cambridge, Mass.: Belknap Press of Harvard University Press.

Lewis, A. R., D. R. Marchant, A. C. Ashworth, L. Hedenäs, S. R. Hemming, J. V. Johnson, M. J. Leng, M. L. Machlus, A. E. Newton, J. I. Raine, J. K. Willenbring, M. Williams, and A. P. Wolfe. 2008. Mid-Miocene cooling and the extinction of tundra in continental Antarctica. *Proceedings of the National Academy of Sciences USA* 105(31): 10676–10680.

Long, J. A., G. C. Young, T. Holland, T. J. Senden, and E. M. G. Fitzgerald. 2006. An exceptional Devonian fish from Australia sheds light on tetrapod origins. *Nature* 444: 199–202.

Lutz, R. A. and D. C. Rhoads. 1977. Anaerobosis and theory of growth line formation. *Science* 198: 1222–1227.

MacNaughton, R. B., J. M. Cole, R. W. Dalrymple, S. J. Braddy, D. E. G. Briggs, and T. D. Lukie. 2002. First steps on land: Arthropod trackways in Cambrian–Ordovician eolian sandstone, southeastern Ontario, Canada. *Geology* 30(5): 391–394.

Marynowski, L. and P. Filipiak. 2007. Water column euxinia and wildfire evidence during the deposition of the Upper Famennian Hangenberg event horizon from the Holy Cross Mountains (central Poland). *Geological Magazine* 144(3): 569–595.

Marynowski, L., P. Filipiak, and A. Pisarzowska. 2008. Organic geochemistry and palynofacies of the Early-Middle Frasnian transition (Late Devonian) of the Holy Cross Mountains, southern Poland. *Palaeogeography, Palaeoclimatology, Palaeoecology* 269: 152–165.

Marynowski, L., P. Filipiak, and M. Zatoń. 2010. Geochemical and palynological study of the Upper Famennian Dasberg event horizon from the Holy Cross Mountains (central Poland). *Geological Magazine* 147(4): 527–550.

Maziane, N., K. T. Higgs, and M. Streel. 1999. Revision of the late Famennian miospore zonation scheme in eastern Belgium. *Journal of Micropaleontology* 18: 17–25.

McGhee, G. R. 1981. The Frasnian–Famennian extinctions: A search for extraterrestrial causes. *Bulletin of the Field Museum of Natural History* 52(7): 3–5.

——. 1982. The Frasnian–Famennian extinction event: A preliminary analysis of Appalachian marine ecosystems. In *Geological Implications of Impacts of Large Asteroids and Comets on the Earth.* ed. L. T. Silver and P. H. Schultz. Boulder, Colo.: Geological Society of America Special Paper 190, pp. 491–500.

——. 1984. Tempo of the Frasnian–Famennian biotic crisis. *Geological Society of America, Abstracts with Program* 16(1): 49.

——. 1988. The Late Devonian extinction event: Evidence for abrupt ecosystem collapse. *Paleobiology* 14: 250–257.

——. 1996. *The Late Devonian Mass Extinction.* New York: Columbia University Press.

——. 1997. Late Devonian bioevents in the Appalachian Sea: Immigration, extinction, and species replacements. In *Paleontological Events: Stratigraphic, Ecological, and Evolutionary Implications.* ed. C. E. Brett and G. C. Baird. New York: Columbia University Press, pp. 493–508.

——. 2001a. Late Devonian extinction. In *Palaeobiology II.* ed. D. E. G. Briggs and P. R. Crowther. Oxford, England: Blackwell Science, pp. 223–226.

——. 2001b. The "multiple impacts hypothesis" for mass extinction: A comparison of the Late Devonian and the late Eocene. *Palaeogeography, Palaeoclimatology, Palaeoecology* 176: 47–58.

——. 2005. Modelling Late Devonian extinction hypotheses. In *Understanding Late Devonian and Permian–Triassic Biotic and Climatic Events.* ed. D. J. Over, J. R. Morrow, and P. B. Wignall. Amsterdam, The Netherlands: Elsevier B.V., pp. 37–50.

——. 2011. *Convergent Evolution: Limited Forms Most Beautiful.* Cambridge, Mass.: Vienna Series in Theoretical Biology; Massachusetts Institute of Technology Press.

McGhee, G. R., M. E. Clapham, P. M. Sheehan, D. J. Bottjer, and M. L. Droser. 2013. A new ecological-severity ranking of major Phanerozoic biodiversity crises. *Palaeogeography, Palaeoclimatology, Palaeoecology* 370: 260–270.

McGhee, G. R., J. S. Gilmore, C. J. Orth, and E. J. Olsen. 1984. No geochemical evidence for an asteroidal impact at Late Devonian mass extinction horizon. *Nature* 308: 629–631.

McGhee, G. R., C. J. Orth, L. R. Quintana, J. S. Gilmore, and E. J. Olsen. 1986. The Late Devonian "Kellwasser Event" mass extinction horizon in Germany: No geochemical evidence for a large–body impact. *Geology* 14(9): 776–779.

McGhee, G. R., P. M. Sheehan, D. J. Bottjer, and M. L. Droser. 2012. Ecological ranking of Phanerozoic biodiversity crises: The Serpukhovian (Early Carboniferous) crisis had a greater ecological impact than the end-Ordovician. *Geology* 40(2): 147–150.

McKay, C. 1996a. Oxygen and the rapid evolution of life on Mars. In *Chemical Evolution: Physics of the Origin and Evolution of Life*. ed. J. Chela-Flores and F. Raulin. Amsterdam, The Netherlands: Kluwer Academic Publishers, pp. 177–184.

——. 1996b. Time for intelligence on other planets. In *Circumstellar Habitable Zones*. ed. L. R. Doyle. Menlo Park, Travis House Publications, pp. 405–419.

McKay, D. S., E. K. Gibson, K. L. Thomas-Keprta, H. Vali, C. S. Romanek, S. J. Clemett, X. D. F. Chillier, C. R. Maechling, and R. N. Zare. 1996. Search for past life on Mars: Possible relic biogenic activity in Martian meteorite ALH84001. *Science* 273: 924–930.

McKinney, M. L., and J. L. Lockwood. 1999. Biotic homogenization: A few winners replacing many losers in the next mass extinction. *Trends in Ecology and Evolution* 14(11): 450–453.

Meyer-Berthaud, B., A. Soria, and A.-L. Decombeix. 2010. The land plant cover in the Devonian: A reassessment of the evolution of the tree habit. In *The Terrestrialization Process: Modelling Complex Interactions at the Biosphere-Geosphere Interface*. ed. M. Vecoli, G. Clément, and B. Meyer-Berthaud. Geological Society, London, Special Publications 339: 59–70.

Mii, H.-S., E. L. Grossman, and T. E. Yancey. 1999. Carboniferous isotope stratigraphies of North America: Implications for Carboniferous paleoceanography and Mississippian glaciation. *Geological Society of America Bulletin* 111(7): 960–973.

Miller, H. 1858. *The Old Red Sandstone*. New York: Hurst and Company, Publishers.

Mojzsis, S. J., G. Arrhenius, K. D. McKeegan, T. M. Harrison, A. P. Nutman, and C. R. L. Friend. 1996. Evidence for life on Earth before 3,800 million years ago. *Nature* 384: 55–59.

Morgan, J. P., T. J. Reston, and C. R. Ranero. 2004. Contemporaneous mass extinctions, continental flood basalts, and "impact signals": Are mantle plume-induced lithospheric gas explosions the causal link? *Earth and Planetary Sciences Letters* 217: 263–284.

Murphy, A. E., B. B. Sageman, and D. J. Hollander. 2000. Eutrophication by decoupling of the marine biogeochemical cycles of C, N, and P: A mechanism for the Late Devonian mass extinction. *Geology* 28(5): 427–430.

——. 2001. Eutrophication by decoupling of the marine biogeochemical cycles of C, N, and P: A mechanism for the Late Devonian mass extinction: Reply. *Geology* 29(5): 470–471.

Newell, N. 1967. Revolutions in the history of life. *Geological Society of America Special Paper* 89: 63–91.

Niedźwiedzki, G., P. Szrek, K. Narkiewicz, M. Narkiewicz, and P. E. Ahlberg. 2010. Tetrapod trackways fron the early Middle Devonian period of Poland. *Nature* 463: 43–48.

Pagani, M., J. C. Zachos, K. H. Freeman, B. Tipple, and S. Bohaty. 2005. Marked decline in atmospheric carbon dioxide concentrations during the Paleogene. *Science* 309: 600–603.

Paton, R. L., T. R. Smithson, and J. A. Clack. 1999. An amniote-like skeleton from the Early Carboniferous of Scotland. *Nature* 398: 508–513.

Poty, E. 1999. Famennian and Tournaisian recoveries of shallow water Rugosa following late Frasnian and late Strunian major crises, southern Belgium and surrounding areas, Hunan (South China) and the Omolon region (NE Siberia). *Palaeogeography, Palaeoclimatology, Palaeoecology* 154: 11–26.

Prave, A. R. 2002. Life on land in the Proterozoic: Evidence from the Torridonian rocks of northwest Scotland. *Geology* 30(9): 811–814.

Prestianni, C. and P. Gerrienne. 2010. Early seed plant radiation: An ecological hypothesis. In *The Terrestrialization Process: Modelling Complex Interactions at the Biosphere-Geosphere Interface*. ed. M. Vecoli, G. Clément, and B. Meyer-Berthaud. Geological Society, London, Special Publications 339: 71–80.

Pritchard, G., M. H. McKee, E. M. Pike, G. J. Scrimgeour, and J. Zloty. 1993. Did the first insects live in water or in air? *Biological Journal of the Linnean Society* 49: 31–44.

Prokop, J., A. Nel, and I. Hoch. 2005. Discovery of the oldest known Pterygota in the Lower Carboniferous of the Upper Silesian Basin in the Czech Republic (Insecta: Archaeorthoptera). *Geobios* 38: 383–387.

Prothero, D. R. 2007. *Evolution: What the Fossils Say and Why It Matters*. New York: Columbia University Press.

Racki, G. 2005. Toward understanding Late Devonian global events: Few answers, many questions. In *Understanding Late Devonian and Permian–Triassic Biotic and Climatic Events*. ed. D. J. Over, J. R. Morrow, and P. B. Wignall. Amsterdam, Netherlands: Elsevier B.V., pp. 5–36.

Racki, G., M. Racka, H. Matyja, and X. Devleeschouwer. 2002. The Frasnian–Famennian boundary interval in the South Polish–Moravian shelf basins: Integrated event-stratigraphic approach. *Palaeogeography, Palaeoclimatology, Palaeoecology* 181: 251–297.

Racki, G. and P. Wignall. 2001. Eutrophication by decoupling of the marine biogeochemical cycles of C, N, and P: A mechanism for the Late Devonian mass extinction: Comment. *Geology* 29(5): 469–470.

Raup, D. M. and J. J. Sepkoski. 1982. Mass extinctions in the marine fossil record. *Science* 215: 1501–1503.

Raymo, M. E. and W. F. Ruddiman. 1992. Tectonic forcing of late Cenozoic climate. *Nature* 359: 117–122.

Raymond, A., P. Gensel, and W. E. Stein. 2006. Phytogeography of Late Silurian macrofloras. *Review of Palaeobotany and Palynology* 142: 165–192.

Raymond, A., and C. Metz. 1995. Laurussian land-plant diversity during the Silurian and Devonian: Mass extinction, sampling bias, or both? *Paleobiology* 21(1): 74–91.

Redecker, D., R. Kodner, and L. E. Graham. 2000. Glomalean fungi from the Ordovician. *Science* 289: 1920–1921.

Reimold, W. U., S. P. Kelley, S. C. Sherlock, H. Henkel, and C. Koeberl. 2005. Laser argon dating of melt breccias from the Siljan impact structure, Sweden: Implications for a possible relationship to Late Devonian extinction events. *Meteoritics and Planetary Science* 40(4): 591–607.

Retallack, G. J., R. R. Hunt, and T. S. White. 2009. Late Devonian tetrapod habitats indicated by palaeosols in Pennsylvania. *Journal of the Geological Society of London* 166: 1143–1156.

Rimmer, S. M., J. A. Thompson, S. A. Goodnight, and Thomas L. Robl. 2004. Multiple controls on the preservation of organic matter in Devonian–Mississippian marine black shales: Geochemical and petrographic evidence. *Palaeogeography, Palaeoclimatology, Palaeoecology* 215: 125–154.

Riquier, L., N. Tribovillard, O. Averbuch, X. Devleeschouwer, and A. Riboulleau. 2006. The late Frasnian Kellwasser horizons of the Harz Mountains (Germany): Two oxygen-deficient periods resulting from different mechanisms. *Chemical Geology* 233: 137–155.

Rothwell, G. W., S. E. Scheckler, and W. H. Gillespie.1989. *Elkinsia* gen. nov., a Late Devonian gymnosperm with cupulate ovules. *Botanical Gazette* 150: 170–189.

Rubinstein, C. V., P. Gerriene, G. S. de la Puente, R. A. Astini, and P. Steemans. 2010. Early Middle Ordovician evidence for land plants in Argentina (eastern Gondwana). *New Phytologist* 188: 365–369.

Sageman, B. B., A. E. Murphy, J. P. Werne, C. A. Ver Straeten, D. J. Hollander, and T. W. Lyons. 2003. A tale of shales: The relative roles of production, decomposition, and dilution in the accumulation of organic-rich strata, Middle-Upper Devonian, Appalachian basin. *Chemical Geology* 195: 229–273.

Sallan, L. C. and M. I. Coates. 2010. End-Devonian extinction and a bottleneck in the early evolution of modern jawed vertebrates. *Proceedings of the National Academy of Sciences USA* 107(22): 10131–10135.

Sallan, L. C., T. W. Kammer, W. I. Ausich, and L. A. Cook. 2011. Persistent predator-prey dynamics revealed by mass extinction. *Proceedings of the National Academy of Sciences USA* 108(20): 8335–8338.

Saltzman, M. R., L. A. González, K. C. Lohman. 2000. Earliest Carboniferous cooling triggered by the Antler orogeny? *Geology* 28(4): 347–350.

Sandberg, C. A., J. R. Morrow, and W. Ziegler. 2002. Late Devonian sea-level changes, catastrophic events, and mass extinctions. *Geological Society of America Special Paper* 356: 473–487.

Scheckler, S. E. 1986. Floras of the Devonian–Mississippian Transition. In *Land Plants: Notes for a Short Course.* ed. T. W. Broadhead. Knoxville, Tenn.: University of Tennessee Department of Geological Sciences Studies in Geology Number 15, pp. 81–96.

Schieber, J. 2009. Discovery of agglutinated benthic foraminifera in Devonian black shales and their relevance for the redox state of ancient seas. *Palaeogeography, Palaeoclimatology, Palaeoecology* 271: 292–300.

Schindler, E. 1990. Die Kellwasser-Krise (hohe Frasne-Stufe, Ober-Devon). *Göttinger Arbeiten zur Geologie und Paläontologie* 46: 1–115.

——. 1993. Event-stratigraphic markers within the Kellwasser Crisis near the Frasnian–Famennian boundary (Upper Devonian) in Germany. *Palaeogeography, Palaeoclimatology, Palaeoecology* 104: 115–125.

Schopf, J. W. 1993. Microfossils of the Early Archean Apex Chert: New evidence of the antiquity of life. *Science* 260: 640–646.

Scott, A. C. and I. J. Glasspool. 2006. The diversification of Paleozoic fire systems and fluctuations in atmospheric oxygen concentration. *Proceedings of the National Academy of Sciences USA* 103(29): 10861–10865.

Sepkoski, J. J. 1986. Patterns of Phanerozoic extinction: A perspective from global data bases. In *Global Events and Event Stratigraphy*. ed. O. H. Walliser. Berlin, Germany: Springer-Verlag, pp. 35–51.

——. 2002. A compendium of fossil marine animal genera. *Bulletin of American Paleontology* 363: 1–560.

Shear, W. A. and G. D. Edgecombe. 2010. The geological record and phylogeny of the Myriapoda. *Arthropod Structure and Development* 39: 174–190.

Shear, W. A. and P. A. Selden. 2001. Rustling in the undergrowth: Animals in early terrestrial ecosystems. In *Plants Invade the Land*. ed. P. G. Gensel and D. Edwards. New York: Columbia University Press, pp. 29–51.

Shubin, N. H. 2009. *Your Inner Fish*. New York: Vintage Books.

Shubin, N. H., C. Tabin, and S. Carroll. 1997. Fossils, genes and the evolution of animal limbs. *Nature* 388: 639–648.

Sierwald, P. and J. E. Bond. 2007. Current status of the myriapod Class Diplopoda (millipedes): Taxonomic diversity and phylogeny. *Annual Review of Entomology* 52: 401–420.

Snitting, D. 2008. *Morphology, Taxonomy and Interrelationships of Tristichopterid Fishes (Sarcopterygii, Tetrapodomorpha)*. Uppsala, Sweden: Acta Universitatis Uppsalensis (PhD dissertation, 54 pp.)

Stanley, S. M. 2007. An analysis of the history of marine animal diversity. *Paleobiology Memoir* 4: 1–55.

Steemans, P., C. H. Wellman, and P. Gerrienne. 2010. Palaeogeographic and palaeoclimatic considerations based on Ordovician to Lochkovian vegetation. In *The Terrestrialization Process: Modelling Complex Interactions at the Biosphere–Geosphere Interface*. ed. M. Vecoli, G. Clément, and B. Meyer-Berthaud. Geological Society, London, Special Publications 339: 49–58.

Stein, W. E., C. M. Berry, L. V. Hernick, and F. Mannolini. 2012. Surprisingly complex community discovered in the Mid-Devonian fossil forest at Gilboa. *Nature* 483: 78–81.

Stein, W. E., F. Mannolini, L. V. Hernick, E. Landing, and C. M. Berry. 2007. Giant cladoxylopsid trees resolve the enigma of the Earth's earliest forest stumps at Gilboa. *Nature* 446: 904–907.

Stigall, A. L. 2010. Invasive species and biodiversity crises: Testing the link in the Late Devonian. *PLoS ONE* 5(12): e15584. doi: 10.1371/journal.pone.0015584

Stössel, I. 1995. The discovery of a new Devonian tetrapod trackway in SW Ireland. *Journal of the Geological Society of London* 152: 407–413.

Streel, M. 2009. Upper Devonian miospore and conodont zone correlation in Western Europe. In *Devonian Change: Case Studies in Palaeogeography and Palaeoecology*. ed. P. Königshof. Geological Society, London, Special Publications 314: 163–176.

Streel, M., M. V. Caputo, S. Loboziak, and J. H. G. Melo. 2000. Late Frasnian–Famennian climates based on palynomorph analyses and the question of the Late Devonian glaciations. *Earth-Science Reviews* 52: 121–173.

Streel, M., K. Higgs, S. Loboziak, W. Riegel, and P. Steemans. 1987. Spore stratigraphy and correlation with faunas and floras in the type marine Devonian of the Ardenne-Rhenish regions. *Review of Palaeobotany and Palynology* 50: 211–229.

Strother, P. K., T. Servais, and M. Vecoli. 2010. The effects of terrestrialization on marine ecosystems: The fall of CO_2. In *The Terrestrialization Process: Modelling Complex Interactions at the Biosphere-Geosphere Interface*. ed. M. Vecoli, G. Clément, and B. Meyer-Berthaud. Geological Society, London, Special Publications 339: 37–48.

Taylor, T. N., H. Hass, W. Remy, and H. Kerp. 1995. The oldest fossil lichen. *Nature* 378: 244.

Taylor, T. N. and J. M. Osborn. 1996. The importance of fungi in shaping a paleoecosystem. *Review of Palaeobotany and Palynology* 90: 249–262.

Taylor, T. N. and E. L. Taylor. 1993. *The Biology and Evolution of Fossil Plants*. New Jersey: Prentice Hall.

Toon, O. B., K. Zahnle, R. P. Turco, and C. Covey. 1994. Environmental perturbations caused by asteroid impacts. In *Hazards due to Comets and Asteroids*. ed. T. Gehrels. Tucson, Ariz.: University of Arizona Press, pp. 791–826.

Veizer, J., D. Ala, K. Azmy, P. Bruckschen, D. Buhl, F. Bruhn, G. A. F. Carden, A. Diener, S. Ebneth, Y. Godderis, T. Jasper, C. Korte, F. Pawellek, O. G. Podlaha, and H. Strauss. 1999. $^{87}Sr/^{86}Sr$, $\delta^{13}C$ and $\delta^{18}O$ evolution of Phanerozoic seawater. *Chemical Geology* 161: 59–88.

Versteegh, G. J. M. and A. Riboulleau. 2010. An organic geochemical perspective on terrestrialization. In *The Terrestrialization Process: Modelling Complex Interactions at the Biosphere-Geosphere Interface*. ed. M. Vecoli, G. Clément, and B. Meyer-Berthaud. Geological Society, London, Special Publications 339: 11–36.

Voigt, S. and M. Ganzelewski. 2010. Toward the origin of amniotes: Diadectomorph and synapsid footprints from the early Late Carboniferous of Germany. *Acta Palaeontologia Polonica* 55(1): 57–72.

Wade, N. 2006. *Before the Dawn: Recovering the Lost History of Our Ancestors*. New York: Penguin Press.

Walker, J. D. and J. W. Geissman. 2009. Geologic time scale. Geological Society of America, doi: 10.1130/2009.CTS004R2C.

Walliser, O. H. 1996. Global events in the Devonian and Carboniferous. In *Global Events and Event Stratigraphy*. ed. O. H. Walliser. Berlin, Germany: Springer-Verlag, pp. 225–250.

Ward, P. D. 2005. *Life as We Do Not Know It*. New York: Viking Penguin.

Ward, P., C. Labandeira, M. Laurin, and R. A. Berner. 2006. Confirmation of Romer's Gap as a low oxygen interval constraining the timing of initial arthropod and vertebrate terrestrialization. *Proceedings of the National Academy of Sciences USA* 103(45): 16818–16822.

Webb, G. E. 2002. Latest Devonian and Early Carboniferous reefs: Depressed reef building after the Middle Paleozoic collapse. In *Phanerozoic Reef Patterns*. ed. W. Kiessling, E. Flügel, and J. Golonka. Tulsa, Okla: SEPM Special Publication Number 72: 239–269.

Wei, F., Y. Gong, and H. Yang. 2012. Biogeography, ecology and extinction of Silurian and Devonian tentaculitoids. *Palaeogeography, Palaeoclimatology, Palaeoecology* 358–360: 40–50.

Wellman, C. H. 2010. The invasion of the land by plants: When and where? *New Phytologist* 188: 306–309.

Wellman, C. H., P. L. Osterloff, and U. Mohiuddin. 2003. Fragments of the earliest land plants. *Nature* 425: 282–285.

Wilson, H. M. 2006. Juliform millipedes from the Lower Devonian of Euramerica: Implications for the timing of millipede cladogenesis in the Paleozoic. *Journal of Paleontology* 80(4): 638–649.

Wilson, H. M. and L. I. Anderson. 2004. Morphology and taxonomy of Paleozoic millipedes (Diplopoda: Chilognatha: Archipolypoda) from Scotland. *Journal of Paleontology* 78(1): 169–184.

Wilson, H. M. and J. T. Hannibal. 2005. Taxonomy and trunk-ring architecture of pelurojulid millipedes (Diplopoda: Chilognatha: Pleurojulida) from the Pennsylvanian of Europe and North America. *Journal of Paleontology* 79(6): 1105–1119.

Wootton, R. J., J. Kukalová-Peck, D. J. S. Newman, and J. Muzón. 1998. Smart engineering in the Mid-Carboniferous: How well could Palaeozoic dragonflies fly? *Science* 282: 749–751.

Xiong, J., W. M. Fischer, K. Inoue, M. Nakahara, and C. E. Bauer. 2000. Molecular evidence for the early evolution of photosynthesis. *Science* 289: 1724–1730.

Young, G. C. 2010. Placoderms (armored fish): Dominant vertebrates of the Devonian Period. *Annual Review of Earth and Planetary Sciences* 38: 523–550.

Yuan, X., S. Xiao, and T. N. Taylor. 2005. Lichen-like symbiosis 600 million years ago. *Science* 308: 1017–1020.

Zachos, J., M. Pagani, L. Sloan, E. Thomas, and K. Billups. 2001. Trends, rhythms, and aberrations in global climate 65 Ma to present. *Science* 292: 686–693.

Zhu, M., P. E. Ahlberg, W. Zhao, and L. Jia. 2002. First Devonian tetrapod from Asia. *Nature* 420: 760–761.

Zhu, M., W. Zhao, L. Jia, J. Lu, T. Qiao, and Q. Qu. 2009. The oldest articulated osteichthyan reveals mosaic gnathostome characters. *Nature* 458: 469–474.

Ziegler, W. and C. A. Sandberg. 1990. The Late Devonian standard conodont zonation. *Courier Forschungsinstitut Senckenberg* 121: 1–115.

Zubrin, R. and R. Wagner. 2011. *The Case for Mars* (Revised Edition). New York: Free Press.

Index

Acanthostega, 72–73, 86, 88–91, 161, 163–172, 174–175, 224, color plate 11
Age of Armored Fish, 266
Age of Crinoids, 266
Age of Sharks, 266
Ahlberg, P. E., 81, 85, 160–161, 167
Algeo, T. J., 113–114, 148–149, 201
Alvarez, L. W., 145
Alvarez, W., 145
Amniotes, evolution of, 49, 62, 70, 235–237, 249–262
Anoxia, 115–117, 120–124, 142–143, 194
Anthoceros, 35
Archaeopteris, 48–49, 51, 110, 114–115, 121, 123, 148–149, 151, 156, 179–180, color plate 5
Arthropod invaders, 54–67; convergent evolution in, 58; Famennian Gap and, 106, 177; first invaders, 57–67, 98; second invaders, 177; third invaders, 223; Tournaisian Gap and, 223; victorious invaders, 52, 237–249
Asteroid impact, 137, 144–148, 154
Averbuch, O., 150
Aysheaia, 20

Balanerpeton, 220, 228–232
Banded-iron formations, 7–8, 21, 23–24, color plates 2, 3
Berner, R. A., 111–114
Biodiversity loss: by extinction, 135–136; by speciation suppression, 135–137, 154–158
Black shale deposits, 115–124, 193–196, 200, 202
Bogdanov, D., color plate 7
Bothriolepis, 95–97, 174–176, 272, color plate 9
Bottlenecks. *See* End-Famennian Bottleneck; End-Frasnian Bottleneck
Brezinski, D. K., 192

Caputo, M. V., 196–198
Carbon isotopes, 7, 118–119, 126, 151, 194, 197–198, 204–207, 210–211
Carroll, R., 79, 213, 233, 236
Casineria, 221–222, 232–235, 237, 253–254

Charcoal gap, 110–111, 113, 139, 141–142
Charniodiscus, color plate 1
Clack, J. A., 106, 140–141, 160–161, 170–171, 186–187, 214, 227–229, 232–233
Clément, G., 80–81
Coates, M. I., 186–187, 214, 265
Collins, M., 20
Convergent evolution, 6–7, 30, 34, 46, 56, 58, 70, 140–141, 227, 229, 230, 250, 256, 258, 263, 269–271, 272
Cooksonia, 38, 40, 43–44
Copper, P., 144
Cowen, R., 236–237, 254–255
Coxhead, P., 41
Crassigyrinus, 218, 226

Daeschler, E. B., 90
Darwin, C., xi, 11, 38
Dehydration, invasion problem of, 1, 3, 29, 56, 58, 70, 275
Densignathus, 84–85, 90, 95, 164–167, 172–173, 175–177, 191
Devonian/Carboniferous extinction. *See* Famennian biodiversity crisis
Devonian extinctions: due to hypoxia, 137–143, 153–154, 194, 199–202; due to hypothermia, 143–154, 199–212
Diadectes, 253
Dickinsonia, color plate 1
Dimetrodon, 259, 261
DiMichele, W. A., 156, 179–180
Dunkleosteus, color plate 7
Dunlop, J. A., 65

Edgecombe, G. D., 65
Eldeceeon, 231, 233
Elginerpeton, 74–78, 84–85, 88–91, 95–97, 160, 173, color plate 8
Elkinsia, 49–50
End-Famennian Bottleneck: characteristics of, 213; consequences of, 266–273; effect of, 213–214, 235, 264–266
End-Frasnian Bottleneck: characteristics of, 159–161; consequences of, 266–273; effect of, 88–89, 160–161, 173, 178, 182, 263–264
Engel, M. S., 249
Eucritta, 220, 226–228
Eugeropteron, 242–244
Euglena, 4–5
Eusthenopteron, 71–73, 77, 80, 176, 236
Eutrophication, 116–124, 137–138, 149–151, 153–154, 194–195, 202

Falcon-Lang, H., 193
Famennian biodiversity crisis: causes of, 199–212; timing of, 182–184; victims of, 51, 179–182, 184, 187–188
Famennian Gap, 104–107, 109, 111, 114, 118, 124–126, 130, 140–143, 151, 161–168, 177–178, 186, 188, 191, 204, 209, 216, 263
Felis, 72
Fossil *Lagerstätten*, 39; Burgess Shale, 58; East Kirkton Limestone, 39, 222–223, 228, 230–231; Gilboa Shale, 39, 46–47, 66, 177; Joggins tree-stump, 254–255, 257–258; Mazon Creek Ironstone, 238, 241–242; Rhynie Chert, 39, 40, 45, 66–67, 177, 222
Frank, T., 199
Frasnian biodiversity crisis: causes of, 137–158, 202, 212; timing of, 102–104, 155; victims of, 49, 99–105, 114
Frasnian/Famennian extinction. *See* Frasnian biodiversity crisis

Gaps. *See* Charcoal gap; Famennian Gap; Romer's Gap; Tournaisian Gap

Geologic record: fossil *Lagerstätten* and, 39, 190; imperfections of, 38–39, 62–66, 79, 84–85, 164–166, 222
Glaciation: Carboniferous, 196–199; Cenozoic, 128, 144, 152, 190–191, 203–212; Famennian, 130–131, 188–191, 193, 195–196, 200, 203–212; Frasnian, 115, 137, 144, 153, 203–212; Proterozoic, 14–15, 19, 21
Glasspool, I. J., 110–116, 156
Gravity, invasion problem of, 1–2, 3, 58, 275
Great Devonian Interchange, 156. *See also* Invasive species
Greererpeton, 218, 224–225
Grimaldi, D. A., 249
Grinspoon, D., 23–24

Habitat homogenization, effect of, 137, 154–158
Hallam, A., 149
Hangenberg extinctions. *See* Famennian biodiversity crisis
Harfoot, M. B., 143
Homo, 72, 159–160, 273, 275
Hook, R. W., 156, 179–180
Horneophyton, 40–41
Hyneria, 175–176, color plate 5
Hynerpeton, 164, 166, 171–173, 175–177, 191, color plates 5, 11

Ichthyostega, 86, 88–90, 161, 163–168, 170–172, 174–175, 178, color plate 11
Ingall, E., 113–114
Insects, evolution of, 52, 67, 238, 240–241, 244; wing evolution in, 52, 237, 242–249, 262
Invaders, terrestrial. *See* Arthropod invaders; Plants; Vertebrate invaders
Invasive species, effect of, 137, 156; Great Devonian Interchange and, 156
Isbell, J. L., 196, 198

Jakubsonia, 161, 164–168, 173–174, color plate 11
Janvier, P., 80–81
Joachimski, M. M., 128–130
Johanson, Z., 82
Johnson, J. G., 132–134, 138

Kellwasser extinctions. *See* Frasnian biodiversity crisis
Kent, D. V., 152
Kill mechanisms: hypoxic, 137–143, 153–154, 194, 199–202; hypothermic, 143–154, 199–212
Kimsey, L., color plate 6
Kopp, R., color plates 2, 3
Kump, L. R., 142–143

Labandeira, C. C., 236
Land invaders. *See* Arthropod invaders; Plants; Vertebrate invaders
Land invasion, problems of: dehydration, 1, 3, 29, 56, 58, 70, 275; gravity, 1–2, 3, 58, 275; radiation poisoning, 3, 16, 21, 27, 275; temperature fluctuation, 2–3, 275
Late Devonian extinction. *See* Frasnian biodiversity crisis
Long, J. A., 80–81
Lycaenops, 261

Manoblatta, 247
Marchantia, 33
Mars: invasion of, 273–275; life on, 8–11, 24–25
McGhee, G. R., 119, 136, 145–147

McGhee, M., v, xi
Metaxygnathus, 74–76, 84–85, 91–92, 96–97, 160–161, 172, 174, color plate 8
Metz, C., 114
Mii, H.-S., 198–199
Monoyios, K., xi, 72, 88, 91, 170, 171, 225, 226, 228, 232, 233, 253, 256, 257, 259
Murphy, A. E., 122, 128
Muttoni, G., 152

Newell, N., 214
Niedźwiedzki, G., 79

Obruchevichthys, 74–76, 84–85, 89–91, 95, 160–161, 173, color plate 8
Olsen, E. J., 145–147
Ophiacodon, 259, 261
Orth, C. J., 145–147
Oxygen, atmospheric: arthropod gigantism and, 247–249, 262; evolutionary rates and, 23–25; in the Carboniferous, 247–249, 262; in the Devonian, 110–114, 137–142; in the Precambrian, 8, 13–15, 19, 21, 23–25
Oxygen, marine: eutrophication and, 115–117, 120–124, 137–138, 149–151, 153–154, 194–195, 202; hypoxic kill mechanism and, 143–154, 194, 199–202
Oxygen isotopes, 126–132, 195, 198–199

Paleothyris, 255–256
Paleuthygramma, 246
Parallel evolution, 81–82, 140–141, 264, 270
Pederpes, 214–215, 218, 224–225
Peripatus, color plate 6
Petrolacosaurus, 256–257
Pikaia, 20
Plants, terrestrial, 19, 21–22, 27–51; convergent evolution in, 30, 34, 46; eutrophication and, 120–121, 137, 148–150, 193, 201–202; evolution of, 6, 14, 17, 27–51; Famennian Gap and, 49, 106, 124–125, 130–131, 141–143; monocultures of, 156–158; Tournaisian Gap and, 187–188
Poty, E., 106–107
Prokop, J., 243
Psilophyton, 45

Radiation poisoning, invasion problem of, 3, 16, 21, 27, 275
Raup, D. M., 211
Raymond, A., 114
Retallack, G. J., 123, 191, 195
Rhizobium, 4–5
Rhynia, 40, 42
Romer's Gap, 187

Sallan, L. C., 214, 265
Saltzman, M. R., 198
Sartenaer, P., 146
Scheckler, S. E., 136
Scott, A. C., 110–116, 156
Sea-level fluctuations, 120, 122–124, 132–135, 144, 153, 190–191, 195–196, 198
Selden, P. A., 223
Sepkoski, J. J., 136, 211
Shear, W. A., 65, 223
Sinostega, 74–76, 84–85, 90–91, 160–161, color plate 8
Speciation suppression: habitat homogenization and, 137, 154–158; invasive species and, 137, 156
Sporogonites, 36
Stanley, S. M., 211
Stenodictya, 245
Stigall, A. L., 136, 156
Streel, M., 125

Stromatolites, 11–14
Strontium isotopes, 108–109, 126, 150–151

Temperature, sea-surface: brachiopods and, 127, 130, 198–199; conodonts and, 128–132; oxygen isotopes and, 126–132
Temperature fluctuation, invasion problem of, 2–3, 275
Terrestrial invasion, problems of: dehydration, 1, 3, 29, 56, 58, 70, 275; gravity, 1–2, 3, 58, 275; radiation poisoning, 3, 16, 21, 27, 275; temperature fluctuation, 2–3, 275
Tetraceratops, 260
Tetrapodomorphs, convergent evolution in, 70; limb evolution in, 68, 71–73, 140–141, 169–172, 269–271; parallel evolution in, 81–82, 140–141. *See also* Vertebrate invaders
Thrinaxodon, 72–73, 261
Tiktaalik, 71–73, 77, 80, 90, 99, 269
Tournaisian Gap, 184–188, 192–193, 205–206, 209, 213–216, 218, 220–221, 223–227, 264, 266
Tribrachidium, color plate 1
Tulerpeton, 80, 164, 166, 172, 176, color plate 11

Ventastega, 84–85, 88, 90–91, 95, 161, 164–167, 169, 174, color plate 11
Vertebrate invaders, 54–55, 57; convergent evolution in, 70, 140–141, 227, 229, 230, 250, 256, 258, 263, 269–271, 272; Famennian Gap and, 104–107, 140–141, 161–168; first invaders, 68–98; parallel evolution in, 81–82, 264, 270; second invaders, 161–178; third invaders, 215–237; Tournaisian Gap and, 184–187; victorious invaders, 52, 249–262; wing evolution in, 237–238, 270–271
Volcanism, mantle-plume, 109–110, 127, 147–148, 154

Wattieza, 47–48, 121
Weathering, continental, 15–17; biological, 120–121, 137, 148–150, 201–202, 212; chemical, 108–109, 121, 126, 137, 148, 150–154, 201–202, 212; strontium isotopes and, 108–109, 126, 150
Westlothiana, 231–232
Wignall, P. B., 149
Woese, C., 3
Wooton, R. J., 243

Ymeria, 163–167, 169, 172, 174, color plate 11

Ziegler, W., 147
Zosterophyllum, 44

Printed in the USA
CPSIA information can be obtained
at www.ICGtesting.com
JSHW011520221024
72172JS00014B/116

9 780231 160575